T0295084

Sustainable Operations Management

This book includes concepts, methodologies, and practices for achieving sustainability in business operations. The underlying concept is explained from two perspectives—organizational level and policy level. In the former, all principles, techniques, and decision-making issues relevant to sustainability at unit level, management of product recovery processes, and sustainability at integrated level are captured. Content on policy-level perspective includes policies, norms, guidelines, and regulatory measures both at global and national levels. The primary goal of this book is the creation of an integrated and value-rich platform for the initiation and management of sustainable operations.

Features:

- This book offers a comprehensive overview of environmental sustainability from the operations and supply chain perspective.
- This book proposes an understandable and innovative viewpoint in explaining sustainable operations management comprehensively as managing operations sustainably at organizational level.
- Readers will learn the concepts, techniques, and the core factors relating to managing operations, keeping in view various dimensions of sustainability and the macro-level guidelines, norms, regulatory measures, and the like in this context.
- This book adds to the knowledge on design, planning, and management of sustainable operations, including the specific management approaches.
- This book includes summary and review questions at the end of each chapter.

This book is aimed at graduate students and researchers in industrial, production, and mechanical engineering, including operations management. It is also recommended as a textbook for courses such as sustainable development, sustainable operations management, and environmental management.

Sustainable Operations Management

Kampan Mukherjee

CRC Press
Taylor & Francis Group
Boca Raton London New York

CRC Press is an imprint of the
Taylor & Francis Group, an **informa** business

Designed cover image: shutterstock

First edition published 2024
by CRC Press
2385 NW Executive Center Drive, Suite 320, Boca Raton FL 33431

and by CRC Press
4 Park Square, Milton Park, Abingdon, Oxon, OX14 4RN

CRC Press is an imprint of Taylor & Francis Group, LLC

© 2024 Taylor & Francis Group, LLC

ISBN: 978-1-498-79652-1 (hbk)
ISBN: 978-1-032-62616-1 (pbk)
ISBN: 978-0-429-19560-0 (ebk)

DOI: 10.1201/9780429195600

Typeset in Times
by Apex CoVantage, LLC

*The book is dedicated to those who really care to hand
over a cleaner earth to our future generations*

Contents

PART I *Sustainability—An Introduction*

PART II *Sustainability at the Policy Level*

PART III   *Sustainability at the Operations Level*

Foreword

The endeavor of Dr. Kampan Mukherjee in writing this book titled *Sustainable Operations Management* is surely a praiseworthy attempt when the policy-makers of the whole world are thriving to make the planet cleaner, greener, and healthier for this generation and for generations in future.

It is a pity that to date there is only a very limited number of textbooks that cover all the basic aspects of sustainability in operations management. This book, perhaps, will become the single literary source that captures the qualitative legislative policy issues and analytical models as two extremes, and similarly, the book elaborates on concepts and philosophical issues on the one hand and hard-core industrial examples on the other.

I know Kampan both personally and professionally. We also successfully collaborated in joint research projects. I am sure that the book will contain enough noteworthy and meaningful materials with technical, academic, and application-worthy values, and the academic community and industrial practitioners will be immensely benefitted by these.

My best wishes to Kampan for the successful launching of this book and his other similar intellectual projects in future.

Dr. Karl Inderfurth
Former Professor and Chair of Production and Logistics
Faculty of Economics and Management
Otto-von-Guericke-University Magdeburg
Magdeburg, Germany

Foreword

Classically defined by the UN's Brundtland Commission in 1987 as the ability to fulfill current generation's needs while ensuring that future generations will also be able to meet their own needs, sustainability continues to be a daunting challenge for humanity. Recent extreme weather events worldwide continue to underscore this topic's importance to all who inhabit this planet. All of this is evidence that despite many years (indeed decades) of attention paid to sustainability, much more work and attention will be needed to fulfill the promise to future generations.

In business, operations management fulfills the need for goods and services by purchasing, producing, storing, and transporting products around the globe. In order to fulfill the promise to future generations, it is imperative that we increase our ability to do these operations in an environmentally friendly manner with as close to zero impact as possible. It is precisely here that this book authored by Dr. Kampan Mukherjee makes an impressive and innovative contribution.

The first part of the book affords the reader an accessible but comprehensive introduction to sustainability, beginning with an examination of the core issues that are necessary to understand the paradigm of sustainability. Diverse viewpoints, concepts, and models are presented, forming a perfect basis for understanding the contribution as a whole.

The second part of the book examines sustainability at the macro or policy level, an essential component of encouraging firms and industries to become more environmentally conscious through curbing their ecological and carbon footprints. After discussing global directives, norms, policies, and perspectives, the work thoroughly examines the Indian government's regulations and guidelines.

The third part of the book examines the operational (or micro) level of sustainability and comprises two chapters. The former deals with sustainable manufacturing (forward logistics, if you will) and producing the products sustainably in the first place. This entails designing the product sustainably, as well as designing sustainable processes and facilities. The latter deals with the exciting area of product recovery management, starting with a solid introduction to cradle-to-cradle issues before examining remanufacturing in detail and incorporating reverse logistics and closed-loop supply chain models and principles.

The book's final part delves into the measurement and assessment of sustainability, something necessary to improve the sustainable performance of operations. The core of this is the two complementary approaches of carbon footprint analysis and life cycle analysis, the former attempting to capture the impact on the carbon footprint specifically while the latter takes a temporally broader cradle-to-cradle view. As there are more than one metric to consider, the part culminates in an interesting scorecard-based approach to analyze integrated business performance in an innovative new way.

Several noteworthy features are especially unique to this work. First, this is the first textbook providing a comprehensive glimpse into sustainability from international and global frameworks, to Indian national policy and legislation, through firm-level strategy and operational issues, including performance measurement. Second, it is peppered with a plethora of short industrial examples that explain the concepts and methodologies in an incredibly accessible and effective manner, as well as tools like the house of sustainability that empower the reader to put the knowledge immediately into practice.

This textbook is well suited both for introducing students to this critical business area and for more advanced study of specific areas of sustainable practice. The key learning and discussion questions allow for the reinforcement of learning, while the introduction of more sophisticated topics like remanufacturing, reverse logistics, and green product design will encourage the more advanced reader to examine these topics in greater detail using the afforded references.

Dr. Ian M. Langella
Professor of Supply Chain Management and
Chair of Finance and Supply Chain Management
John L. Grove College of Business
Shippensburg University
Pennsylvania, USA

Foreword

The term sustainability is a popular keyword that is being used in various contexts, such as agriculture, energy, tourism, and business. Although the concept of sustainability is understood in its general meaning of "long-term" and "durable," the meaning of sustainable development is far more complex and provides specific objectives for organizations, nations, and society. Realizing the need to study sustainability in business, Elkington, in 1997, provided an organization-level perspective and introduced the accounting framework called triple bottom line, which stresses on three dimensions of sustainability: environmental, social, and economic. Thus, organizations are expected to control the negative social and environmental outcomes of business while enhancing their financial performance. The concept of sustainability is still in its research phase although organizational interests have developed over the last decade, and the markets are transitioning toward sustainable products in the wake of shifting consumer preferences. Due to such high levels of customer interest, business organizations are under extreme pressure to adopt sustainable practices. However, organizations feel that fulfilling consumers' demand for sustainable products is a joint responsibility of the entire supply chain. This book addresses the key issues of sustainability in operations management.

I am indeed privileged to write a foreword to this book. I have the pleasure of knowing the author, Dr. Kampan Mukherjee, for nearly two decades through our interactions. This book is the outcome of his long experience in teaching, research, and consultancy in the area of operations and sustainable management. He has taken serious efforts to explain the important areas of sustainability in operations in a lucid style with key learning points covering sustainability at the policy level and operational level, then explaining product recovery management through reverse logistics and closed-loop supply chain management perspectives. Toward the end, he covers measuring sustainability performances, especially about carbon footprint and life cycle analysis, with relevant and interesting examples. He has attempted to explain even the complexities in sustainability in a simple style so that anyone can understand and implement them in his/her workplace.

I am quite sure that both the academic community and practitioners will get a good exposure on various concepts, industrial examples, methodologies, policies, and models that are presented in the book with great details. The book also gives directions for research scholars to get inspirations in terms of research leads for their research especially on topics like green product, process, and facility designs; product recovery management and remanufacturing; reverse logistics; and the like.

This book is perhaps the first textbook that captures all aspects (other than the social dimension) of operational sustainability, both at national and global policy levels, conference of parties, other macro-level initiatives or legislative efforts, strategic issues, modeling approaches, and performance management methodologies. The book includes various short industrial examples while explaining concepts and discussing methodologies. Further, readers may also be aware of some new concepts like the house of sustainability and various new indices of effectiveness and efficiency in buyback planning of used products in product recovery management.

I recommend this book to every professional, teacher, and student of management, particularly operations management and supply chain management.

Dr. T.A.S. Vijayaraghavan
Professor
Production, Operations Management and
Decision Sciences (PODS)
Member, Board of Governors, XLRI
Chairperson, Centre for Logistics and
Supply Chain Management
XLRI–Xavier School of Management
Jamshedpur, Jharkhand, India

Preface

BACKGROUND

Years ago, the first time when I offered sustainable operations management (SOM) as an elective course for the MBA program in Indian Institute of Management Kashipur, the main challenge that I faced was the nonavailability of a textbook on that subject that would cover all relevant topics under SOM. I had to manage with lot of edited books or research papers, research reports, case studies, and other related literature available on the internet. That actually motivated me to take up this book-writing project.

Management professionals of this century are expected to expand their perspectives and domain of considerations beyond sheer profitability, covering both ecological surroundings and society at large and incorporating inclusiveness as the main slogan for success.

As new world order changes, success in both societal and technological development perhaps significantly lies on understanding the appropriate trade-offs and compromises in capturing the totality for the benefit of any producer organization or the user of the product in a broader perspective.

ABOUT THE BOOK

This book, in fact, covers all aspects of sustainability in operations management, except the issues on social dimension. Its unique features are as follows:

- Readers will be exposed to various extremely opposing concepts, like qualitative (policy-related) versus quantitative (analytical models) approaches, concepts/theories versus industrial applications, and sustainable new product development versus remanufacturing of used products.
- The book has four distinct but related parts, each of which is meant for a specific interest area of a reader: concepts/theories/definitions (Part I), macro-level policy-making (Part II), strategies and planning at the operational level (Part III), and performance analysis with sustainability (Part IV).
- Not only is this book an appropriate source for exposure to sustainability in operations, but it may also be treated as a rich reference book for higher studies or research because it contains outcomes of extensive research studies like the house of sustainability, effectiveness indices on buyback decisions, and a lot of references to research publications. The research scholars and research-minded professionals will surely be benefitted by this book.
- The readers will be benefitted by a summary at the end of each section (sub-chapter) titled "Key Learning."
- Key concepts and definitions are highlighted in bubble boxes in bold font at convenient locations in each chapter, which are the readymade records of those definitions.

- There is no exclusive section on case studies. All the industrial examples and applications have been included along with the relevant text materials. This makes the flow smoother and reading simpler. Of course, some industrial examples exclusively on product recovery process are included in Sections 4.1.3 and 4.2.5.
- The ideas, methods and concepts discussed in all five chapters are somewhat interrelated. So there is every possibility of repetition of some terms or some concepts in more than one sections or sub-chapters. Please note that these are unavoidable occurences.

I shall be very happy if the academic and industrial communities are benefitted by this book and if some practitioners or researchers can use some of the concepts or ideas to implementation-worthy strategies or meaningful research projects.

About the Author

Kampan Mukherjee, after teaching for 35 years at the Indian Institute of Technology (IIT ISM) Dhanbad and 7 years at the Indian Institute of Management (IIM) Kashipur as Professor in Operations Management and Decision Sciences, joined Institute of Management Technology (IMT) Ghaziabad as Professor in Operations Management in 2021. He earned his PhD from Moscow State University of Economics, Statistics, and Informatics (formerly MESI) as an Indo-USSR government research scholar in 1988, and in 1998, he was invited for advanced research in LAMSADE, Paris Dauphine University, as Senior Visiting Fellow of the Government of France. He published research papers in several reputed journals. He is the recipient of Silver Medal and Best-Case Study Award from the Indian Institution of Industrial Engineering in 1992. He was a regular Visiting Professor in Otto-von-Guericke-University Magdeburg, Germany. He taught there in 2000–2001 on sabbatical and in 2004, 2007, 2009, 2011, and 2013 during summer sessions. As Visiting Professor, he also taught in Curtin Business School, Curtin University of Technology, Australia; Lappeenranta University of Technology, Finland; Vienna University of Economics and Business, Austria; ESDES Lyon Business School, Catholic University at Lyon, France; and several other universities in Europe and the UAE and also in XLRI, IIM Shillong, and so on in India. He was Dean in both IIT and IIM and established the Department of Management Studies in IIT Dhanbad in 1997. He was conferred the Life Time Achievement Award in Operations Management by the Society of Operations Management in 2012 jointly organized by IIT Delhi, IIM Lucknow, and Grand Valley State University of the USA.

Acknowledgment

It is my pleasure to acknowledge all the help and support extended by people for successful completion of this book writing project.

At the very outset, I am to acknowledge the sacrifice and tolerance of my wife during the days of the book writing, when hardly I could pay attention to the essentials of domesticity in the family.

The world of sustainability in industrial operations was introduced to me by my German friend and acclaimed researcher Dr. Karl Inderfurth (former Professor, Otto-von-Guericke-University Magdeburg) during my first visit as Visiting Professor in 2000–2001. Till 2013, I learned the whole domain of sustainability extensively, with special reference to product recovery management more intensely, during my several visits to Germany and to other European countries through academic and research interactions.

I am thankful to the administrators (Director and Deans) and colleagues in Indian Institute of Management Kashipur and Institute of Management Technology Ghaziabad for extending necessary support and encouragement.

I am happy to mention the technical help from my young colleague Dr. Kaustav Chakroborty of IMT Ghaziabad during this book writing project, particularly in the digitization stage.

It is my immense pleasure to share the credit of this book writing with all the well-wishers mentioned here and those whose names have been missed inadvertently.

Part I

Sustainability—An Introduction

1 Sustainability—An Inclusive Paradigm

1.1 UNDERSTANDING SUSTAINABILITY AND ITS CORE ISSUES

The task of introducing "sustainability" seems to be difficult enough, which perhaps becomes more difficult if I accept the challenge of making people perceive its implication, the pitfalls of "non-sustainability," and the meaning of living with a sustainable lifestyle. Interestingly, this difficulty persists in the organizational level, in the policy-making level, and also in the social community or NGO level. Perhaps, response to some fundamental queries may facilitate the process of understanding this phenomenon.

In what way do our habits, activities, and lifestyle affect nature? How do human habitats, habitats of other living beings, and nature coexist? Do our economic activities endanger the scope for our further development and the existence of the living environment around us? Does it jeopardize the existence or even the growth of our future generations? Why is understanding sustainability, its core elements, and its framework so important for ever-lasting existence of this beautiful planet?

In this pursuit, let me try to explain this complex idea with simple examples. At the outset I would prefer to explore the role of sustainability in our day-to-day life. The following discussions address some issues, which are expected to make us understand this phenomenon more simply and appropriately.

When we buy branded jeans, we usually focus on its fit, stitching, color, texture, and so on. Do we really show our concern regarding how the cotton was harvested or water contamination caused by toxic chemicals used in the textile industry? Are we aware of the fact that harvesting of pesticide-intensive cotton degrades the quality of the soil or even biodiversity? Do we know the working environment of the employees who stitched the clothes? If the answer is no in each of the questions, then perhaps we are not accepting (knowingly or unknowingly) the fact that our existence depends on the existence of other members of the society and the planet as a whole. Can the contaminated soil be used for production of agricultural products? Are you sure that these will not lead to sickness of individuals? Don't you agree that a bad work environment causes unhealthy future life of employees? Textile factories continuously emit air pollutants and water effluents to surroundings. What is their impact?

I just recall a small story, but as I cannot exactly remember its source, I cannot cite it here, but let me recognize and praise the story writer and its underlying meaning. It is a story of a school teacher who explains in a class of small children the development of mankind. In the prehistoric era, people in groups or clans used to live on fruits collected from trees and animals hunted locally. Once these resources are exhausted, they moved somewhere else in search of new sources for food, and thus,

DOI: 10.1201/9780429195600-2

they were nomads. Then, thanks to evolving technologies, human beings have created shelters, learned agriculture, and created facilities required for the satisfaction of their needs. All basic and other extra resources are now available around. Urbanization is growing. Now they need not travel for food or shelter or for the satisfaction of any other basic physiological needs like nomads did. The teacher explains this part of the history of mankind. But one of students stands up and raises an objection. She argues that in that case, we are still as nomads. Then, she explains the reason behind this hypothesis. She says that people are felling trees and making use of all natural resources in one place, and once they are exhausted, people look for some other location with plenty of natural resources. The same is true in construction activities in the process of urbanization. Once a location becomes overcrowded by civil construction or becomes a pure concrete jungle, we move to some empty space with an abundance of natural resources like trees and bushes. Now we occupy the new space and try our utmost to convert it to another concrete jungle with all the resources required to maintain our amenities. Thus, villages are converted to towns and towns to cities. Urbanization goes on unhindered. So the student is absolutely correct. Mankind could not develop itself, although there is no dearth of man-made inventions, which have apparently made our life easy and comfortable. But the weight and pressure of these inventions on this beautiful planet are growing day by day. Can we still restart our journey for the real development of this intelligent race? Can this urbanization be redefined and reengineered and implemented in a different way?

We cannot show indifference to the impact of our activities on the surroundings. So what we learn is the fact that the textile industry apparently may be confined to the factory, but it significantly affects the ecosystem and society at large. Purchasing a branded, well-designed pair of jeans is not only represented as a number in the accounts in the supply chain but also has an enormous impact on the environment where we live and the quality of life of the people working in the manufacturing of this piece of apparel.

The consequence of intense and unplanned urbanization leads to the exhaustion of natural resources and maintenance of the prehistoric nomadic lifestyle. This is not at all an intelligent way of developing mankind.

Let us take another example. When people do pant and sweat outside during summer, we enjoy the coolness and comfort in our sweet home or office with powerful air conditioners humming steadily. We are happy, and the AC manufacturer is happier. But what are the other consequences? ACs consume massive energy to function, and the source of that energy is most likely the burning of fossil fuels. Thermal power generation pollutes the atmosphere and also depletes the scarce mineral resources. The hot air pumped out of the room creates heat bubbles locally. Urban heat islands disrupt the climate, leading to abnormal rain showers, cloud formations, and so on. Very recent news mentions the interesting fact that Delhi, as an infamous urban heat island, represents a hole, being surrounded by foggy neighborhoods during January. The most crucial fact is that ACs use refrigerants, which puncture the ozone layer, causing diseases like skin cancers. We thus find the following three consequences because of cooling ourselves in comfortable enclosures:

- Depletion of non-renewable fossil fuels, which is already scarce and going to be scarcer in near future

- Major air pollution or carbon emissions because of thermal power plant operations
- Refrigerants of ACs, which emit pollutants that are likely to puncture the ozone layer, causing diseases like skin cancer

So our effort of keeping cool through the AC is actually making the earth hotter, and our comfort inside may lead to intensive discomfort outside in near future.

These and other similar other examples, like room-heating systems, electric ovens, roads, and air or water transports, contribute local or global environmental degradations in varied degrees. We may perhaps accuse industrialization for creating this nuisance. Industrialization is in fact both cause and effect of economic development. The process is essential and also inevitable, although its pace may fluctuate over time and space. Industries strive to grow, and this growth symbolizes economic development. What may be the consequences? An intelligent decision on selecting location of a power plant or steel manufacturing plant is primarily governed by various economic and socio-political parameters. First, this often results in deforestation. By photosynthesis, trees absorb carbon dioxide, which is the primary cause of global warming and climate change. Moreover, forests also moderate diurnal range of air temperature, maintain atmospheric humidity, and control rainfall. These are the benefits of the forests (most colorful component of nature), in addition to their key role as the primary source of food and other essentials for human life. Besides, establishing a manufacturing plant or power plant along with its other facilities in a non-urbanized location results in a lot of relocations and displacements of human habitats. These directly impact human livelihood and lifestyle. Affected families may experience discomfort and displeasure due to changing environments. These occur in addition to the generation of various types of pollutants and destruction of the landscape.

Industries are becoming more aware of this phenomenon and very often take steps opting for some forms of compensation—physically or monetarily. Interestingly, financial returns and implementation of control mechanisms on environmental degradation need not always be conflicting in nature. If energy consumption process is efficient, energy cost is reduced and simultaneously there is reduction of pollution due to lower consumption of energy (primarily generated from thermal or non-renewable resources). Redesigning of a product with less material content (or with more reusable materials) lowers the material cost and at the same time maintains the conservation of some natural resources used as materials for the finished products.

Further, can this pollution demon be tamed and used for our benefit? The generation of solid waste occupies space on earth and/or incineration of these wastes further pollutes the atmosphere. Waste management seems to be one of the important decision areas in managing urban life. This is true in managing manufacturing process of chemical plants, the steel and power sectors, in particular, which are infamous for being huge waste generators. Managing wastes involves collecting, storing, transferring, and reprocessing waste products.

The effectiveness of managing waste products primarily depends on how these are reprocessed or recycled and whether these can be converted to some usable products.

I have seen in Vienna such a plant that converts the biomass waste products of the city to electricity. This is also popular in some other big cities of the developed countries. Actually, municipal waste landfills are large sources of methane emissions in most of the populated cities of the world. When burned under controlled conditions, rather than letting it escape into the atmosphere, methane becomes a rich renewable energy source for generation of electricity, heat, or fuel.

Pollution damages health, creates disorder in the climate, and affects the organic cycles of plants and agricultural products, and it is the most undesirable element of global and local economic activities. Moreover, pollution is the generation of wastes, which degrade the environment. These wastes (in solid, liquid, or gas form) are undesirable but inevitable as byproducts of any process. Now issue is how to manage it. Practically there are four logical ways to tackle the problem:

1. Reduce the generation of wastes at the process stage
2. Control or arrest the outflow of wastes by creating barriers so that the harmful components do not contaminate the environment
3. Reduce the harmful effect of the wastes in the environment by injecting some agents, once the pollutants or wastes had already been mixed with the environment
4. Develop some mechanism for reusing the wastes, leading to the production of some beneficial products for the society

These strategies are applicable at various stages of waste generation and waste flow chain. We may use the following terms to represent these four possible strategies.

1. Reduce
2. Control outflows by filtration
3. Treat
4. Reuse

Strategy 1 focuses on process improvement and even redesign of product or use of substitute materials in the process, so that least wastes or pollutants may be generated. Strategy 2 includes use of devices for filtration at the outflow stream of pollutants, so as to restrict their discharge to environment. Implementation of various treatment mechanisms of the wastes already discharged is represented by strategy 3. Perhaps, one of the most interesting attempts is represented by strategy 4. This is the endeavor of converting negative elements to positive ones, a harmful impact to a beneficial one.

Typical examples are use of biomass or even municipal solid wastes in cities for generation of energy and beneficial uses of fly ash generated at thermal power stations as pollutants. Fly ash can be used for road construction, brick manufacturing, cement manufacturing, making support systems in underground mines, or even agricultural uses.

> *Pollution can be controlled by taking any or a combination of the four generic strategies—reduce, screen unwanted items from outflows, treat, and reuse.*

Now, let us try to summarize what we have learned from the previous discussions thus far:

1. Manufacturing of an apparel or activities in the textile industry at large is likely to affect environment and soil quality. Besides, a poor working condition in the textile sector surely adds to the negative effect and reflects the lack of our responsibility toward the society where we are living. This may even be true in other industries with various intensities.
2. Extensive use of some products like ACs makes the environment warmer (increasing the need for ACs in future, leading to a vicious cycle) and may cause significantly adverse impacts on the climate and the human body.
3. Serious consideration is required for maintenance of forests while intensifying the industrialization and urbanization for economic development of a country. Proper schemes of afforestation can only compensate loss of forest covers due to rapid growth of industrial activities. Moreover, care should be taken for societal disruption caused by these activities.
4. Reduction of material content in a product or control on generation of energy from non-renewable sources (e.g., coal, oil) would be a wise strategy in sustainable use of limited natural resources of the planet.
5. Unwanted waste products or pollutants may be further treated and reused for the benefit of the people instead of creating waste heaps or incineration.

All these five learnings point toward a single phenomenon—creation of a healthy, comfortable, and pleasant abode for human beings, which will continue to exist for a long period in future (hopefully permanently).

Here, we see the implication of the world sustainability. The meaning of sustainability per English dictionary is the ability to maintain something for some time at some level. However, sustainability is to be understood along with its essential components and framework in a little more comprehensive manner.

1.1.1 Proposed Paradigm of Sustainability

Let us first look at the key terms associated with the word "sustainability." The following two terms seem to explain the core meaning of sustainability.

- Continuity

    It means maintenance of something at its present state or trying to achieve and maintain some defined state. Of course, it may not mean any growth expectation, if not specifically mentioned as the achievement of sustainable growth.

- Futuristic

   Sustainability does not mean any achievement of proposed or desired outcome in the short term. It has an inbuilt focus on long-term and futuristic achievement.

Putting these words together, sustainability means the *creation of some capability or striving for the maintenance of some state of existence for years to come or, ideally, forever.*

If we look for comprehensive meaning of sustainability, the domain of considerations expands and its primary focus will be the continuity of existence of this planet and its inhabitants. This may be explained by a unique concept of eternal triangle of influences, proposed in this book. It is interactions among some human-influenced activities, societal activities, and natural activities, as shown in Figure 1.1.

Natural activities (activities of nature) are the activities involving the environment and living beings of the planet other than human race, or *Homo sapiens*. Environmental actors of this set of activities may be classified under three groups—nonliving entities (glaciers, water bodies, hills, climate, etc.), animals, and plants. The impact of these activities is reflected on climate change, various natural catastrophes, melting of glaciers, forest fires, and the like.

By societal activities, we mean initiatives, planning, policy-making, and administration of communities, formal groups, nations, or even global organizations like the United Nations. These activities lead to formulation of new plans, policies, or guidelines. Implementations, monitoring, and controlling are also included under this set of activities.

By man-made activities,[1] we mean all economic activities and consumption behavior of people. So this set of activities covers the exploitation of natural resources, conversion and related processes (production, transportation, etc.) for creating products, mode and intensity of product use, and other activities at the end of the useful life of products, including their disposal.

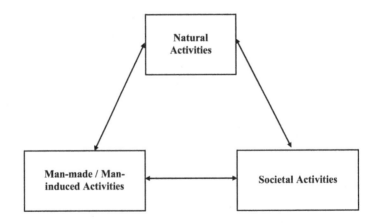

FIGURE 1.1 Eternal triangle of influences.

Each set of activities of Figure 1.1 influences and is influenced by other two sets of activities. National policy relating to economic growth, urbanization, or industrializa- tion may motivate people to create new business ventures and triggers the necessary value-additive processes. For example, the current Make in India policy surely motivates people to invest in startups, particularly in manufacturing sector. Societal activities (like national policy) thus trigger man-made activities. It may, on the other hand, lead to excessive loss of forest covers in the country, causing more air pollution and low rainfall in several places (the impact on nature). A conscious community may put pressure on the government to formulate new guidelines, ordinances, or restrictions on environment management. This shows how resultant natural activities make society active in taking corrective measures. Subsequently, man-made activities (business-related activities) of some sectors like power, chemical, and textile may be financially affected because of these restrictions. However, this may ultimately improve the condition of nature and thus reduce harmful effects like climate change and other natural catastrophes. Efforts on reforestation help regain some loss of forest covers and thus contribute positively to the environment or climate. On the other hand, a conscious society is expected to create customers who may demand environment-friendly products and processes. As a logical consequence, factories will be compelled to orient their managerial activities accord- ingly. The existence of an environment-conscious society may even enable creating new sectors of economy enriching man-made activities. For example, sectors producing pol- lution controllers like electrostatic precipitators or cyclone separators are the new addi- tions to national economies. Consultancy services on environmental impact assessment, environmental management planning, carbon footprint analysis, life cycle analysis, and so on have become popular sectors now-a-days in service industry. This description shows how each set of activities influences others based on the Figure 1.1.

Thus, the concept of sustainability originates from the holistic perception of inter-influences among these three sets of activities. Interactions among these sets of activities may be termed as the eternal triangle of influences (ETI). Dynamism of this ETI determines the healthy existence of the human race, society, and nature, which is expected to continue forever. This expectation can be realized only if this dynamism is a positive contributor to societal development, ecological balance, and favorable existence of living and nonliving beings in nature. The appropriate func- tioning of ETI thus contributes to sustainability of the human race, society (essential for the existence of mankind as a group or as groups), and nature (complete ecosys- tem covering nonliving and living beings).

> *The eternal triangle of influences (ETI) depicts the mutual interactions of influences among man-made activities, societal activities, and natural activities, the dynamism of which is essential for the existence of mankind and nature.*

Man-made or man-induced activities encompass the set of activities striving for the creation, maintenance, and use of physical products meant for the survival of the human race in this planet. In a value chain framework, we may conclude that these originate from the extraction of natural resources and end at the disposal of materials (after their use). The whole value chain is created, managed, maintained, and controlled by human beings. So we are to take the responsibility of any resulting consequence because of the activation of this value chain. Figure 1.2 shows the generic value chain of human activities—extraction of natural resources, production, movement or transportation, use, and ultimately disposal. Both society and nature affect and get affected by the chain of the man-made activities in different forms. For example, society supplies resources to humans, and all economic activities are guided and controlled by societal norms, governmental guidelines, and legislative restrictions. Customers, who are the consumers of the end products, are also part of society. Nature not only supplies natural resources; it also takes in disposed materials. The balance of the ecosystem may be perturbed by various activities of human beings. These *anthropogenic* effects trigger dysfunctions and malfunctions in nature, which may further affect society. Society may create new guidelines, norms, and mechanisms for better monitoring and control. Activities in the ecological system are primarily reactive to the impacts of human activities. Society functions both reactively and proactively. Nature tries to maintain an order and balance. It only reacts when the balance is disturbed by human activities, and the results are global warming, natural disasters, and change in seasonal cycles. On the other hand, the establishment and maintenance of a stable society demands the creation of appropriate frameworks, guidelines, and regulations. Otherwise, it would be chaotic, leading to imbalance of power, conflict of interest, crimes, poverty, war, and so on. Societal guidelines often direct human activities proactively for the creation of nice living environments and the maintenance of nature. However, any disruptive event caused by human activities (also reflected on nature) may be handled by society reactively by designing a new guideline or formulating legislative measures.

A generic value chain is proposed in this context as extension to ETI and pictorially depicted in Figure 1.2. The extraction of natural resources, their conversion to usable products, and their transportation for delivery to customers are the business activities in the generic value chain. Each of the activities is having its implication on the components of ETI. All the activities under I, II, III, IV, and V are likely to

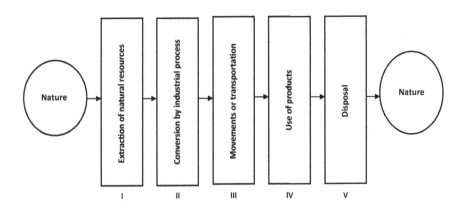

FIGURE 1.2 Generic value chain of economic activities.

generate wastes, in the form of pollutants in various scales. The processes involved in these activities are guided by societal and governmental legislative norms and regulations.

The activities shown as IV relate to use of products by intermediate or end users. Intermediate users (in case of industrial products) may reprocess the products, mostly applicable for industrial products for further value additions. In that case, activities of types II and III may be repeated. However, end users intend to dispose products at the end of their useful life or when their perceived value expires. At the dump yard, the life of a product usually ends by incineration, landfilling, and use of garbage in road construction. However, in some other cases, the products are recycled or remanufactured by extracting their reusable parts. That means once again the activities of type II and III will be repeated. Incidentally, the use of primary products (type IV) also emits pollutants at the end of their use. If disposals lead to incineration, there is air pollution. In case of landfilling, there is chance of soil degradation, depending on the biodegradability and toxic content of the products. So all the activities in types I, II, III, IV, and V affect the environment, and type I activities deplete the limited natural resources of the earth. However, all these are influenced by societal guidelines and laws. So although the dynamism of our existence along with the planet is an outcome of interactions of the influences in ETI, society and institutions have the primary responsibility of restricting the negative impact of the man-made activities.

> *A generic value chain may be created depicting man-made activities, which originates in nature and ends in nature. The generic set of activities under the chain cover five sets of activities, each of which affects the environment and nature.*

Key Learning

- Although industrialization reflects economic development, management of the eventual generation of unwanted wastes is crucial for maintaining a healthy environment for the present and future generations.
- Trees and forests are not only the prime sources of food and essential resources for our day-to-day living; trees protect us from environmental pollution and absorb unwanted carbon content of air pollutants. Unplanned urbanization often causes deforestation and danger to humans.
- Sustainability means long-term, healthy, and comfortable existence of human beings in this planet. This essentially demands both industrial development and non-conflicting coexistence of humans with animals and living beings. Meaningful compromises may be sought for the maintenance of the balance of the ecological system.
- Sustainability thus refers to long-term existence of nature and ecological system along with their influencers and influences. The synergy may be explained by interactions of activities in nature and by human beings and society, depicted as the eternal triangle of influences.

1.2 CONCEPTS, VIEWPOINTS, AND MODELS OF SUSTAINABILITY

Since the "Adam and Eve" days, when nature got the first footprint of mankind, the human habitat has been maintaining an irrefutable bonding with nature functionally, technically, and emotionally. Prehistoric human existence reflects the humanity's dependence on nature or, rather, interdependence between earth (along with its relevant content) and *Homo sapiens*. Nature, as a huge entity, covers the physical environment (oceans, mountains, deserts, etc.), plant life (trees, shrubs, etc.), and other living beings (animals, marine life, microorganisms, etc.). Activities of nature is reflected by seasonal cycles, rainfalls, snowfalls, and also natural calamities, like floods, storms, and landslides. Functionally, this connection can be explained by thousands of human activities, such as agriculture and the exploitation of natural resources for the existence and growth of mankind. Prehistoric evidence also supports the fact that human beings can exist only by forming into groups and societies. This group formation and group dynamics are outcomes of both the natural processes and purposeful societal process. So the interaction among humans, nature, and society is a naturally driven phenomenon, as depicted in the eternal triangle of influences.

Human beings, right from the beginning, accepted the fact that their livelihood depends on the contribution of nature in various forms. However, population and thus consumption went on increasing, leading to exploitation, and thus generates pressure on the limited natural resources (particularly, nonrenewable ones) of the earth. Moreover, the consumption of resources is associated with waste generation and disposal. The process of resource consumption and waste generation becomes intense with speedy technology development, which results in high rate of product obsolescence. On the other hand, free and competitive market economy, perhaps, could not pay all necessary attention toward the reduction of global poverty and malnutrition. A balance of economic growth and equitable need-based consumption is yet to be achieved.

In last few decades, sustainable development has become a popular area of focus, rather a buzzword in both academic and business worlds. Scientific journals and conferences usually have sustainability as a common theme of discussion. Relevant articles on sustainability or sustainable development, environmental management, product recovery management, and the like are being added for the enrichment of course curricula and teaching materials. Boardrooms of corporations and public service organizations are abuzz with related discussions, and most of the national and international forums have deliberations and debates (often intense ones) on various issues of sustainability.

But the realm of sustainability, as we perceive it today, has emerged from a long evolution. Scores of hypotheses, theories, concepts, and viewpoints have been proposed, argued, and established over time. Proper understanding of sustainability thus requires revisiting the history and evolution of its current understanding.

The most common mode of depicting the milestones is to follow the timelines. However, the origin of the modern understanding emerges from the debate on whether or not we are going to face the *doomsday*. On the one hand, there was the initial belief on complete exhaustion of resources because of population explosion; on the other hand, subsequent skepticism emerged, which is supported by technology development and exploitation of newly discovered resources.

Although people had been aware of impact of human activities on the environment, the debates continue, time and again, on whether high economic growth of the few affects the living conditions of the many. Sustainability took its formal shape much later. The global community created a unified phenomenon called sustainability only at the later part of last century.

1.2.1 HISTORY AND EVOLUTION OF CONCEPTS AND VIEWPOINTS

Desta Mebratu (1998) and Jacobus A. Du Pisani (2006) explored the history and evolution of sustainable development concepts meaningfully with the critical study of literature. Here, these two excellent reports are referred extensively, and a modified form of classification is proposed on views and ideas leading to emergence of current concept of sustainability. Thus, a set of *five perspectives* is proposed, which are expected to cover the historical viewpoints and concepts. Figure 1.3 depicts how contemporary sustainability evolved from its philosophical perspective in prehistoric era.

1.2.1.1 Perspective I: Launching Pad for the Growth of Civilization— Prehistory and History till Industrial Revolution

Since time immemorial, human beings had been caring, fearing, and perhaps loving the three primary components of nature. These are biotic factors (all living organisms—animals, microorganisms, plants, etc.), abiotic factors (all nonliving entities of nature—mountains, rivers, the air, oceans, etc.), and regular and irregular (because of imbalance of interactions among other components and human activities) climatic factors (seasonal cycles, rainfall, floods, storms, etc.).

Several historians, sociologists, and philosophers studied the prehistoric stage of human history and concluded the initial trend of human civilization in various ways. Reports by Mebratu (1998) and Pisani (2006) contain detailed discussion and relevant sources on this topic.

Two primary directions emerged from this trend in prehistoric days, which enable the creation of foundation for the contemporary framework of sustainable development. These are *harmony with nature* and the process of *industrial development*.

Harmony with nature is the understanding and belief that harmony needs to be maintained for meaningful existence of human beings.

Interestingly, this belief existed in prehistoric days as religious beliefs, cultural norms, and philosophy of life in different corners of the globe.

Let us start with Indian philosophical notions written as Vedic scriptures sometime during 1500 BC to 500 BC or even before that. All the four Vedas—the *Rig Veda*, *Sama Veda*, *Yajur Veda*, and *Atharva Veda*—are supposed to be the oldest books of knowledge and philosophy. These are expected to contain the viewpoints of the early Indus Valley civilization and Hinduism. These books and other related scriptures emphasize the fact that human activities and philosophy of life are not independent of nature and celestial bodies. Nature is to be treated well, adored, and even worshipped. The principle of replenishment has been explained in the form of harmony with earth in *Rig Veda* by the statement *"You give me and I give you"*

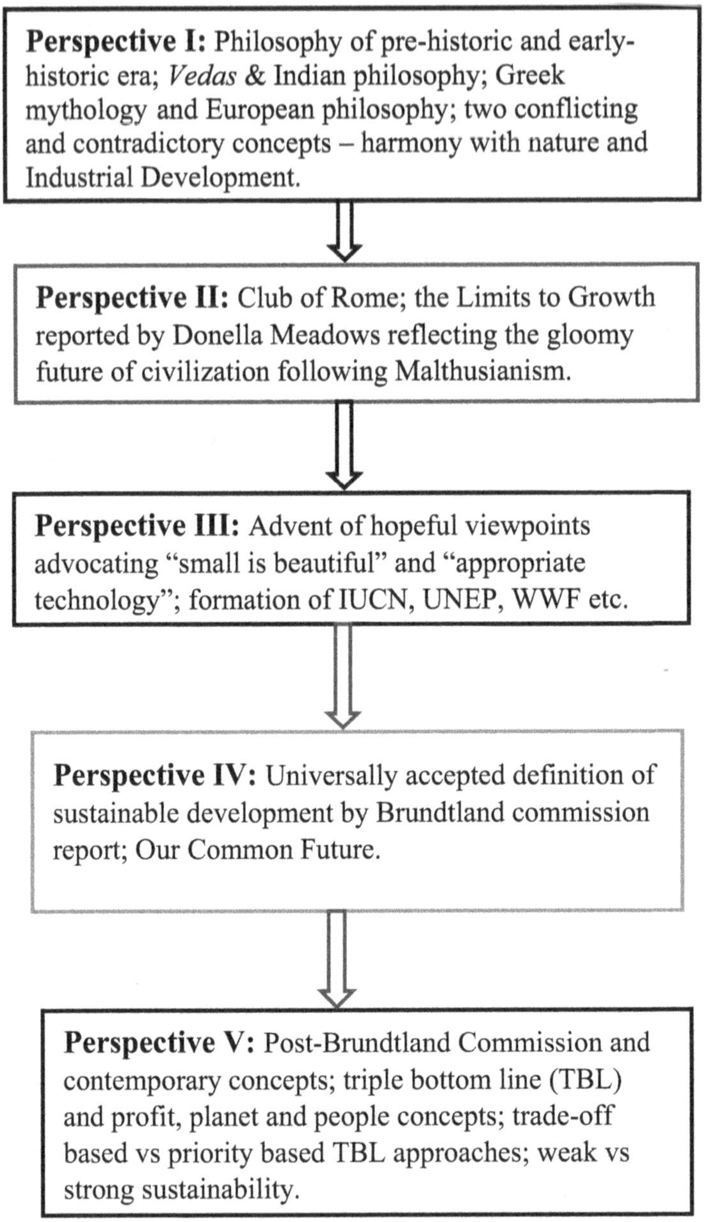

Perspective I: Philosophy of pre-historic and early-historic era; *Vedas* & Indian philosophy; Greek mythology and European philosophy; two conflicting and contradictory concepts – harmony with nature and Industrial Development.

Perspective II: Club of Rome; the Limits to Growth reported by Donella Meadows reflecting the gloomy future of civilization following Malthusianism.

Perspective III: Advent of hopeful viewpoints advocating "small is beautiful" and "appropriate technology"; formation of IUCN, UNEP, WWF etc.

Perspective IV: Universally accepted definition of sustainable development by Brundtland commission report; Our Common Future.

Perspective V: Post-Brundtland Commission and contemporary concepts; triple bottom line (TBL) and profit, planet and people concepts; trade-off based vs priority based TBL approaches; weak vs strong sustainability.

FIGURE 1.3 Evolution of perspectives, concepts, and viewpoints on sustainability.

as written by Nimisha Sarma (2015). It is further hypothesized that life is primarily controlled by five elements of nature, or *pancha mahabhoota*—namely, *Akasa* (sky), *Vayu* (air), *Tezas* (fire), *Apaha* (water), and *Prithvi* (earth). All these are to be worshipped and properly maintained; otherwise, seasonal cycles, rainfall pattern, agricultural productivity, and so on are likely to be badly affected.

Renu Tanwar (2016) also cited quotes from Vedas, like *"Do not cut trees, because they remove pollution"* (Rig Veda, 6:48:17) and *"Do not disturb the sky and do not pollute the atmosphere"* (Yojur Veda, 5:43). Moreover, elements of nature are also worshipped as deities, namely *Surya* (sun), *Agni* (fire), *Pawana* (air), and the like. Worshipping of nature was also prevalent among our great ancestors in continents like Asia, Africa, and Australia, deities being the trees, mountains, rivers, and the like. Most often, this worshipping used to be because of fear.

Similar perception prevails in Greek mythology as well. *Gaia* is Mother Earth, the goddess responsible for maintaining harmony, wholeness, and balance within the environment. *Demeter* (by Greek mythology), also known as *Ceres* by Romans, was one of the prominent deities looking after harvesting and grains, very similar to the god of seasons and climate. In almost all civilizations, "environmental problems, such as deforestation and salinization and loss of fertility of soil occurred, which we would today refer to as sustainability problem" (Pisani, 2006). In Hawaiian tradition, nature is considered as a living entity. So like another living being, nature needs proper care for its existence. "For African tradition, man is not master in the universe; he is only the center, the friend, the beneficiary, the user" (Mebratu, 1998). Germans had shown their concerns and worries, even before 17th century, for the negative impacts of excessive deforesting and mining activities, particularly on surrounding wildlife. Thus, our forefathers popularized and implemented the essential concept of living, which is **maintenance of harmony with nature**.

The second concept that emerged at that time is the role and essence of *industrial development* in the process of civilization and improvement of living standard of mankind. Initially human beings were nomadic and continued their mobility, searching for better hunting grounds and sources of food and other essentials. Slowly they domesticated animals and cultivated plants. Farming and agriculture became primary activities for sustenance. Essentials for living started getting traded among them, crossing the geographical boundaries. Concepts like money, trade, business, and power became popular in society.

Forests used to be the primary source of foods, fuel, and other living necessities (house building, furniture making, wheel making, etc.). With expansion of human population, consumption of these resources went up manifold. Scarcity of natural resources was apprehended. At that time, Britain took the lead in initiating *Industrial Revolution*, which took place sometime during 1760 to 1840. People started using iron and steel as raw materials for new industrial models. Coal became the key fuel. The contributions of the Industrial Revolution significantly enriched the industrial sectors through the entries of new technologies like, steam engine, IC engine, electricity, and power loom, as efficient replacement of human power by machines. Machine-operated activities improved the productivity, and people started mass production and exploiting economies of scale. People could understand the concept of factory system and its elements—namely, the division of labor and specialization of functions. Standard of living began to improve consistently for the first time in history. This resulted in a remarkable upsurge of consumption. History witnessed new world order with huge depletion of natural resources and degradation of environment because of intensified mining activities, electricity generation, use of automobiles, steel manufacturing, and so on. Economy took an upward trend of growth, and simultaneously perhaps it was the beginning of inequity in the pattern of consumption, deepening the fragmentation of society into various economic classes across the globe. Therefore, although the Industrial Revolution positively contributed to the economic development of mankind, it has its negative impact on sustainability. **Practically, industrial development does not support living in harmony with nature.**

Out of the five perspectives, which are supposed to explain various concepts derived from understanding and activities of mankind in prehistoric and early historic era, perspective I considers two critically conflicting concepts— living in harmony with nature and Industrial Revolution.

1.2.1.2 Perspective II: Alarm Bells for Doomsday—Limits to Growth and Exploitation of Resources

The sound of alarm bells was first heard way back in 1798, with manifestation of its indication described by an English political economist Thomas Robert Malthus in his book *An Essay on the Principle of Population*. The picture was quite gloomy, perhaps reflecting the detrimental effect of Industrial Revolution. "Population, when unchecked, increases in geometric ratio and subsistence for man in an arithmetic ratio" (Rogers et al., 2008). This was an outcome of a study based on concept of classical economics—population growth (i.e., consumption) leading to limiting the supply of agricultural land and resulting in diminishing returns to agricultural production. So the scarcity of food supply and natural resources is inevitable. The same proposition is also applicable for exhaustion of coal as the primary source of non-renewable energy resources, argued by Pisani (2006).

An Italian businessman, Aurelio Peccei in 1968, formed the *Club of Rome*. It conducted its first study involving eminent scientists and economists from MIT to predict future possible living conditions of human beings, considering resources and other economic issues. The outcome of this computer-simulation-based study reinforced the gloomy Malthusianism, and it was reported as a popular book *The Limits to Growth*, written by the team leader Donella Meadows in 1972. It indicates that if the growth rate of industrialization and population continues unabated, then the limits will surely be reached within next 100 years. Consequently, this limit will trigger a mechanism of decline. Many Malthusian supporters like Lester Brown became quite anxious and vocal about conservation of natural resources and the creation of global policies on industrialization and environmental issues. Institutes like the *World Watch Institute* and *Earth Policy Institute* were established in late or mid-1970s.

The apocalyptic conclusion of Malthusianism and the report of the Club of Rome infused a sense of alertness because of limitless growth and industrialization. This represents a perspective of warning and a need for restructuring the process of economic development for existence of mankind in this planet.

Perspective II covers the Malthusian proposition of a doomsday in near future, which was reinforced by the model-based testing of the Club of Rome. Subsequently, the **Limits to Growth** *theory of Donella Meadows pointed out the ringing of alarm bells and the need for restricting the growth of consumption and the necessity for conservation of natural resources.*

1.2.1.3 Perspective III: Emergence of Rays of Hope—Propositions for the Avoidance of Impending Dooms

The critics of *The Limits to Growth* soon could bring out the shortcomings of its theory, which is primarily the fixed production curve as the basis for the study. The study also failed to consider the future possibilities of creating or exploring new or alternative resources. Further, the technological innovations could enhance productivity or pollution control more intensively than what was perceived during the report of the Club of Rome.

In 1973 British economist E.F. Schumacher published a book titled *Small Is Beautiful: A Study of Economics as if People Mattered*, a collection of essays. A new concept emerged, termed as appropriate technology, which concerns rapid depletion of natural resources and the corresponding destruction of environment. This technology, perhaps, is meant for its need-based use opposing the existing perception of "bigger is better."

In different forums people started using the word "sustainability" during 1970s, although little loosely, as till then there was no standardized and universally accepted definition. Earth Day rallies had been witnessed in USA. Some global environmental organizations were established to fight against environmental destructions worldwide like *Friends of the Earth (1971)* and *Greenpeace (1972)*.

The international conference of *United Nations on Human Environment* at Stockholm was held in 1972, with the participation of 113 states, which actually addressed the inherent conflict between environmental and economic priorities. The *United Nations Environmental Program (UNEP)* was created to facilitate partnerships among nations for the better care and maintenance of environmental parameters and improvement of the quality of human life. The Stockholm Declaration, however, primarily focused on trade-off issues. This seemed to be a meaningful step for the formulation of a complete concept and definition of sustainable development. The UNEP review created the new term, "eco-development," in 1978. Conservation of flora and fauna emerged as another meaningful concept, which was the outcome of strategy formulation by the *International Union for the Conservation of Nature (IUCN)*, along with UNEP and World Wildlife Fund (WWF, also known as World Wide Fund for Nature).

This perspective is the background for creating the unified definition of sustainable development, which encompasses the understanding of limitation of growth, development without environmental destruction and with conservation of natural resources, and need for combining economic and environmental priorities.

Perspective III refers to the viewpoints of sensing some rays of hope advocating the concept of "small is beautiful" in place of "bigger is better" and the concept of appropriate technology. The popularity of green consciousness can be seen by the formation of global institutions and organizations like IUCN, UNEP, and WWF.

1.2.1.4 Perspective IV: The Ultimate Definition of "Sustainable Development"—Report of the Brundtland Commission

At this stage, people could already understand that economic and technological development is either inevitable or necessary, and at the same time, development most often degrades nature and badly affect the surroundings. So there is the inherent conflict. The UN General Assembly took a serious step for integrating environmental, societal, and economic development issues by forming a group of around 20 leaders and experts from all over the world in early 1980s. This *World Commission on Environment and Development (WCED)* was constituted under the chairpersonship of Ms. Gro Harlem Brundtland, the Norwegian prime minister. In 1987, the outcome of several deliberations of the commission was published as *Our Common Future*, a landmark report, popularly known as the *Brundtland Report*. A meaningful concept and definition of **sustainable development** was born. It is defined as a process of *development that meets the needs of the present without compromising the ability of future generations to meet their own needs* (WCED, 1987).

This definition actually leads to creation of three derivative concepts required for formulation of unit-level, national-level, and global development strategies.

1. *Needs*—This is particularly applicable for understanding the essential needs of the world's economically poor community, which demands overriding priority. This calls for creating any mechanism for intra-generational distribution of resources judiciously and logically keeping the primary focus on human values.
2. *Limitations*—It involves limitations on current state of technology, natural resources (particularly the non-renewable ones), and also present and future consumption patterns. Perhaps, its detailed analysis is mandatory to meet present and future needs simultaneously.
3. *Intergenerational need satisfaction*—Appropriate consumption and use of resources is to be assessed between generations while maintaining their required development.

The report of the Brundtland Commission is, in fact, a sign of hope and, to some extent, positivism. However, the implementation of all the three concepts is really a herculean task for the world leaders. We are yet to achieve them in reality. Some critics opine that Mrs. Brundtland intended to communicate a slogan of idealism and that is why it was deliberately formulated as a definition of some sort of vagueness. Others commented that this definition could unite the "first world politicians with green electorates to appease and third world politicians with economic deprivation to tackle" (Benton, 1994).

Despite the criticisms and some lack of clarity, the definition proposed by the Brundtland Report, along with the related concepts, seems to be the most accepted definition and understanding of sustainable development. Experts and practitioners have already accepted it as the only possible standard definition of sustainable development. This seems to be the most prominent milestone in achieving a synergy between sustainability and development process of mankind.

> *Under Perspective IV, we include the standard definition of sustainable development, based on the Brundtland Report of 1987, which explains this as a process of development that meets the needs of the present without compromising the ability of future generations to meet their own needs.*

1.2.2 CONTEMPORARY CONCEPTS AND THEORIES ON SUSTAINABILITY DURING THE POST-BRUNDTLAND COMMISSION PERIOD

This section includes the discussion on modern theories depicting the possibilities of achieving sustainability and bringing in the harmony among economic or business endeavors, nature, and the community or society at large. This is presented as Perspective V of the various stages of understanding sustainability.

1.2.2.1 Perspective V: Post-Brundtland Commission Era— Contemporary Concepts and Viewpoints

While accepting the inherent conflict between economic growth and environmental protection, the Brundtland Report expressed the possibility of economic development without environmental degradation and societal deterioration. Thus, the needs of social equity, economic growth, and environmental maintenance are to be met simultaneously through an integrated approach.

Logically, three fundamental components emerged from the sustainable development definition—environment, economy, and society, which subsequently became formalized as the *triple bottom line (TBL)* concept (Elkington, 1997). TBL was first coined in 1994 by John Elkington, founder of the British consultancy firm Sustain-Ability. This is also popularly known as the 3Ps concept—*profit*, *planet*, and *people*. However, issues still remained in strategizing the three components both at organizational and global levels.

As the purpose, role, set of activities for goal achievement, and stakeholders of the three bottom lines are complexly different, it is really a difficult task to integrate them under a single sustainability-targeted strategy. However, achievement of TBL-based sustainable development may be possibly explained by following two viewpoints per published reports.

1.2.2.1.1 Trade-Off–Focused TBL

This viewpoint proposes simultaneous consideration of all the three components of sustainability. The conflict in simultaneous achievements of the three components is managed by the acceptable trade-offs. This is most popularly represented as a Venn diagram

(Rosen and Kishaway, 2012). Sustainable development is achieved, if the development process includes caring for the economic (financial) viability, environmental stewardship or concern, and the social responsibility. Figure 1.4 shows the common areas covering the three domains of concern. Although sustainable development (SD) is intersection of all the components, the intersection of any two components also bears some meaning.

1.2.2.1.1.1 Zone A: Bearable Process of Development It implies the formulation of national/global/societal policies for protection of environment and natural resources in activities of society, perhaps emphasizing more on consumption and disposal. It includes, for example, restriction on pollution or waste generation and water consumptions regulation. It may also include lifestyle changes of people in society, like carpooling and reducing plastic consumption. Naturally both environment and society components are taken care of.

1.2.2.1.1.2 Zone B: Viable Development Process Here, corporate viability implies focus on both environmental and economic issues. Policies relating to subsidies or incentives, legislative requirements, payment of taxes or penalties, environmental stewardships, and the like are included in this zone, which shows monetary assessment of credit or discredit on environmental concerns in businesses. At the strategic level, it may include inclusion of sustainability as strategic goal, Euro Certification (Bharat Series in India) in automobile designs, and the like. At the operational level, it also covers implementation of energy-efficient systems, lean manufacturing, Six Sigma, and the like, which essentially address wastage reduction and resource conservation.

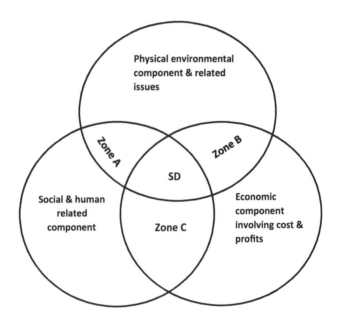

FIGURE 1.4 Trade-off–based TBL.

1.2.2.1.1.3 Zone C: Social-Equity-Based Development It includes corporate social responsibility (CSR) schemes, consideration of business ethics, worker participation, and gender equity in business policies and strategies. It may also cover all philanthropic activities, contributions to healthcare services, sports activities, and the like.

The main challenge a policy-maker faces while implementing TBL is the existence of conflict among the three components. Dis-incentivizing investment for a thermal power plant is of course beneficial in terms of achievement of environmental sustainability, but it costs a lot on the national economy, particularly for developing countries. It may remain true so long the technology is not matured enough for economically exploiting the renewable energy sources. This tri-criteria strategic decision is to be taken based on trade-offs among the three components and their acceptable tolerance limits. This compensation or trade-off function(s) is (are) to be developed for a specific business process and the type of resources. For example, an organization needs to estimate the profit reduction for investment on reduction of every tonne of CO_2e, or carbon dioxide emission, or for providing vocational training of an unemployed person in the society. On the other hand, the management of the organization is also to decide on the acceptable or tolerance limit of such profit reduction. So both the trade-off and the tolerance limit of acceptable loss are important while implementing TBL under the trade-off–focused viewpoint.

1.2.2.1.2 Priority-Focused TBL

This viewpoint is more understandable, conceptually less complex, and devoid of issues like compensation or trade-offs. This concept proposes pre-emptive prioritization. In other words, it means that we prioritize the importance of the three dimensions or components of sustainability. Once the need of the topmost dimension is fully satisfied, *only then* will the need of next-level dimension be considered. And this continues till we reach the bottommost dimension of sustainability. It is neither a case of simultaneous consideration of dimensions nor that of any trade-off between two dimensions.

Any meaningful activity by human beings or the continued existence of society depends on existence of this planet and the health of the nature and the physical environment. So the environment (i.e., the planet) bears the top priority in any sustainability-focused activity. Once environmental concerns and protection get utmost attention, focus should be directed toward maintaining society with excellent living environments and quality of life. Thus, society or people are to be considered in the next level of prioritization. Being the key stakeholder, society is the supplier of human resources and the users (customers) of the output of any business or economic activity. Per this viewpoint, the focus on economic or financial return is given the least importance. It means that once a corporation extends the maximum possible support for environmental protection and social stewardship, it is now eligible to orient its strategy or planning to profit-making. Figure 1.5 depicts this interdependence among the three components or dimensions in priority-based TBL.

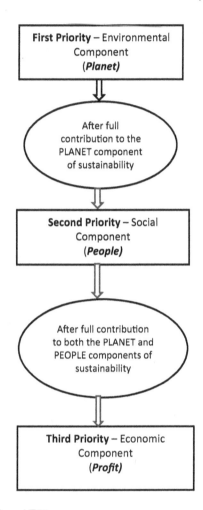

FIGURE 1.5   Priority-based TBL.

Is it so easy to convince any profit-making organization that profit-making should be of secondary or even tertiary interest to the organization, if it becomes sustainability-conscious? Are the nations ready to primarily attach more importance in caring for the environment than in GDP growth? On the other hand, we also put our argument supporting this TBL that nothing can exist without the well-beings of the earth and the natural environment. If Mother Earth does not exist in healthy condition, where will the existence of profit-making business organizations be? Also, we may think of the disaster and loss in managing a business due to the impact of a hostile, crippled, and antagonistic community or society. However, although its practicability at the organizational level is questionable, this viewpoint is quite useful at the national and international policy levels. The viewpoint may be depicted as Russian nesting dolls, or *Matryuska*, similar to cosmic interdependence model (Mebratu, 1998).

> *The realization of TBL may be made following any of the two approaches—trade-off–based or priority-based approach. The former proposes simultaneous consideration of the three components with sacrifice of one for achievement of some other component optimally. However, the latter considers them separately, and the consideration of components will be done with pre-emptive prioritizations.*

1.2.2.1.3 Weak and Strong Sustainability Paradigms

Weak and strong sustainability seems to be the two possible mutually exclusive paradigms, any of which may be considered as the base or direction for achieving sustainability. However, each of these paradigms has the supporting and opposing arguments. The debate between the trade-off– and priority-focused viewpoints of sustainable development is somewhat similar to the conflicts between these two paradigms of sustainability.

These two paradigms emanate from the concept of *capital*, which may be defined as the stock (asset) that provides current and future utility (Neumayer, 2003). Natural capital is the system representing the totality of nature, which comprises evolving and continuously interacting biotic and abiotic elements. Performance of human functions and services entirely depend on the complete ecosystem capacity. On the other hand, human beings, because of necessity, convenience, and amenities, create man-made capital. Man-made capital is the result of manufacturing processes, like machineries, factories, or other facilities. Conversion or value-additive processes by this capital create consumable goods or services. Physical objects like electronic goods or buildings used in day-to-day life are also included under man-made capital. Incidentally the man-made capital is built and developed utilizing the natural capital. The conflict emerges in accepting the justification of destroying one type of capital for the generation of the other.

Weak sustainability is the acceptance of possible *substitution* of natural capital by manufactured capital so long the total value of aggregated capital remains unchanged. It is also based on the assumption that there does not exist any essential difference in roles of these two types of capital. "It does not matter whether the current generation uses up non-renewable resources or dumps CO_2 in the atmosphere as long as enough machineries, roads and parts are built in compensation" (Neumayer, 2003).

However, a manufactured product may be reproduced if it is destroyed. But once a glacier is melted, it cannot be remade, and a changed climate cannot be reversed, and so does the extinction of species. Although reforestation can somewhat compensate for deforestation, it is one of some rare reversible cases. However, some lead times in this process of reforestation will be the loss along with some loss in terms of quantity of green cover. Moreover, the role of the natural capital is multidimensional in terms of its service to human beings—it is a source of security, health, food, and basic materials. Further, we do not have the right to ask the future generation to breathe polluted air and face water crisis in exchange for creating huge capacity for production of goods and services.

Strong sustainability, in contrast, rejects the assumption that substitution of one type of capital by other is possible. It emphasizes the fact that it is essential to maintain proper ecological functioning of natural system for survival of mankind. Profit potential in no way can compensate environmental degradation and exhaustion of non-renewable natural resources, neither the societal sufferings.

Nevertheless, a middle path is more practicable, as profitability and national income are the key elements for sustainability of a corporation and a nation, respectively. It focuses on the identification of some core elements or critical components of environment or a society. For example, average temperature of earth's surface or essential coverage of forestry on the globe needs to be maintained at any cost for existence of species. Of course, the exact assessment of these critical elements and their threshold value is both crucial and difficult.

The two viewpoints of capturing the interdependencies of three components of sustainability in TBL emanate from these two concepts of sustainability. Weak sustainability accepts the compensation of loss of one component by gaining another, and thus, the retention of sustainability and development may be managed by feasible trade-offs. On the other hand, strong sustainability considers pre-emptive prioritization of the environment over economic returns. Most of the policies, however, are formulated for meeting the target threshold value of critical elements.

> *Weak sustainability accepts capital substitution so long as the total capital in the earth remains unchanged. So the loss of natural capital is permissible if the new man-made capital compensates it. Strong sustainability does not allow any destruction of natural capital, even for the sake of its conversion or substitution by other type of capital.*

1.2.3 INCLUSIVENESS FOR SUSTAINABILITY

Policy-making and management of sustainability are expected to address the following questions.

- What are the impacts of each economic activity (exploitation of natural resources, conversion processes, transportation, consumption, and disposal) on the surrounding environment and society? **Assessment orientation**
- How do we minimize, if not eliminate, the negative impacts? **Plan orientation**
- How do we design operations strategies and plans that judiciously balance the achievement of 3Ps? **Strategy orientation**
- How do we motivate an organization, community, or nation to implement sustainability-friendly policies and practices? **Policy orientation**
- How do we review, monitor, and control the environmental degradation and depletion of scarce resources of the earth? **Control orientation**

These issues may be addressed during *macro-level* and *micro-level* management efforts. Macro-level policies, strategies, and plans reflect the decisions at the global or national level, whereas micro-level ones are the management efforts at the organizational or supply chain level. Let us propose a phenomenon for easy elaboration of inclusiveness for achieving sustainability.

1.2.3.1 "WE" Phenomenon

Let us propose a simple "WE" phenomenon, which emphasizes on *inclusiveness* of primary stakeholders for managing the businesses to achieve sustainable development.

The proposed paradigm is the expansion of perception on inclusiveness of processes, actors, and key stakeholders, which may be explained by three phases (as shown in Figure 1.6) describing the maturity of sustainability. Here we consider the case of a manufacturing organization.

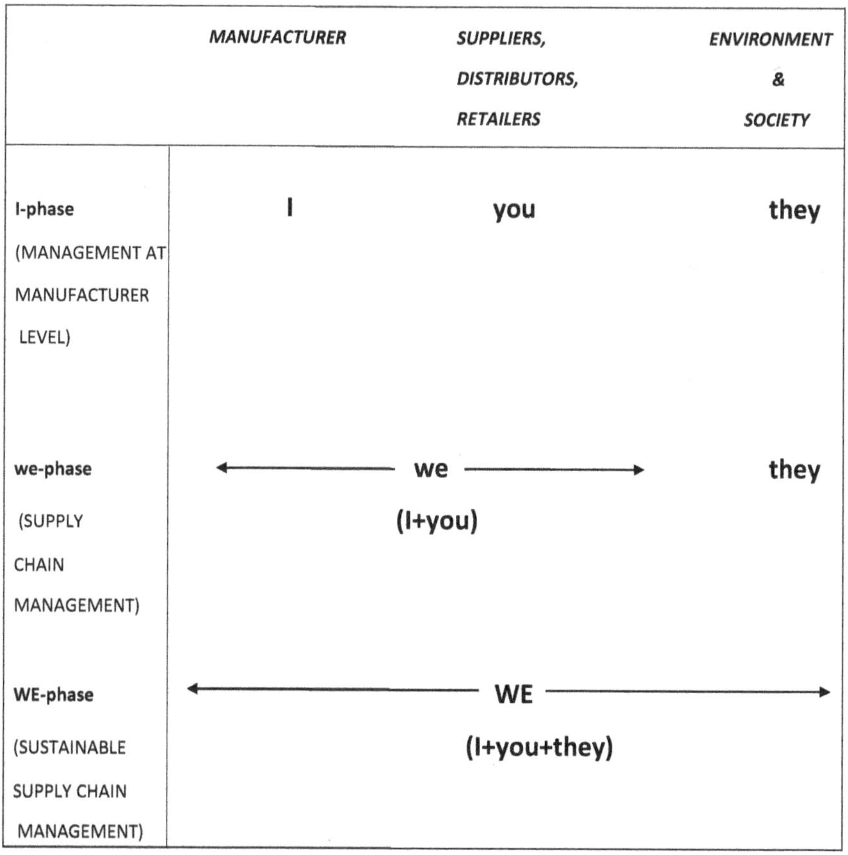

	MANUFACTURER	SUPPLIERS, DISTRIBUTORS, RETAILERS	ENVIRONMENT & SOCIETY
I-phase (MANAGEMENT AT MANUFACTURER LEVEL)	**I**	**you**	**they**
we-phase (SUPPLY CHAIN MANAGEMENT)	←	**we** ——————→ **(I+you)**	**they**
WE-phase (SUSTAINABLE SUPPLY CHAIN MANAGEMENT)	←	—————— **WE** ——————→ **(I+you+they)**	

FIGURE 1.6 "WE" phenomenon.

The figure is self-explanatory and depicts the achievement of ultimate goal of attaining "WE" phase of maturity by inclusion of interests and relevant parameters pertaining to all stakeholders beyond the organizational boundary covering the eco-system and society at large. The "WE" paradigm is meant for framing policies at the national or global level or formulating strategic plans at the organizational level. In both the cases, due considerations are to be given to human habitat at large, all living beings, climate, conservation of natural resources, and so on. All the global forums, primarily initiated by United Nations, do strive for effective planning and execution of the "WE" concept.

The "WE" phenomenon describes the journey of inclusiveness. A business initially keeps its focus on the operations management, considering only the interests of its manufacturing and related operations. In the second phase, other direct stakeholders in business environment (suppliers, distributors, wholesalers, etc.) are included by supply chain management. Now the interests of other members of supply chain are also included in business goals. The ultimate "WE" phase extends its wing further and includes non-economic and non-business entities, like the environment and society, and we name it as sustainable operations or sustainable supply chain.

In the present world order, the role of global intervention on this issue cannot be overemphasized because of the following facts.

- Economic success in any business activity is mostly dependent on collaboration and coordination among nations, crossing the boundary of a specific location.
- Environmental issues cannot be localized, as environmental degradation in one corner of the globe may have some effect on another corner.
- Society is becoming multicultural because of intense mobilization of people and efficient information processing and communication.

Key Learning

- A study on the evolution of viewpoints and practices of sustainable development shows the emergence of two historical phenomena playing the role of prime movers. These are harmony with nature and accelerated development after the Industrial Revolution.
- The first alarm bell for possible doomsday was rung by Robert Malthus, indicating the danger of unchecked population growth, way back in the end of 18th century, which was subsequently supported by Donella Meadows in the book *The Limits to Growth* after almost two centuries. The formation of the *Club of Rome* is a reflection of the global sense of alertness.
- The formal and most accepted definition of sustainable development originated from the publication of the book *Our Common Future*, the report of

the Brundtland Commission. This proposes the fact that while planning for development process, every generation should take into account the scope for meeting the needs of future generation.

- In 1997, the *triple bottom line (TBL)* concept became one of the popular ways of operationalizing sustainable development by integrating the three components of development—environment, society, and financial returns.
- Five perspectives have emerged showing the evolution of various viewpoints and concepts of sustainable development.
- Two viewpoints are proposed to integrate the three components of TBL—*trade-off-focused viewpoint* supporting the *weak sustainability* paradigm and *priority-focused viewpoint* supporting the *strong sustainability* paradigm. Sustainable development can be achieved by integrating all the three components using appropriate compensation policy per the former viewpoint, whereas the latter pre-emptively prioritizes environmental issues over others in any development process.
- The "WE" phenomenon opines the role of inclusiveness and its expansion capturing ecosystem and society as key stakeholders during policy-making and management of business processes for achieving sustainability.

Prior to the discussion session, it is expected that student groups will be formed. Now each of these questions may be discussed among the group members. The objective of the discussion session is to encourage students to think threadbare and explore all related issues, not arriving at the answer or solution to the problem,

---

Discussion Questions

1. Is "small is beautiful" still valid in world order of today? Does it represent two advantages simultaneously—cost and sustainability? Is it not the fact that large corporations are more inclined toward sustainability-based strategies?
2. Obsolescence rate of most of the products (particularly innovative products) is very high nowadays. Does it reflect a trend of counter-sustainability? Should there be retardation of NPD to solve this problem? Is this strategy equally applicable across all the industries?
3. How can the trade-off–based TBL model be translated to practice? How optimization may be achieved? What would be the model formulation for achievement of sustainability?
4. Why the implementation of need-based consumption in the market will remain as an unfulfilled dream? What are issues supporting and opposing this phenomenon?

NOTE

1 Readers are requested to consider "man" as a gender-neutral term. These are nothing but activities made or influenced by mankind.

REFERENCES

Benton, D. (1994). *The Greening of Machiavelli: The Evolution of International Environmental Politics*. London: Royal Institute of International Affairs/Earthscan.

Elkington, J. (1997). *Cannibals with Forks, the Triple Bottom Line of the 21st Century*. Oxford: Capston Publishing Ltd.

Meadows, D. (1972). *The Limits to Growth: A Report of the Club of Rome's Project on the Predicament of Mankind*. New York: Universe Books.

Mebratu, D. (1998). Sustainability and sustainable development; Historical and conceptual review. *Environmental Impact Assessment Review*, *18*, 493–520.

Neumayer, E. (2003). *Weak versus Strong Sustainability: Exploring the Limits of Two Opposing Paradigms*. Northampton: Edward Elgar.

Pisani, J. du. (2006). Sustainable development—historical roots of the concept. *Environmental Science*, *3*(2), 83–96.

Rogers, P., Jalal, K., & Boyd, J. (2008). *An Introduction to Sustainable Development*. London: Earthscan.

Rosen, M. A., & Kishaway, H. A. (2012). Sustainable manufacturing and design. *Sustainability*, *4*(2), 154–174.

Sarma, S. (2015). Environmental awareness at the time of vedas. *Veda-Vidya*, *26*, 221–224.

Tanwar, R. (2016). Environmental conservation in ancient India. *IOSR Journal of Humanities and Social Sciences (IOSR—JHSS)*, *21*(9), 1–4.

WCED, World Commission on Environment and Development. (1987). *Our Common Future*. Oxford: Oxford University Press.

Part II

Sustainability at the Policy Level

2 Macro-Level Initiatives for Curbing Ecological and Carbon Footprints

2.1 POLICIES, NORMS, AND DIRECTIVES AT GLOBAL PERSPECTIVES

The *Millennium Ecosystem Assessment (2003)*, while describing the possible frameworks for the evaluation of the state of the ecosystem and its close interactions with the human beings, clearly indicates the possible threats. Its warning mentions the strains on natural functions of the earth because of unwise activities and sometimes unwanted human interventions in natural processes.

The existence of the ecosystem, along with the continuity of its natural functions, is the prime cause of existence of living organisms, including the human race. If ecosystem fails, the sustainability of future generations will be at stake. The Millennium Ecosystem Assessment and other similar reports reiterate the fact that important ecosystem services have been impaired because of global warming, air pollution, diminishing natural resources, deforestation, depletion and deterioration of water resources, and so on. Incidentally, all these are anthropogenic impacts or, in other words, outcomes of mindless or selfish deeds of mankind. It is also true that fully or partially, immediately or in near future, human beings themselves can control these. As we can degrade ecosystem, we can also retard the process of this degradation. In some cases, we may even reverse the process. This is the fundamental understanding, the belief, and the real motivation for taking up the initiatives at global level or national level. If the group members accept these policies or the members get involved in formulating the policies, the implementation becomes effective. On the other hand, a group (the UN, the EU, or a nation) consists of various stakeholders with diverse interests, individual goals, and even socio-cultural setups. This poses challenges in getting the consensus in policy formulations.

2.1.1 CHARACTERISTICS OF MACRO-LEVEL POLICIES ON SUSTAINABLE DEVELOPMENT

Attempts at formulating any macro-level policy (national or global level) on effective sustainable development face challenges due to the following characteristics of the implementation process of sustainable development.

1. **Multiple activities**

   Sustainable development demands consideration of all activities relevant to the exploitation of natural resources, conversion processes, product

DOI: 10.1201/9780429195600-4

designing, and the use of products and their disposal. Treatment of each type of activities requires a specific set of parameters and their priorities. For example, sustainable resource exploitation (as raw materials for business process) requires consideration of conservation issues along with environmental degradation (pollution effects) and social impacts. On the other hand, sustainable product design is expected to consume less energy during its production, its use at customer end, and also in some cases, during remanufacturing or recycling of the product or its components and parts.

2. **Hierarchical framework**

Success of sustainable development lies on seamless activities of a hierarchical process. Top- or macro-level activities include policies, guidelines, and norms formulated at the global or national level, which are influenced by global condition of ecosystem and overall social environment. These policies are implemented at the micro level—that is, organizational or business level—which affect and subsequently modify the state of physical and social environment. Thus, the top-down approach is very applicable here. Policies, as a result, need modification or reformulation in the changed environment, and the smooth dynamicity should be maintained. Any undesired lag in this cycle may lead to disaster or it may require radical change in policies.

3. **Multidimensionality**

Sustainable development represents a multidimensional phenomenon both in the cause and effect of universal activities. Broadly, two sets of cause factors are primarily applicable for enabling or resisting this process of sustainability. These are man-made cause factors and causes created by elements of nature or ecosystem. On the other hand, (un)sustainable activities affect the surrounding physical environment and/or society and/or economic or business units. Each relevant factor may be assessed by several dimensions, which are used for measuring various ecological parameters considering degradation of land, marine areas, and environmental resources due to their scarcity and environmental pollution. The dimensions of the impact assessment may also be in terms of lifestyle impact, social impact, health-related parameters, and so on.

4. **Uncertainty and imprecision**

Most of the issues relating to planning and control of sustainability essentially need exhaustive knowledge on inherent mechanism of nature and complex functioning of society. Ironically, acquisition of exhaustive knowledge is quite difficult or, rather, impossible. So the existence of uncertainty and imprecision makes the decision-making quite complex and the subsequent implementation of the decisions quite challenging.

5. **Conflict of interest**

Conflict of interest is prevalent in implementing and maintaining sustainable development because of the involvement of multiple stakeholders. This may be explained in both vertical and horizontal perspectives. A vertical conflict may exist because of the difference in goals between the United Nations and different nations or between a national policy and a policy of a particular organization/sector. On the other hand, there may be difference of opinions between a developed and a developing nation or between one sector (high-tech, such as the IT sector) and another sector (traditional, such as the agricultural sector).

> *Macro-level policy-makers often encounter challenges while implementing sustainable development because of multiple activities, hierarchical framework, multidimensionality, uncertainty and imprecision, and conflict of interest.*

In this backdrop, global initiatives had been taken to establish various institutions in the form of bureaucratic organizations, conventions, or even forums for discussion. Numerous non-government organizations (NGOs) had been created at the global and national levels to motivate, negotiate, or make people aware of the current situation of degradations. Quite often these NGOs play the role of enablers as the representatives of affected people. These institutions formulate policies, ensure implementation, and monitor and evaluate the process and outcomes.

So the effectiveness of sustainable development policies needs consideration of certain elements, although debates still continue on what the exact characteristics of an ideal policy and how to implement it are. Challenges do exist, but the following elements surely help in enhancing the effectiveness.

1. **Long-term planning horizon**

   Because of essential nature of sustainable development, this policy should enable the creation of a framework, applicable for long-term planning with due consideration to its trade-offs with short-term planning.

2. **Modification of economic analysis**

   Market-based pricing and costing mechanism is not in the assessment of policies for sustainability. The cost of environmental degradation and social costs are supposed to be matched with the related benefits along with localized market profitability. Environmental accounting or sustainability-based accounting is going to be the appropriate accounting systems in near future.

3. **Uniformity of cost-effectiveness**

   Minimization of total cost should be the aim of policies, where some cost elements reflect the impact of hazards of environmental degradation and social disorders. It is as if society and ecosystem demanded the compensation for using every unit of their resources and for the suffering inflicted by us for our financial gains. Moreover, there should be equity on cost-benefit among all nations for every proposed intervention.

4. **Principles of environmental sustainability**

   Some basic principles of environmental sustainability are to be considered in policy formulation.

   - *Assimilation*: Ecosystem has its inherent assimilation capacity and rate for handling pollutants. The release of polluting substances should be limited to that assimilation capacity for maintenance of sustainability.
   - *Regeneration*: In order to maintain the resource conservation goal, policies should ensure the consumption of renewable natural resources limited to regenerative or renewable capacity of nature.

- *Substitutability*: Whenever there is a need for using non-renewable natural resources, it should be done very economically and efficiently. Moreover, sustainability is maintained if the rate of consuming a non-renewable resource is limited to rate of generation of any substitutable renewable resource or any other artificially created or man-made resource.
- *Risk of irreversibility*: The adverse effect of human activities sometimes may create an irreversible impact on the ecosystem. The limit of the impact should surely be taken into consideration, and policies should strictly include these limits while formulating guidelines on emission of waste and unwanted pollutants. This principle also explains extinction of some animals and critical thresholds of the regenerative capacity of nature.

Any policy formulation on environmental sustainability needs to consider principles like assimilation, regeneration, substitutability, and risk of irreversibility.

---

5. **Integration**

Policies are supposed to maintain integration across various processes and sectors crossing national boundaries. Essentially, these call for international cooperation for their execution.

6. **Transparency and accountability**

Information on the criteria of assessment of sustainability, prioritization, and action-taking reports and their impacts are to be made transparent through a participatory approach accessible to others. Every policy, guideline, or direction is expected to have a built-in component of accountability.

7. **Implementation mechanism**

A policy may be implemented as a mandatory rule with some mechanism of penalty for non-compliance or as a voluntary effort or pledge for achieving a target decided by an organization or nation.

2.1.2 INTERNATIONAL ORGANIZATIONS OR INSTITUTIONS FOR POLICY-MAKING AND SUPPORT IN IMPLEMENTING SUSTAINABLE DEVELOPMENT

2.1.2.1 Role and Objectives of International Institutions

As mentioned earlier, these institutions were created at different stages of timeline, as people went on striving for evolving mechanisms to control environmental degradation with changes in their perspectives. Some of them are mainstream sustainability-oriented (with overall objectives of improvement), while others are focused on specific issues of concerns. These are listed in the following.

1. **Mainstream institutions**

- International Union for Conservation of Nature (IUCN)
- United Nations Environmental Program (UNEP)
- United Nations Framework Convention on Climate change (UNFCCC)

- Commission for Sustainable Development (CSD)
- World Commission on Environment and development (WCED)
- Intergovernmental Panel on Climate Change (IPCC)
- Conference of Parties (COP)
- World Wildlife Fund (WWF)
- World Nature Organization (WNO)

2. Institutions with focused issues

- Convention of International Trade on Endangered Species (CITES)
- Convention on the Conservation of Migratory Species (CMS)
- Global Environment Facility (GEF)
- Food and Agriculture Organization (FAO)
- Convention on Long-Range Transboundary Air Pollution (LRTAP)
- Convention on Biological Diversity (CBD)
- United Nations Forum on Forests (UNFF)
- International Renewable Energy Agency (IRENA)
- The Economy of Ecosystem and Biodiversity (TEEB)
- World Meteorological Organization (WMO)
- International Energy Agency (IEA)
- Greenpeace

There are also major groups or stakeholders directly associated with environmental issues for a nation or the globe as a whole. These groups may not be formal organizations but do act significantly from various countries. *UNEP* (www.unenvironment. org) has a process of accreditation for these groups through the *United Nations Environment Assembly (UNEA)*. Most of these groups are non-government organizations. The list gets updated at regular intervals, and 23 Indian groups have been accredited till 6 June 2018.

Let us primarily look into the roles and activities of these global institutions, which are considered to be the institutions in policy-making and which primarily represent the voice of common people in this planet.

2.1.2.2 International Union for Conservation of Nature (IUCN)

The **IUCN** is among the **oldest environmental organizations** and was founded as a membership-based organization in 1948. Currently it has a large membership base of 216 governmental agencies, along with more than 1,100 NGOs and 160 member countries. Various scientists and other business communities can also get memberships if they show serious interest and are involved in addressing and studying global natural resources. There are more than 13,000 experts worldwide now (www.iucn.org, retrieved on 25 August, 2018).

The primary aim of the IUCN is to encourage, enable, and sensitize world communities for the conservation of both integrity and diversity of nature and for the equitable and sustainable use of natural resources. Its activities include almost 14 diverse themes, ranging from business and biodiversity to world heritage. Governmental agencies representing various nations, multilateral agencies, foundations, and member organizations take care of funding the IUCN.

The IUCN has created nine categories of threatened species of the earth, which indicate the degree of criticality in initiating the measures for saving them from extinction. These are extinct, extinct in the wild, critically endangered, endangered, vulnerable, near threatened, least concern, data deficient, and not evaluated.

2.1.2.3   United Nations Environmental Program (UNEP)

The **formation of the UNEP** is an outcome of the UN Conference on the Human Development (UNCHD), popularly known as **Stockholm Conference, held in June 1972**. In fact, the UNEP was proposed and founded by the UN General Assembly in December 1972. It primarily influences almost all global activities of the UN relating to environmental management. Broadly, its aim is enabling cooperation among countries and international organizations for global, regional, and national policy-making and implementing appropriate practices in environmental protection and achievement of sustainability. One of its crucial roles is assisting developing countries on these activities.

The mission of the UNEP includes providing leadership, enabling partnership among nations, and motivating them to improve the quality of life of present generation without compromising that for the generations to come.

The goals of the UNEP may be summarized as follows:

- Promoting awareness on environmental threats and their impacts
- Assessing, analyzing, and monitoring the current environmental status
- Establishing mechanisms for better cooperation, participation, and partnership on environmental protection
- Extending support for formulating policies, legislation, and also consultancy to national organizations and NGOs

2.1.2.4   United Nations Framework Convention
for Climate Change (UNFCCC)

By the end of 1980s, people became concerned on climate change and deterioration of air quality due to various man-made activities. In the **Earth Summit**, or the **UN Conference on Environment and Development (UNCED), held in 1992 at Rio de Janeiro, Brazil**, a declaration was made to set a framework for international agreements on global environmental issues. One of the major **outcomes of the Earth Summit is Agenda 21**, which is a powerful or somewhat daring program for achieving sustainable development in the 21st century. The framework, which is supposed to be an overview of these actions, is known as the **UN Framework Convention for Climate Change (UNFCCC)**, and it was opened for signature by various states. By 1992, 158 states had signed the framework. Subsequently, an international treaty for stronger commitment among developed countries was made through the Kyoto Protocol in 1997.

The primary objective of the UNFCCC is to stabilize greenhouse gas concentration in the atmosphere at a level that would prevent dangerous anthropogenic interference in affecting the climate system. It was operationalized by assessing emissions and removing greenhouse gases by nations who signed the Kyoto Protocol. This was obligatory for developed countries. Under this convention, most of the industrialized countries expressed their willingness to extend their support to developing countries for their actions on preventing climate change. The UNFCCC is also supposed to evaluate and monitor the implementation of the treaty. Developed countries are to report regularly the current status of their emission and action plans for its reduction. It is voluntary for developing countries.

2.1.2.5 Commission for Sustainable Development (CSD)

The CSD was primarily formed to monitor the implementation of action plans per Agenda 21 of the Earth Summit. It was proposed that the achievement of sustainable development requires addressing the nine key areas of concern: critical activities and elements of sustainability; funds and financial mobilization; science and technology; decision-making structure; role clarification of critical interest groups; rehabilitation and resettlement, along with human health issues; deforestation and biodiversity; atmosphere, ocean, and marine life; and toxic and hazardous waste management.

The roles and activities of the CSD may be summarized as follows:

* To monitor and review progress per Agenda 21 guidelines
* To analyze relevant information collected from reliable or competent sources and to communicate the report to appropriate institutions or agencies
* To provide recommendations on these issues
* To encourage and attract national policy-makers for greater involvement
* To create platforms for cooperation and mutual support among nations for closer interaction

2.1.2.6 World Commission on Environment and Development (WCED)

This commission, perhaps, is the most well-known and famous among all in the history of sustainable development. The **WCED, more commonly known as the Brundtland Commission**, of about 22 members, published its report *Our Common Future* in 1987, and we got a near-perfect definition of sustainability. Although a lot of criticisms erupted subsequently, the people, by and large, accepted the definition by this commission.

Impact of the WCED has been discussed in detail in the previous chapter. It emphasized environmental protection, economic growth, and social equity for designing developmental policies of the existing generation, keeping in mind the need for the development of future generations as well. Most of the popular theories and concepts of sustainability have been derived from this definition only.

2.1.2.7 Intergovernmental Panel on Climate Change (IPCC)

In 1988 the UNEP and the World Meteorological Organization (WMO) explored the possibility of establishing a **specialized international body** that will be primarily meant for carrying out **scientific studies on environmental issues** involving global experts from relevant domains of knowledge. The **IPCC** was thus created in **1988–1989**, as a joint venture of the WMO and the UNEP to be funded by a trust fund. This is a panel with representatives from various countries, organizations, and individuals, and it is supposed to provide an objective (more politically neutral) view of the process and mechanisms involved in climate change. It operates under the purview of the United Nations.

The role of the IPCC includes production of updated information on climate change on the basis of objective assessment involving thousands of scientists all over the world. The **IPCC was awarded the Nobel Peace Prize in 2007** for its effort in disseminating appropriate knowledge on climate change due to man-induced activities.

2.1.2.8   Conference of Parties (COP)

The **COP is a forum** created by holding annual conferences, representing the supreme decision-making body with the objective of resolving conflicting issues through discussions. The COP is organized by the UNFCCC. It reviews the implementation of convention directives, which also includes the use of legal instruments for effective execution. Parties submit current status, emission inventory, and implementation reports to the COP. Its presidency changes in all five recognized UN regions—Africa, Asia, Latin America and Caribbean, Central and Eastern Europe, and Western Europe and others.

2.1.2.9   World Wildlife Fund (WWF)

WWF (a.k.a. World Wide Fund for Nature) was established in 1961. With its headquarters at Switzerland, it works in more than 100 countries, with the primary objective of building and maintaining a healthy living environment of this planet. So its role is enabling nations and people to conserve biodiversity and non-renewable natural resources and promoting reduction of pollution and wasteful consumption.

2.1.2.10   World Nature Organization (WNO)

The WNO is an intergovernmental organization. Its treaty came into force in the recent past (2014). It is treated as a center of competency in the protection of nature and the environment. Its activities include offering support and consultancy in managing the environment and scientific and technology transfer.

2.1.2.11   Convention of International Trade in Endangered Species (CITES)

In 1975 CITES was established mainly for better conservation of wild flora and fauna. It primarily ensures fair international wildlife trade to prevent reduction of wildlife populations. It helps member countries in the management and control of wildlife trade. With its headquarters at Geneva, Switzerland, CITES listed endangered species in three categories: Appendixes I, II, and III.

2.1.2.12   Convention of the Conservation of Migratory Species (CMS)

Enforced in 1983, with its initial adoption in 1979 at Bonn, Germany, the CMS was established based on the agreements among member countries for the conservation of migratory terrestrial, marine, and avian species. The stringency of agreement depends on the degree of extinction of the species.

2.1.2.13   Global Environment Facility (GEF)

The GEF, being an independent financial organization established in 1991, takes care of funding and cooperation among nations (on financial and economic issues) for projects related to biodiversity, climate change, ozone layer, management of pollutants, and so on, which are directly contributing to betterment of environment.

2.1.2.14   Convention on Long-Range Transboundary Air Pollution (CLRTAP)

Established in 1979 by UNEP Governing Council under the Earth Summit program, the CLRTAP or simply LRTAP is supposed to assess and monitor the quantity of air pollutants moved long-distance crossing geographical boundaries. Its ultimate goal is to limit this movement and develop strategies to combat discharge of air pollutants by scientific studies, collaborations, and policy negotiations among concerned parties (or nations).

2.1.2.15   Convention on Biological Diversity (CBD)

The CBD is a multilateral international body (supposed to be legally binding), opened for signature in the Earth Summit of 1992 and enforced by end of 1993. In tenth COP of 2010, the Nagoya Protocol was adopted for implementation. Its focus was on three specific goals related to biodiversity: conservation of biodiversity, sustainable use of biodiversity, and fair and equitable sharing of benefits arising out of the use of genetic resources.

2.1.2.16   United Nations Forum on Forests (UNFF)

The main goal of the UNFF, established in 2000 by the Economic and Social Council, is to conserve and sustainably develop forest reserves at national, regional, and global levels. The role of the UNFF includes facilitation of assessment and monitoring and policy implementation through cooperation and political negotiations across the globe.

2.1.2.17   International Renewable Energy Agency (IRENA)

The first conference of IRENA was organized in 2000. Its statute was formalized in 2010. IRENA primarily strives for promoting transition of energy use from nonrenewable resources to renewable ones through knowledge sharing, enabling and sharing of policies, and technology transfer.

2.1.2.18   The Economy of Ecosystem and Biodiversity (TEEB)

The TEEB is an organization created to study the environmental accounting in a little unconventional way, considering the benefit by conserving biodiversity against the cost of biodiversity loss and loss due to ecological degradation. G8 + 5 Environment Ministers at Potsdam, Germany initiated the first phase of this study in 2007. UNEF took over the second phase. TEEB predicted that around 18% of global economic output would be lost (as cost of environmental degradation) in 2050.

2.1.2.19   World Meteorological Organization (WMO)

In 1951, the WMO came into existence as a specialized agency for carrying out meteorological, hydrological, and geophysical studies. Its report keeps track on climate change and predicts the future meteorological data.

2.1.2.20   International Energy Agency (IEA)

The IEA, being an intergovernmental autonomous organization, was established in Paris in 1974. Its role is primarily advisory in nature to the member nations, along with some non-member countries, like China and India, on energy security, economic development, and environmental protection. It supports the countries in generation and management of clean energy.

2.1.2.21   Greenpeace

Greenpeace is essentially a nongovernmental organization, set up in 1971 for the conservation and protection of environment. Its primary goal is to ensure the ability of the earth to nurture life in all its diversity. It meets its objective by campaigning and initiating protests or actions, if necessary, against global warming, deforestation, overfishing, genetic engineering, and similar issues.

In the preceding sections, popular institutions have been shown in boldface. Table 2.1 systematically exhibits the list of global organizations striving to propagate the idea of sustainability by making sustainability policies, supporting developing countries to implement such policies, and informing the people worldwide.

The most critical and popular international institutions, primarily associated with the UN in managing, implementing, and maintaining sustainable development worldwide, are the IUCN, the UNFCCC, the UNEP, the IPCC, the WCED, annual COPs, and the IEA.

2.1.3   COPs and Important Global Summits

It is quite apparent that the effectiveness of any worldwide initiative entirely depends on cooperation, mutual support, motivation, and sharing of scientific ideas, technology, and funds. These are really hard tasks due to wide variations of development, political status, possession of natural resources, and policy level prioritizations among nations. The role of the Conferences of Parties (COPs) is creating the right forums for the exchange of ideas and for discussion on these issues for formulating any meaningful policy or guidelines applicable for all nations of the globe. Annual COPs are organized regularly. The first COP, COP1, was organized in Berlin in 1995 and the latest one, COP27, was held in Sharm El Sheikh, Egypt in November 2022. With its primary focus on climate change issues, any COP intends to enable agreements among nations and strives for conflict resolution in this context. Some COPs certainly concluded with successful and feasible recommendations, while some resulted in inconclusive outcomes or non-agreements. Nevertheless, famous global summits may be treated as important milestones in this pursuit, and their impacts do play significant

TABLE 2.1
List of World Organizations or Institutions at a Glance

Series Number	Name	Year of Establishment	Main Objectives, Roles, and Status
I. World organizations/forums/conventions for general environmental concerns			
1	International Union for Conservation of Nature (IUCN)	1948	Conservation of integrity and diversity of nature; equitable and sustainable use of natural resources
2	United Nations Environmental Program (UNEP)	1972	Key influencer of environmental management of the UN; cooperation among countries for policy-making and implementation of practices
3	United Nations Framework Convention for Climate Change (UNFCCC)	1992	Creation of international treaties for a stronger commitment for stabilizing and removal of GHGs from the atmosphere
4	Commission for Sustainable Development (CSD)	1993	Monitoring implementation of action plans per Agenda 21 of the Earth Summit
5	World Commission on Environment and Development (WCED)	1987	Report of the Brundtland Commission, *Our Common Future*, with its famous definition of sustainable development
6	Intergovernmental Panel on Climate Change (IPCC)	1988	Specialized international body to carry out scientific studies on environmental issues and their impact on climate change
7	Conference of Parties (COP)	1995 (COP 1)	Annual conferences for discussions and decision-making resolving conflicts among nations on sustainability.
8	World Wildlife Fund (WWF)	1961	Primary objective of maintaining biodiversity, conservation of natural resources, and pollution control
9	World Nature Organization (WNO)	2014	Intergovernmental organization with its treaty on environmental protection and support in scientific and technology transfer

(Continued)

TABLE 2.1 *(Continued)*
List of World Organizations or Institutions at a Glance

Series Number	Name	Year of Establishment	Main Objectives, Roles, and Status
II. World organizations/forums/conventions for specific areas of concern			
1	Convention of International Trade in Endangered Species (CITES)	1975	Keeping control of international trading of endangered or near-extinct species per the agreement
2	Convention on the Conservation of Migratory Species (CMS)	1983	Monitoring agreements and supporting conservation of migratory terrestrial, marine, and avian species
3	Global Environment Facility	1991	Management of funding and cooperation among countries on projects relevant to global environmental protection and maintenance of the overall ecosystem
4	Convention on Long-Range Transboundary Air pollutant (LRTAP)	1979	Assessing and monitoring air pollutants moving across long-range geographical boundaries
5	Convention on Biological Diversity (CBD)	1993	International agreement on biodiversity
6	International Renewable Energy Agency (IRENA)	2010	Main goal of enabling nations to use energy from renewable sources through a formal statute
7	The Economy of Ecosystem and Biodiversity (TEEB)	2007	Study and application of environmental accounting, incorporating benefits derived from conservation of biodiversity
8	World Meteorological Organization (WMO)	1951	Specialized agency for meteorological studies
9	International Energy Agency (IEA)	1974	Advisory role to nations on energy security, economic development, and environmental protection
10	Greenpeace	1971	Non-governmental organization globally famous for its activities on conservation and protection of nature

roles in global policy-making on sustainability. Among various conventions, treaties, or summits, four seem to play most significant role in policy-making: the *Rio Earth Summit*, the *Kyoto Protocol*, the *Paris Summit*, and *Agenda 2030* and SDGs.

2.1.3.1   Rio Earth Summit

The United Nations Conference on Environment and Development (UNCED) was held at Rio de Janeiro, Brazil, during first two weeks of June 1992, famously known as the Rio Earth Summit of 1992. A short document known as the **Rio Declaration** was prepared on environment and development. A non-binding action plan known as **Agenda 21** was the outcome of the discussions for formalization of global action plan for the sustainable development into the 21st century, which comprises recycling, energy efficiency, and conservation issues. Subsequently, after ten years, **in 2002** the **World Summit on Sustainable Development (WSSD)** took place in Johannesburg, South Africa. On the 20th anniversary of the Earth Summit, the **United Nations Conference on Sustainable Development (UNCSD)** was organized in the same city of Brazil, which is popularly known as **Rio+20 Summit, in 2012.** The conference published a non-binding document called **"The Future We Want."** The decision also emphasized launching a set of measurable targets for assessing the achievement of sustainable development globally.

2.1.3.2   Kyoto Protocol

The Kyoto Protocol is hailed as the most widely debated global treaty, negotiated in Kyoto, Japan, during COP 3 of UNFCCC in 1997. The agreement was primarily meant for the control of six greenhouse gases (GHGs): carbon dioxide, methane, nitrous oxide, perfluorocarbons, hydrofluorocarbons, and sulfur hexafluoride.

GHG emissions are responsible for an increase in global average temperature of the earth's surface (i.e., global warming). According to the IPCC, the long-term effect of global warming is the overall rise in sea levels, resulting in inundation, melting of glaciers and Arctic permafrost, extreme climate-related events (like floods, typhoons, and droughts), and an increased risk of extinction of 20–30% of plant and animal species. The treaty calls for meeting a target reduction of 5.2% of global GHG emission levels, with reference to the 1990 level, within the commitment period of 2008–2012. Each nation is supposed to fix its own target accordingly. It is applicable to two groups of countries—Annex I and non–Annex I countries. Annex I countries are primarily developed countries, and the commitment is mandatory for them, whereas the attempt for meeting the targets for non–Annex I countries is considered to be a voluntary exercise. Signatories formally accepted the protocol in 2005.

The parties in the COP judiciously proposed the following mechanisms for motivating Annex I countries to contribute to the global GHG emission reduction.

2.1.3.2.1   Clean Development Mechanism (CDM)

Annex I countries may invest in technology and infrastructure on **CDM** projects in less-developed countries and earn the credit of emission reduction measured as **certified emission reduction (CER) units**, which may be used in place of actual emission reduction in the investor country.

2.1.3.2.2   *Emission or Carbon Trading*

CER units may be traded in specific emission-trading markets. The **joint implementation (JI)** mechanism may be executed between Annex I countries through earning of **emission reduction units (ERUs)** by investing in technology and infrastructure for emission reduction in another country. Both CERs and ERUs may be traded in specific markets.

There were challenges in implementing this protocol because of non-acceptance of some countries, which were quite infamous in terms of GHG emission. In COP18 in Doha, Qatar, in 2012, delegates agreed to extension of commitment deadline to 2020.

2.1.3.3   Paris Summit

COP21, held in Paris in 2015, is considered to be a successful summit with agreement achieved among 196 signatories on issues like curbing global warming and providing financial support to developing countries. With current average temperature of earth surface being around 15 degrees Celsius, it was decided to restrict the temperature increase to 2 degrees only. If possible, countries will attempt to restrict it to 1.5 degrees. A 32-page document was signed, known as the **Paris Agreement**, which may be treated as the replacement of the Kyoto Protocol. The relevant actions are supposed to be reviewed in every five years, and it was also decided to create an annual Green Climate Fund of USD 100 billion to support the not-so-developed countries for replacing existing technology with clean technologies. The Paris Agreement came into force in November 2016, and the signatory nations should strive to achieve the GHG reduction target through **nationally determined contributions (NDCs).** Actually, once a country on its own comes forward to reduce emission it is included under **intended nationally determined contributions (INDCs).** Later on, its contribution is treated to be an NDC, which has some special clauses under the Paris Agreement.

Unfortunately, the withdrawal of USA from its commitments in 2017 and its resultant impacts on commitments of others tarnished the success story of the Paris Summit (Zhang et al., 2017).

2.1.3.4   Agenda 2030 and the 17 SDGs

During the UN Sustainable Development Summit in September 2015 at New York, a decision was made to implement the mission of "Transforming our world: the 2030 Agenda for Sustainable Development." Participants of the summit could identify the areas of concern like climate change, economic inequality, innovation, sustainable consumption, peace, and justice for the transformation of our world by 2030. Emphasis was on people, planet, prosperity, peace, and partnership. Agenda 2030 included 17 Sustainable Development Goals (SDGs) associated with related 169 targets. The United Nations Development Programs (UNDP) is expected to provide support to nations to integrate SDGs in their national development plans and policies. The following is the list of SDGs. All goals are meant and applicable for all people across the globe.

SDG 1: End poverty
SDG 2: End hunger with food security and improved nutrition

SDG 3: Ensure healthy lives and well-being

SDG 4: Ensure inclusive and equitable quality education and lifelong learning opportunity

SDG 5: Achieve gender equality and women empowerment

SDG 6: Ensure sustainable management of water and sanitation

SDG 7: Ensure access to affordable, reliable, and sustainable energy

SDG 8: Promote sustained and inclusive economic growth and productive employment

SDG 9: Build resilient infrastructure and promote inclusive and sustainable industrialization

SDG 10: Reduce inequality within and among countries

SDG 11: Make human settlements inclusive, safe, resilient, and sustainable

SDG 12: Ensure sustainable consumption and production patterns

SDG 13: Take urgent actions to combat climate change and its impacts

SDG 14: Conserve and sustainably use the oceans, seas, and marine resources

SDG 15: Protect, restore, and promote sustainable use of terrestrial ecosystems and forests and combat desertification, land degradation, and biodiversity loss

SDG 16: Promote peaceful and inclusive societies, provide access to justice, and build effective and accountable institutions

SDG 17: Strengthen the means of global partnership for sustainable development.

Every year, the COP discusses relevant issues on sustainable development, and its outcome, if conclusive, proposes a global policy on sustainability. Among others, four such summits seem to be good contributors to achievement of sustainability—the Rio Earth Summit (Rio Declaration, Agenda 21, and Rio+20 Summit), the Kyoto Protocol (CDM, JI, and carbon trading using CERs and ERUs), the Paris Summit (Paris Agreement and INDCs/NDCs), and the UN Sustainable Development Summit (Agenda 2030, 17 SDGs, and 169 targets).

Key Learning

- Global endeavors for providing leadership to nations along with support in policy-making on sustainability and for monitoring the implementation of action plans gave rise to creation of various global institutions, organization of conventions, and formulation of agreements.
- The UNEP, by and large, influences all activities, and the UNFCCC primarily deals with managing activities relating to climate change. COPs are organized annually to discuss various relevant issues and resolve conflicts.

- The most acceptable functional definition of sustainable development emerged from the WCED report.
- The Kyoto Protocol represents the first successful attempt in quantifying the acceptable targets for controlling GHG emissions.
- The goal of the Paris Summit was the refinement of nationwide emission targets for national policy-making and the creation of a fund to support developing countries in implementing green technologies.
- Agenda 2030 established 17 goals, the achievement of which is expected to transform the world to a decent and living-worthy planet. These goals capture all elements of sustainability, including issues like justice and equity.

2.2 GUIDELINES AND REGULATORY ISSUES AT THE INDIAN GOVERNMENTAL LEVEL

Initiatives at the global level had their resultant impact on framing of rules and regulations by government of India, which are meant for guiding the national policy-making of the country. As mentioned in previous chapter, historically Indians had already shown their concerns on environmental degradation and manifested their deep understanding of the intrinsic relationship between humans and nature. Mythological accounts also reflect the existence of this perception of Indians since time immemorial.

2.2.1 INITIATIVES OF THE INDIAN GOVERNMENT IN POLICY-MAKING AND CREATING LEGISLATIVE SUPPORT

Incidentally, in initial years after independence of India (1947), no policy formulation or formalization of this concern was explicitly visible. After the Stockholm Conference of 1972, India had first created the *National Committee on Environmental Planning and Coordination (NCEPC)* within the Department of Science and Technology as an advisory body for the study of the environment and its possible degradation. Its functions also include guidance on project appraisal, taking into account the environmental factors. In the fifth Five-Year Plan, the environmental concerns are more emphasized, and in the sixth Five-Year Plan, the Environment and Development was explicitly considered as a planning document. It also provided guidelines for creating institutional structure for environmental management at the central and state levels. In 1980 the Tiwary Committee report of the Planning Commission motivated the government to set up a separate *Department of Environment* (in 1980). Subsequently, the government of India started considering the protection of wildlife and forests an important issue in India, which has become a groundwork for strengthening the jurisdiction of central government to overrule the state decisions on these areas of environmental preservation and protection. Thus, the federal Department of Environment turned into a full-fledged *Ministry of Environment and Forests (MoEF)* in 1985. Initial initiatives of India in this pursuit may be divided into three directions—*environmental pollution, forest management,* and *resource conservation.* In this context, it is to be noted that environmental degradation and carbon emissions impact on wildlife activities, and deforestation contributes to climate change across the globe. But as the priority on climate change (along with global warming) has become more prominent, the MoEF has been renamed and reconstituted as *Ministry of Environment, Forest, and Climate Change (MoEFCC)* in May 2014.

The *Water (Prevention and Control of Pollution) Act* came into existence in 1974 through a parliamentary decision to control water pollution. The government of India established the regulatory bodies federally (i.e., in Delhi) and in different states, the *Central Pollution Control Board (CPCB)* and the *State Pollution Control Boards (SPCBs)*, respectively, to look after the proper implementation of parliamentary acts for proactive and reactive activities in pollution control of the country. These are supposed to guide and advise the government in framing plans for related issues. These boards are also engaged in data collection and analysis of the relevant activities for meaningful studies. Subsequently, in 1981, a similar act was enacted, the *Air (Prevention and Control of Pollution) Act*. The implementation of this act is logically under the scope of the mentioned boards. Like the developed countries, India took its subsequent steps in sustainable development after the Rio Conference, followed by plans for execution of Agenda 21 in 1992. This called for involving all stakeholders in society in a more comprehensive manner to manage sustainable development. The MoEF declared a policy document for abatement of pollution, which expressed more concerns on implementation strategies. It emphasized on following guidelines.

- To reduce and control pollution generation at its *source*
- To use *best available technology* (energy-efficient, resource-consumption-efficient, and green technology) for most of the economic activities
- To adopt *principle of "polluters pay"* (i.e., the polluter bears the responsibility and is liable to pay the penalty)
- To prefer involvement of *public participation* (considering stakeholders and society at large) in decision-making activities related to the prevention and control of environmental pollution

The *National Forest Policy of India* was declared in 1988, with emphasis on the protection of forests and maintenance of ecological balance and environmental stability. The concern areas of this forest policy are summarized as follows:

- Protection of ecological balance and maintenance of biological diversity by proper preservation of biotic components of nature
- Control of soil erosion to mitigate floods and droughts and prevent the extension of sand dunes in desert areas
- Increase of forest covers through extensive afforestation
- Maintenance and increase of forest resources, like fuel wood, fodder, timber, and other forest produce, to meet the requirements of rural and tribal population, as well as national needs
- Increase of awareness among people and public movements for the maintenance of forest sustainability

The National Forest Policy was redrafted in 2018, including its focus on international challenge on climate change.

In June 1992, the MoEF formulated its *National Conservation Strategy and Policy* documents on Environment and Development. This policy is meant for meaningful conservation of biosphere reserves for the protection of biodiversity and of wetlands, mangroves, and coral reefs. Moreover, the documents include combating desertification, conserving forests, and protecting wildlife.

So Indian legislative initiatives for implementing the intention of creating and maintaining a clean environment may be outlined as the following critically relevant acts and national policies.

Incidentally, some of these have been amended later on.

The list of governmental initiatives may be classified under relevant groups as shown here. It is based on the information and facts collected from various online resources and the published report of Priyadarshini (2016).

1. **Environment protection:**
   Environment Protection Act, 1986; Water (Prevention and Control of Pollution) Act, 1974; Water Cess Act, 1977; Air (Prevention and Control of Pollution) Act, 1981; and others.

2. **Management of forests and biodiversity:**
   Indian Forest Act, 1927; Forest (Conservation) Act, 1980; Wild Life (Protection) Act, 1972; Biodiversity Act, 2002; and others.

3. **National policies for environmental management:**
   National Environment Policy, 2006; National Forest Policy, 1988; National Conservation Strategy and Policy Statement on Environment and Development, 1992; Policy Statement on Abatement of Pollution, 1992; and others.

4. **Policies of relevant sectors (with impact on environmental management):**
   National Agriculture Policy, 2000; National Population Policy, 2000; National Water Policy, 2002; and so on.

5. **Rules and regulations for sustainable economic activities:**
   Hazardous Wastes (Management and Handling) Rules, 1989; Manufacture, Storage and Import of Hazardous Chemical Rules, 1989; Restriction of Hazardous Substance (RoHS) Rules 2012; Bio-Medical Waste (Management and Handling) Rules, 1998; Municipal Solid Waste (Management and Handling) Rules, 2000; E-Waste (Management) Rules 2016; Ozone Depleting Substance (Regulation and Control) Rules, 2000; Regulation on Recycling of Waste Oil and Non-Ferrous Scraps, 1999; Batteries (Management and Handling) Rules, 2001; Prohibition on the Handling of Azo Dyes, 1997; Noise Pollution (Regulation and Control) Rules, 2000; Prohibition against Open Burning of Waste Oil, 1997; Prevention of Dumping and Disposal of Fly Ash, 1999; Recycled Plastics Manufacture and Uses Rules, 1999; and so on.

6. **Additional measures for effective execution of regulations:**
   Environmental Impact Assessment Notifications I, II and III, 1992 and 1994; Public Hearing Notifications I and II, 1997; Taj Trapezium Zone Pollution (Prevention and Control) Authority Order, 1998; Public Liability Insurance Act and Rules, 1991; National Environment Appellate Authority Act, 1997; National Environment Tribunal Act, 1995; and others.

2.2.2 IMPACT OF THE PARIS SUMMIT ON INDIAN POLICIES

The Paris Agreement is not meant for adherence to legally binding emission reduction target; rather, it is a case of voluntary commitment to climate change actions made by the *intended nationally determined contributions (INDCs)* (or sometimes also shortened as NDCs) of 195 countries.

Being an emergent national economy with expected growth rate of around 7% till 2030, India is striving for global leadership within next couple of decades or so. With its intensified industrialization and accelerated economic activities, it is becoming the third largest energy consumer and GHG emitter. However, thanks to the high population growth, per capita emission rate of India is still very low.

In order to meet the global target of limiting the increase in the warming of the earth's surface by 1.5 degree Celsius (per the Paris Summit), India took a pledge by INDC for comprehensive planning in environmental and ecological improvement. Its specific commitments are as follows.

- *Reduction of GHG emissions* by 33–35%, back to the 2005 level, by 2030.
- Achievement of the *target of 40% energy generation from non-fossil-fuel sources* (i.e., renewable energy sources) by 2030 through proper technology transfer and green technology. This may be achieved, if necessary, with the support of the Green Climate Fund (as expected by the implementation of the Paris Agreement).
- *Creation of additional carbon sinks* for 2.5 to 3 billion tonne of carbon dioxide equivalent by 2030 through intensive afforestation and meaningful urban planning, including the creation of additional forest covers.

The excellent note jointly prepared by the National Resources Defense Council (NRDC), Administrative Staff College of India (ASCI), SEWA (Self-Employed Women's Association), Indian Institute of Public Health (IIPH), Council of Energy, Environment and Water (CEEW), and Energy and Resources Institute (TERI) includes detailed description of Indian initiatives post–Paris Summit era.

These initiatives are outlined, which are classified under some categories.

2.2.2.1 Sustainable Energy (Generation and Use)

In order to meet the renewable energy goal of Paris Summit, India has fixed its target of installing power plants of 175 GW from renewable energy sources, including 100 GW of solar, 60 GW of wind, 10 GW of biogas and 5 GW small hydro-power energy. The country intends to achieve this ambitious target by 2022. The following are some specific endeavors in this direction.

- *National Solar Mission* represents the formalization of this initiative for expansion of solar energy uses for electricity generation. Kurnool Ultra Mega Solar Park of Andhra Pradesh is also an example of growth of solar energy market, which is also catching up as a mode of power generation. With plenty of sunlight available in the land, it is estimated that India's solar energy potential may even be as high as 750 GW. Simultaneously, the importing of cheaper solar cells from China and better management of techno-economics have resulted in selling the solar power at a price as low as INR 3.0 per kilowatt-hour (kWh), as recorded in 2016.

- India is also growing in wind energy generation. It is now almost fourth largest wind energy producer in the globe with 32 GW installed capacity, which amounts to 10% of the total installed power capacity of the country. In one single year 2016–2017, the country could add 5.4 GW of wind power. Price of wind power has also reduced to INR 3.42 per kWh as recorded in October 2017.
- *National Mission of Enhanced Energy Efficiency (NMEEE)* scheme enables improvement in energy efficiency in various sectors. It could achieve a reduction of 8.67 million tons of oil equivalent between 2012 and 2015.

2.2.2.2 Green Building Construction

The *Bureau of Energy Efficiency (BEE)* designed the *Energy Conservation Building Code (ECBC)* in 2007, which has been updated in 2017, and perhaps it is to be included as amendment in Energy Conservation Act in 2018. Various states are in the process of implementing ECBC—2017 in infrastructure development. The *Leadership in Energy and Environmental Design (LEED)* certification is also becoming popular in Indian civil construction sector. Per the report on US Green Building Council in 2016, buildings on a 15-million-square-meter space have already been awarded LEED certification. India has also created its own indigenous certification scheme, named *Green Rating for Integrated Habitat Assessment (GRIHA)*. It is expected that these endeavors would lead to a saving of around 3453 terawatt-hour (TWh) of electricity by 2030.

2.2.2.3 Manufacture and Use of Green Appliances

The BEE could popularize *the star-rating scheme* among power consuming electrical products and white goods like tube lights, air conditioners, refrigerators, transformers, and so on. Of course, air conditioners need special treatments, as making an air conditioner sustainable not only requires energy efficiency but also a design that *replaces the HFC emission* to avoid its effect on the ozone layer. In this context, India also has an ambitious goal of stopping the use of HFC refrigerants by 2024, per the Kigali amendments of the Montreal Protocol. This leads to formulating a *National Cooling Action Plan* for the manufacturing and use of zero-global-warming refrigerants. In the energy efficiency front, India could replace the incandescent lamps by 3.4 million LEDs by 2017.

2.2.2.4 Sustainable Transport Sector

India has already instituted its own standards for keeping a check on air pollution on road first time in 2000 based on European norms (Euro), which is known as the *Bharat Stage (BS) Emission Standard*. BS III norms (for all over India) and BS IV (in some cities) had already been enforced in 2010. In 2016 the government of India decided to directly adopt BS VI (equivalent to Euro VI emission standard) norms by 2020, skipping BS V. Per the Corporate Average Fuel Consumption standards, passenger vehicles are expected to improve fuel mileage by 15% by 2022 for better fuel efficiency in transportation. In 2013 the *National Electric Mobility Mission Plan 2020* aims at selling only electric vehicles by 2030 with the provisions of subsidies in production of electric and hybrid vehicles. Of course, the net improvement can only be achieved if India can transform its electricity generation process to clean power generation mostly using renewable energy sources. The creation of a mass transit system in urban areas by metro rail network is another initiative for discouraging use of individual vehicles.

2.2.3 INDIAN INITIATIVES TOWARD SDGS UNDER AGENDA 2030

India has been comfortably maintaining its journey of development keeping sustainability in focus, by addressing some key SDGs in various policies and schemes. One of the most meaningful slogans or phrases often quoted by India's prime minister is *Sabka Saath Sabka Vikas*, which literally means "Development of All through Collective Efforts." In fact, this actually emphasizes the fact that India's development process should be activated by inclusive development. So this all-inclusiveness technically reflects the wish of the land to involve all stakeholders in this journey. Thus, beneficiaries and actors of national development programs are supposed to include society, businesses, and nature.

The NITI (National Institution for Transforming India) Aayog, for submission to the *United Nations High Level Political Forum on Sustainable Development*, New York, in 2017, prepared a *Voluntary National Review (VNR)*. In this report, seven SDG goals were emphasized as contribution of the country toward sustainable development (VNR, 2017). Out of these seven goals, India's concern on five SDGs needs mentioning.

2.2.3.1 Goal 1: End Poverty

Since the early '90s, the poverty level has declined considerably in India, perhaps because of the sustained growth pattern of the country. The *Mahatma Gandhi National Rural Employment Guarantee Act* substantially enabled creation of new employments. Schemes like this also help poor people get access to life insurance and financial supports in house building. Various programs have also been launched for availing access to education, health, and nutrition security. Moreover, better water management programs enabled around 70% of the population to get 40 liters of drinking water per capita per day. There is also some visible improvement in sanitation in villages by governmental supports and better campaigning or awareness programs. The *Pradhan Mantri Ujjwala Yojona* scheme supports the poor population use clean fuel (liquified petroleum gas) instead of direct fossil fuel for household purpose.

2.2.3.2 Goal 2: End Hunger and Achieve Food Security

India seems to have progressed in this direction during last decade or so. The reported cases of diseases relating to retardation of growth rate of children under five years was reduced from 48% (2005–2006) to 38.4% (2015–2016), and those of underweight children have also shown a dip from 42.5% to 35.7%. *India's Public Distribution System* of food grains and *Mid-Day Meal programs* in primary schools had shown some positive result in enhancing general nutrition level in food particularly among rural community. Agricultural sector has already introduced organic and sustainable farming by governmental initiatives. Perhaps the thrust on organic food production is an outcome of demand from environmentally conscious consumers as well.

2.2.3.3   Goal 3: Ensure the Health and Well-Being of People

India has promoted vaccination coverage among children 12–23 months of age. Infant mortality rate in the country had shown a decline trend from 57 (in 2005–2006) to 41 (in 2015–2016) in 1,000 live births. The formalization of health improvement schemes has been made by implementing the *National Health Policy 2017* of India.

2.2.3.4   Goal 4: Achieve Gender Equality and Empower Women

Women literacy rate has improved from 55.1% (in 2005–2006) to 68.4% (in 2015–2016). The *Beti Bachao, Beti Padao* ("Save the Girl Child, Educate the Girl Child") program strives for the reduction of gender inequality and intensification of girls' education. In various professions women's participation is considerably increasing in India, including in global sports event. Schemes like *Sukanya Samriddhi* is another example for giving financial incentives as long-term investment in banks through preferential interests.

2.2.3.5   Goal 9: Build Resilient Infrastructure and
Promote Sustainable Industrialization

Transportation (mass transit) and roadways, internet connectivity, and electrification have been rapidly expanded connecting all parts of India, including villages. The Make in India campaign has remarkably boosted the manufacturing sector. The Indian youth community is showing more interest in entrepreneurship than taking employment, which is resulting to creation of new employments. New startups are coming up almost in every city or town, of course more in service sector. Government is also incentivizing small- and middle-scale enterprises (MSMEs). The growth rate of foreign direct investment (FDI) by 7.5% during 2014 to 2017 shows attractiveness of India as a prospective business hub.

Nevertheless, India is yet to achieve its goal of becoming a country of sustainable economic endeavors, and a nation of greenness and cleanliness. Per WHO global air pollution database released in Geneva (report of TNN on 2 May 2018) India has 14 out of 15 most polluted cities in the globe in terms of harmful particulate matters (i.e., $PM_{2.5}$) (please refer https://sustainabledevelopment.un.org). In recent past Delhi being the national capital has drawn much attention for its polluting environment. Like some other congested metropolitan city in the globe, the air quality index (AQI) in Delhi is often recorded as either "very poor" or "severe" based on its categorization. The AQI has its own categorizations, which has been elaborated in Chapter 5. Sometimes low wind speed may cause this severity. During festival days, it often becomes worse. On the other hand, there is also some rays of hope. Extensive work is taking place in the *Ministry of New and Renewable Energy (MNRE)*, established in October 2006. The *Solar*

Energy Corporation of India (SECI) is promoting for single-location new solar photovoltaic (PV) plant at Ladakh and Kargil. The Ladakh plant is going to be the largest such solar plant in the world. This project is expected to save 12,750 CO_2 equivalent per annum. The projected installed capacity of the Ladakh plant is 5,000 MW and that of Kargil plant is 2,500 MW. Another burning issue of India is e-waste generation from disposals of electronic and electrical gadgets, particularly mobile phones and laptops, which are piling up day-by-day. India is also taking steps in this direction by enacting and modifying e-waste rules as *E-Waste (Management) Rules 2016*, by stringent target-based control of CPCB and by implementing *Extended Producers Responsibility (EPR)*. However, it is not enough. This issue is still remaining a pain point of the government of India.

Key Learning

- Indian first milestone toward achieving sustainability is formation of the MoEF in 1985, and for effective implementation of all the acts, rules, and policies on environmental issues, it has established Central and State Pollution Control Boards.
- Primarily all initiatives were identified in three directions—environmental pollution, forest management, and resource conservation.
- Various acts, rules, and policies came into existence during last three to four decades for managing unwanted waste disposals, which are polluting environment, for managing forests and biodiversity, and for gaining overall sustainability of the nation.
- After the Paris Summit of 2015, India declared its targets under INDC. This includes 33–35% reduction of GHG emissions (back to the 2005 level), creation of sink for absorption of 2.5 to 3 billion tonne of carbon dioxide equivalent, and establishment of power generation plant from renewable resources with 40% share of total national electricity generation. All targets are expected to be achieved by 2030.
- India is showing some progress in achieving at least five SDGs out of 17 SDGs agreed upon as Agenda 2030 of the UN.
- India is yet to show good progress in controlling air pollution, managing urban and electronic wastes, and other social dimensions in terms of happiness of countrymen, although government is giving considerable efforts in doing so.

Prior to the discussion session, it is expected that student groups will be formed. Now each of these questions may be discussed among the group members. The objective of the discussion session is to encourage students to think threadbare and explore all related issues, not arriving at the answer or solution to the problem,

Discussion Questions

1. Most of the UN initiatives in achieving sustainability are based on weak sustainability concept. If the planet is really in danger, why cannot the global body like the UN take any step founded on strong sustainability, even in a modified form?

2. Let us first categorize the countries in terms of an environmental sustainability index. The factors should include all types of pollution, ecological footprint, water use, conservation of scarce non-renewable resources, and generation of unwanted hazardous wastes. Now apply the "polluters pay" concept. The creators of environmental degradation should support (financially and by sharing expertise) the countries affected by the degradation through any mechanism. Is it a utopian scheme? If not, what may be pros and cons of this proposal?

3. Are our legislative mechanisms capable enough to implement all policies and directives at the ground level? If not, what do you suggest to strengthen or modify them? Is it necessary to create any new department under MoEFCC for this purpose or restructure the existing department, particularly for meeting the target of 2030? How do we motivate common Indians toward making India a green, clean, and healthy abode to live in?

REFERENCES

About the Sustainable Development Goals. www.un.org/sustainabledevelopment/sustainable-development-goals. Accessed on 7 October 2018.

https://sustainabledevelopment.un.org. Accessed on 7 October 2018.

http://theenviro.blogspot.com/2012/11/objectivesrole-and-purpose-of.html. Accessed on 24 August 2018.

https://unfccc.int/process-and-meetings/the-paris-agreement. Accessed on 7 October 2018.

Millennium Ecosystem Assessment Report. (2003). *Ecosystem and Human Well-beings: A Framework for Assessment.* Washington, DC: Island Press.

Priyadarshini, G. I. (2016). Environmental policies in India towards achieving sustainable development. *IOSR Journal of Humanities and Social Sciences, 21*(1), 56–60.

Report of TNN on 2 May 2018. https://timesofindia.indiatimes.com/city/delhi/14-of-worlds-15-most-polluted-cities-in-india/articleshow/63993356.cms. Accessed on 9 January 2019.

United Nations Environment Programme. www.unenvironment.org. Accessed on 24 August 2018.

Voluntary National Review on Implementation of Sustainable Development Goals. Prepared by NITI Aayog, Government of India for presentation to UN High Level Political Forum. 2017. https://sustainabledevelopment.un.org/content/documents/15836India.pdf. Accessed on 2 January 2019.

Zhang, H. B., Dai, H. C., Dai, H. X. L., & Wang, W. T. (2017). U.S. Withdrawal from Paris agreement: Reasons, impacts and China's response. *Advances in Climate Change Research, 8*(4), 220–225.

Part III

Sustainability at the Operations Level

3 Key Domains of Sustainable Operations Management in Manufacturing

The previous chapter addresses various parameters, challenges, initiatives at the UN and national levels and the emergence of rules, regulations, and policies in dealing with sustainability. These are mandatory and voluntary frameworks and guidelines that are to be observed or adhered to at the organizational level. Achieving sustainability is the strategic goal of this organization. But it is not so simple a task. Issues are many. How do we operationalize these macro-level initiatives at the unit level (i.e., organizational level)? What should be the areas of concern in managing operations? How do we prioritize them? What activities should be eliminated or at least avoided? There are many others.

Let us consider two types of products at retail stores. Product A is a detergent (produced by P&G, HUL, or a similar FMCG), and Product B is a refrigerator (LG, Godrej, or of a similar make).

During managing operations of Product A sustainably, we are expected to focus on some of the issues:

- Does wastage generated during production create hazard or toxic effect in land, water, or air?
- Is packaging (both industrial and consumer) material made of recycled plastic material?
- Does the company take care of reusing disposed packaging materials?
- Does detergent require lots of water for proper washing and rinsing?

Similarly, for Product B, our attention may be directed toward a set of issues, some of which are as follows:

- Is there any further possibility of overall reduction of materials used in manufacturing the refrigerator?
- Is the generation of wastes and emission of pollutants during the manufacturing process within safe limit?

DOI: 10.1201/9780429195600-6

- Is the manufacturing process not only economically efficient but also energy-efficient?
- Does it consume huge electricity during its use?

Now let us see the activities in the service sector like a restaurant. The following are some of the issues relevant to establishing sustainability in operations.

- How much is the food waste generation per day?
- Are the cooks keeping the ovens burning unnecessarily by an inefficient chain of processes in the kitchen?
- Are there cases of using huge amounts of pesticides during farming to increase the production of vegetables used in cooking?
- Are suppliers of raw materials or farmers located quite far?

This list of issues relating to product or service is showing the areas of concern in making the operations sustainable.

Per the World Commission on Environment and Development (WCED), sustainable development does not compromise with the need of future generation in extending the development further. Sustainable operations management may be defined as a set of skills and concepts that allow a company strategize and manage its business processes in order to obtain competitive returns on its capital assets without sacrificing the needs of the stakeholders and with due regard to the impact of its operations on people and environment. However, it is better to describe its definition considering two perspectives—macro level and micro level.

In the macro-level perspective, sustainable operations management may be conceptualized as a set of international and intranational decisions involving the consumption of natural resources, waste and pollutants introduced into global ecosystems, the maintenance of societal well-being while keeping the balance of source and sink, and the continuity of mankind's development.

In the micro-level perspective, on the other hand, sustainable management of operations is management of a set of business processes or a supply chain for the satisfaction of both the shareholders and the customers, keeping due regard to environmental degradation, ecological balance and conservation of natural resources, and the well-being of the society. In short, the management process is expected to be effective, responsive, and efficient with significant contribution to betterment of surrounding nature, environment, and society at large. The general principles for sustainability in operations broadly cover the following guidelines.

- Generate less pollutants in the air, water, and land and less generation of toxic items as waste materials.
- Achieve energy efficiencies in all stages.
- Use fewer materials both in products and processes.
- Use more reusable materials.
- Create less wastage and reuse byproducts.
- Use green logistics.
- Use of fewer toxic, hazardous, and non-biodegradable materials.
- Design for reuse.

In any industrial organization, this management process is supposed to involve three interrelated areas of decision-making:

1. Sustainable product design
2. Sustainable process design and planning
3. Sustainable facility design

3.1 SUSTAINABLE PRODUCT DESIGN

The designing process of a marketable product is initiated by understanding and studying the needs of customers. This is often known as the voice of customers (VoC), and this is included in the quality function deployment (QFD) model to assess the prioritization of the technical features of the product. The QFD is one of the most popular and useful management tools in product design and development, which helps in translating VoC to design targets and required quality assurance. Product design, in general, is a result of judicious consideration of marketing, technical, and financial factors. Although marketing factors are exogenous in nature, primarily represented by VoC, technical and financial factors are internal parameters. The availability of various resources and technology are technical factors, whereas financial factors are all costs and financial parameters contributing to profitability.

> *The quality function deployment (QFD) is a product design model, which translates the needs of customers or voice of customers (VoC) to design targets and required quality assurance via prioritization of technical features.*

Sustainability is an added dimension in sustainable product design. This dimension is included in **governmental policies and regulations** (mandatory), **corporate sustainability goals** (voluntary), and **VoC**. Several organizations (Sony, McDonald's, Walmart, IOCL, and so on) voluntarily included sustainability in their corporate goals, which is uplifting brand image and values in the market. Customers' awareness on environmental and other sustainability issues also adds new criteria of their needs in VoC. The governmental policies and enactment of regulations get updated regularly matching with the current status of environmental parameters and knowledge on negative impacts of ecological degradation. Once a city is affected by air pollution due to dust particles (e.g., $PM_{2.5}$), the government or even the local authority becomes compelled to introduce restrictions on burning of crop residues or stubbles. Similarly, the emission standard of automobiles is becoming more stringent over the time in terms of carbon monoxide, hydrocarbon, or sulfur dioxide contents of the exhaust. This is because of the global consciousness of their harmful effects on the human body and this planet.

Bharat Stage Emission Standard (BSES) I started in 2000, and India decided to move to BS VI, skipping BS V by 2020. Simultaneously, customers are also becoming aware and educated, and this is reflected on growth of demand for environment-friendly design in vehicles. They are now looking for cars with higher

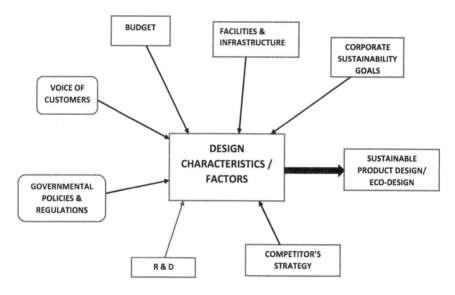

FIGURE 3.1   Basic framework of sustainable product design.

BSES standard, even at the cost of paying higher car price. This led to significant change in design of a vehicle. Sustainable design could be achieved by adding **catalytic converters** in the vehicle and, now, new fitments like **diesel particulate filters** and **selective catalytic reduction modules**. Figure 3.1 shows how the exogenous factors like current customer needs and governmental regulations affect the designing of sustainable products along with the others.

During the late 1980s and early 1990s, both the global producers and consumers started looking for new greenness-focused design criteria beyond functionality, marketability, and cost. Footprint of sustainability needs to be integrated with other conventional factors. Producers had started accepting the fact that these additional factors, which adds costs to production, help create an innovative and attractive product, and hence enhance corporate brand value.

Sustainable design essentially addresses the following concerns.

1. Less material consumption (conservation of natural resources)
2. Less pollutant generation during production, transport, and use (environmental degradation)
3. Energy efficiency during production and use (both an economic and environmental concern)
4. Use of reusable materials (conservation of resources and delaying disposals)
5. Use of recyclable and recycled components/parts/materials (value recovery of used products)
6. Design for remanufacturing (scope for remanufacturing of products/components/parts after primary use)
7. Avoidance of non-biodegradable materials (clean disposal)
8. Avoidance of toxic and hazardous materials (health and safety of human body and society)

In this context, various new terms became popular during last couple of decades. These are *sustainable product design, green design, eco-design, environmentally conscious design, design for environment (DFE)*, and so on. By and large, all these means design of environment-friendly products. However, two other terms are also used for design of reused, recycled, and remanufactured/refurbished products. These are *design for remanufacturing (DFR)* and *design for disassembly (DFD)*, which primarily refer to detachability or separability, modularity, use of recyclable or recycled components or parts, durability of the core, and so on.

Let us share a formal definition of DFE (Billatos and Basaly, 1997). DFE is *"a process that must be considered for conserving and reusing the earth's scarce resources: energy and material consumption is optimized; minimal waste is generated and output waste streams from any process can be used as raw materials (inputs) of another."* Thus, DFE aims at designing a product, taking into account the conservation of limited resources of the planet and the reduction or, preferably, complete elimination of environmental impact due to unwanted waste generation during the production process. **The most important consideration during implementation of DFE is the estimation of environmental impacts during the whole product life and subsequently relating them to various design indicators.**

Design for environment (DFE) is a product design concept that comprehensively takes care of the conservation of scarce natural resources, restriction on the generation of pollutants and unwanted wastes, and the reuse or remanufacture of used products, components, or parts.

3.1.1 Sustainable Product Design Practiced by Factories

A proper green design is expected to balance and integrate the conventional design criteria with eco-design criteria. A smart green design creates products that use less energy and natural resources, products that can be recycled or reused easily, and products that promote energy and material efficiency even among consumers. Reusable napkins, cloth or cotton mesh bags instead of plastic bags, recycled fabrics, precision induction cooktops, reusable water bottles (made of glass, stainless steel, etc.), LED bulbs, and rechargeable batteries are some examples of simple eco-friendly alternatives to traditional products. On the other hand, the use of solar energy in various large and small appliances had started drawing the attention of green-conscious manufacturers, retailers, and customers. Solar appliances fall in two broad categories: (1) solar photovoltaic-based items that produce electricity to power DC-type electric devices and (2) solar thermal systems that capture heat to be further used for cooking food or heating water. We thus get solar-lighting systems, like solar lanterns or flashlights (even street lighting), solar cookers, solar water heaters, and the like. The US Department of Energy suggests a seven-step approach for somebody who intends to power his/her home with solar energy. It is advisable to start the planning from proper requirement analysis and energy audit of home. Let us have a glimpse of practices by some corporations, which are engaged in producing sustainable products.

3.1.1.1   Bosch Home Appliances

Dishwashers are designed for sustainable use with primary focus on conservation of water footprints with eight to ten liters per wash cycle on an average. Moreover, to save water and energy, there is the inbuilt program of "half load" in the machine.

3.1.1.2   Samsung

Samsung's F500 Washing Machine introduces the Ecobubble green technology, which creates water and air bubbles with detergents, which make washing with cold water as good as washing with hot water without Ecobubble technology. Samsung's air-conditioning systems also use energy-saving digital inverter technology with lower consumption of electricity. Because of the digital inverter compressor, Samsung's refrigerators also consume less energy. Of course other white good makers are also doing the same. Further use of R600a refrigerant reduces environmental damage, as it has zero ozone depletion potential and low global warming effect.

3.1.1.3   Daewoo

The Daewoo microwave has the technology of zero-energy consumption. If the oven is not used for ten minutes, it will enter into saving mode with no electricity consumption.

3.1.1.4   Southcrop Whitegoods

Its dishwashers consume less than 18 liters of water in full load. This dishwasher is significantly lighter and has low material use in its manufacturing, and the design is meant for quick disassembly for remanufacturing and recycling (design for disassembly).

3.1.1.5   Harman Miller

Being a well-known multinational furniture manufacturer, Harman Miller formulated its environmental goal in 2004. Soon it could achieve its goal of reduction of operational footprints, and 100% electricity is sourced from renewable resources. Higher durability of items reduces materials consumption, and its chairs are made of recycled materials, amounting to almost 70% of total material content of the chairs. Steel and aluminum contents are also recycled.

3.1.1.6   Xerox

Photocopiers of Xerox Europe are manufactured incorporating almost all green design features, like using less raw materials and more reusable or recyclable parts. Xerox India has established its secondary market as supplier of reconditioned/ refurbished/remanufactured photocopiers.

3.1.1.7   IKEA

IKEA, the Swedish multinational, established its strong footprint in ready-to-assemble furniture and home appliances. It announced to embrace the Planet Positive approach by 2030. The sustainability manager of IKEA announced that all products will be designed from the very beginning to be repurposed, repaired, reused, resold, and recycled. IKEA applied the circular design principle of designing a product, which essentially meet the following commitments.

- Producing products using only renewable and recycled materials
- Removing all single-use plastic products from the IKEA product list by 2020
- Becoming climate-positive and reducing the total IKEA climate footprints by an average of 70% per product
- Achieving zero-emission home deliveries by 2025

IKEA primarily uses materials like aluminum, rough unpolished wood, recycled plastics for kitchen shutters, recycled certified wood for kitchen cabinets, plastic films from recycled plastic bottles, and so on. Almost 77% of the wood used in IKEA products is FSC certified or recycled. Its basic approaches of "more from less" and circular design (reusing, recycling, separability, etc.) are the foundations of its sustainable product design.

3.1.1.8   Background of Sustainable Car Design

Most of the contemporary attempts toward the inclusion of sustainability in car design boil down to manufacturing electric cars. There is a heavy rush in auto sector (Mahindra and Tata Motors in India, for example) to make electric vehicles primarily for the environment-conscious customers in affordable price. Whether it is an all-electric, hybrid, or plug-in hybrid variety, efforts are made to achieve a downward trend of prices for cars and commercial vehicles. It is expected that sales of electric cars will grow to around 7% (i.e., 6.6 million per annum) globally by 2020 (DOE, 2014). "The History of Electric Cars" (DOE, 2014) surprisingly shows that electric cars were introduced more than 100 years ago by a series of breakthroughs in Europe and the US. As reported by Z. Shahan (2015) in his article "Electric car evolution" (available in the Clean Technica site), a Hungarian inventor, Anyos Jedlik, built the first car powered by an early electric motor back in 1828. After a series of improvisations and developments, the marketable electric cars became available to citizens. Subsequently, electric vehicles threw a strong challenge to gasoline-powered vehicles in the US, and this continued for some years. But Ford's highly efficient Model T ultimately made the gasoline-powered automobiles winners in the race, and they completely replaced the electric vehicles in 1930s.

After nearly 40 years, the Arab oil embargo led to the rising cost of oil and its scarcity, which resulted in re-establishment of interests on electric-powered automobile technology worldwide. In the 1990s environmental consciousness started influencing governmental policies and industrial community toward this shifting of focus. The first hybrid electric car was introduced by Toyota by manufacturing its Prius model in 2000. Toyota used a nickel metal hydride battery as source of power. Tesla Motors introduced a luxury electric sports car as the second entry to this car segment.

In this direction, India is striving for achieving its sustainability goal by announcing its mission by 2030 to completely have electric vehicles, signing agreement with Japan on electric vehicle development, and incentivizing electric vehicle battery manufacturers.

Tata Motors' Tigor, Tiago, and e-Vision models are typical examples of its contribution to electric car segment. It has also successfully completed its Starbus hybrid electric bus for public transport. On the other hand, Mahindra's contributions in this pursuit include the e-Verito, e20 Plus, and e-Supro cargo van models. The government of India (Ministry of Power) formed the Energy Efficiency Service Limited (EESL), a large energy service company whose 100% shareholder is the government and which is meant for supporting and facilitating various energy efficiency projects. Tata and Mahindra bagged large orders from EESL for supplying electric vehicles.

3.1.2 PRODUCT DESIGN ON THE BASIS OF PRODUCT VALUE

Any business organization continuously attempts to increase its financial earnings by simultaneously improving the value of a product or service, as determined by its worth perceived by customers. Simultaneously, the organization strives for the reduction of the cost or investment for creating that value. The pricing policy is primarily determined by the cost of products sold and profit expectancy of the shareholders. Of course, there exist a set of other external factors in this business dynamics, like the competitors' strategy, governmental regulations, the system and structure of the organization, and the mission and vision of the organization. The design team is expected to formulate the design strategy for enhancing the product value.

Sustainable design is also related to the extension of the valuable life of a product. Broadly, product life may be classified into two types—usable or useful life and marketable life.

The *usable life* of a product at the customer's end is determined by its functionality (controlled and enhanced by repair and maintenance), obsolescence (difficulty in getting spare parts, components, and after-sales service), unreliability (high frequency of breakdowns, demanding replacements), uneconomic use (incremental cost of use), legal nonviability (legal issues/guidelines or restrictions on driving old diesel-powered vehicles on metro roads, for example), and so on. This is the period of its use by the customers. After the end of this life, the customer will be interested in replacing the product by a new one and dispose the old one. Any activity related to the recovery and/or addition of the product's value at the end of its life enables the extension of its usable life.

The *marketable life* of a product is determined by the demand of the market, as shown in four distinct stages (introduction, growth, maturity, and decline) of a product life cycle. Although this influences new product development decision and the production planning activities, market factors do affect this marketable life.

It may be noted that *usable life* of a product is determined by the technical parameters and work environment at the user's end, whereas the *marketable life* is primarily influenced by market-related factors.

Attempts are made both by customers and manufacturers to add value for the extension of the useful life of a product. The following two ways may extend the useful life of a product:

1. Extension of useful (with usable value) life at the customer's end: This emphasizes the restoration and enhancement of the functionality of the product by repair, maintenance, or upgrade.

2. Value recovery after the end of the useful life of a product: This may be carried out by OEM or other manufacturers by refurbishing, remanufacturing, or even recycling.

Some sustainable design strategies may be adopted for achieving these two goals, as proposed by Giudice (2008). Table 3.1 displays the summarized list of the strategies identified based on the guidelines of Giudice.

TABLE 3.1

Design Variables, Design Strategies, and Impact on Product Life Cycle

Design Level	Design Issues/ Variables	Sustainable Design Strategies	Impact on Product Life Cycle	
			Useful Life Extension	End-of-Life Recovery
System	Designing the product	Minimize the number of components and parts	✓	✓
		Extensive modularity	✓	✓
		Inclusion of multifunctional and upgradable components	✓	✓
		Accessibility to components	✓	✓
		Design for remanufacturing	✓	✓
	Relations between components	Restrictions to limited connections and variety	✓	✓
		Design for disassembly	✓	✓
Component	Parts or materials	Eliminate hazardous materials	✓	✓
		Reduce non-biodegradable materials	✓	✓
		Reduce material variety		✓
		Increase standard materials with high compatibility and recyclability		✓
	Shape	Optimize performance and reliability	✓	✓
		Ease of removal of parts	✓	✓
	Dimensions	Reduce volume and even weight	✓	✓
		Optimize performance and reliability	✓	✓
		Ease of removal of parts	✓	✓

Product life may be classified as having usable life and marketable life. The usable life of a product is determined by the technical parameters and work environment at the user's end, whereas its marketable life is primarily influenced by market-related factors. The extension of usable life is one of the primary objectives of sustainable product design.

3.1.3 Golden Principles of Sustainable Product Design

Designing a sustainable product needs due consideration of the following six dimensions of concern. An effective integration of these six dimensions requires prioritization and/or inclusion of possible trade-offs among the criteria in formulating the final design of the product. Decisions on appropriate priorities for the dimensions are made matching with demands from contemporary business environment, market, society, and governmental policies. The relevant factors impacting on sustainable product design may be summarized as follows:

1. Functionality (customer voice/needs): quality, usability, reliability, product useful life span, disposal issues
2. Marketability and distribution-friendliness (market and downstream factors of supply chain): customer acceptability, competitive positioning in the market, distributor/wholesaler/retailer specific requirements, Exim policy of the government, GST
3. Technical factors (feasibility and availability related): availability of raw materials/parts/components, feasibility and availability of technology and know-hows, infrastructure and layout, usability and maintainability of the product, estimated product life cycle
4. Financial factors (costs and revenues as producer's voice or needs): product and logistics cost, pricing factors in a competitive market, profitability, fund acquisition and management, inventory and other related working capital management factors
5. Legislative compliances (governmental policies/regulations): mandatory restrictions/compliances, need for accreditations/certifications, regulations/ rules
6. Sustainability (call of environment/ecosystem): corporate sustainability goals, materials consumption rate, energy efficiency, pollutant emission, waste generation, use of non-biodegradable or hazardous materials, need for use of recycled parts/materials, recyclability/reusability/remanufacturability

The relevant factors having impact on sustainable product design are functionality, marketability, technical factors, financial factors, legislative compliances, and sustainability issues.

Short et al. (2012) carried out an empirical study to understand the awareness and attitude of large firms of Sweden and the UK on green designing. The result shows that although the corporations are aware of the sustainability concern in product design and they do agree to consider it in design process, most of them are not quite equipped with techniques and approaches. In this pursuit, the concerned actors or stakeholders are advised to implement the following ten golden principles proposed by Luttropp and Lagerstedt (2006) and Luttropp (2017).

Golden Principle 1: The functional value of the product and corresponding components or materials is to be given due importance, along with the environmental impact, so that the reuse program may be subsequently initiated for reusing the remaining value with some value addition.

Golden Principle 2: Human resources should be managed in a sustainable way during the production and distribution of the products.

Golden Principle 3: All materials, including input materials to the manufacturing process of suppliers, are to be studied. Hazardous materials are to be minimized or, if possible, completely removed.

Golden Principle 4: Material resources are to be used efficiently in production. The use of materials in logistics and other material-handling activities should also be efficient. The lean concept is preferred in this context along with implementing all possible activities for waste reduction.

Golden Principle 5: Firms are to remain profitable for maintenance of economic sustainability. Management should strengthen its promotional activities, so that the additional revenues may be earned by enhancing corporate image, which will offset the extra investment required for introducing greenness in product design.

Golden Principle 6: Energy efficiency is an essential consideration in product design. This is to be assessed in all stages of the life cycle (i.e., production, logistics, use at the customer's end, and even at disposal or reuse).

Golden Principle 7: Separability, detachability, and modularity are to be taken into consideration during product design for easy disassembly during upgrading, refurbishing, or remanufacturing.

Golden Principle 8: Each firm should plan for recovery and reverse logistics of used products so as to repair, upgrade, reuse, remanufacture, or recycle them.

Golden Principle 9: Environment-friendliness should be inculcated as a corporate culture in the organization, like TQM implementation.

Golden Principle 10: There should be proper display of all relevant information in the form of appropriate labeling on the product, and its packaging must show the plastic parts, the detachable and recyclable parts, and the like. This reporting mechanism helps in the disposal, recycling, and disassembly after the useful life of the product.

3.1.4   ESSENTIAL FACTORS IN MARKET COMPETITION

Two sets of factors play crucial roles in sustainable product designing and subsequent survival in a competitive business environment—*enablers of competitiveness* and *relevant cost factors.*

3.1.4.1   Enablers of Competitiveness

Manufacturers consider some factors while designing a new product, such as price, quality, delivery speed, response time, and after-sales services. These contribute to competitive advantage to the manufacturers. Although other factors also do play important roles in winning and surviving in the market competition, the primary determining factors for sustainable product design are greenness (legal requirements and environment-friendly demands of customers) and price.

Usually, the existence of this type of products may be explained by the following series of activities occurring sequentially.

1. One of the prospective producers intends to take the lead and invests significantly on research and development for new product green design.
2. Correspondingly, its suppliers invest for relevant component designs and for a new production setup.
3. The production of and market for the new product meets the mandatory governmental regulations and current market demand for greenness. The producer gets the first mover's advantage and enjoys winner's status in the market.
4. As the market of the green product matures, other producers strive to catch up with the new trend and invest in technology adoption or imitation of the design. Incidentally, this investment amount is much lower than the investment required for research and development and new product development.
5. Now competition grows, and subsequently the product price (if not, service level, speed of delivery, etc.) becomes a differentiating factor. So manufacturers put their efforts in making the production process (or even the whole supply chain) more efficient by implementing leanness (leanness and greenness, for example), Six Sigma, process improvement techniques, and so on. This results in reduction of marginal costs.

Thus, greenness (primarily driven by external forces) and product price are the two enablers of competitiveness, which dictates sustainable product designing.

3.1.4.2   Cost Factors

Qian (2011) had shown that the decision on new product development is essentially influenced by cost factors, which ultimately lead to classifying product types as development-intensive products (DIPs), marginal-cost-intensive products (MIPs), and marginal-cost-and-development-intensive products (MDIPs). DIPs and MIPs are supposedly two extreme types of products. Actually, industrial economists identified these classes to incorporate better-quality-focused product design or designs with additional features.

Some design strategies require intensification of research activities with high investment, which substantially add to fixed costs and the resultant products are known as DIPs. The development of new product models or new facility layouts or setups is the representation of DIP. Designing an electric car in place of a diesel-powered one is a typical example of DIP. It is very common in pharmaceutical and software industries. The feasibility of this product design is determined by achieving economies of scale with expectation of a high sales volume. These strategies actually reflect radical change in product designs, which subsequently demands new process design and new supporting infrastructure.

On the other hand, MIPs include products made by adding new parts or materials to an existing design for the enhancement of functionality and other values. It is more like incremental value addition. The white goods sector and electronic industry (covering products like mobiles and laptops) often launch products as MIPs. MIPs do not significantly increase fixed costs in the profit-and-loss accounting document, whereas the marginal or variable costs do increase because of additional material costs and/or labor costs. In reality, DIPs also require some changes in marginal costs, and MIPs may enhance fixed costs slightly. MIDPs, in fact, are the representation of a class of products that leads to some changes on both these costs.

Green or sustainable product (GP) designs are classified into similar groups development-intensive green products (DIGPs) and marginal-cost-intensive green products (MIGPs), adding the greenness factor along with cost factors in product design. Examples of DIGPs include the new design of electric cars in the automotive industry, new battery technology, the inclusion of DFR in designing, closed-loop supply chain design, designing solar-power-operated products as replacements of electric ones, designing air conditioners with higher star rating, and designing washing machines with lower water consumption rate. MIGPs use scrap aluminum for mobile phones or laptops, replace lead-acid lithium-ion batteries, install exhaust emission control devices in manufacturing plants, reuse wastewater by purification process, and so on. As mentioned earlier, these are practically MDIGPs with varied proportions of both fixed costs and marginal costs.

Decisions on sustainable product design thus require judicious consideration of both fixed and marginal cost factors for the achievement of competitive advantage, balancing greenness and pricing.

The crucial factors in sustainable product design are competitiveness and cost. There exist two extreme forms of green product design: development-intensive green products (DIGPs) and marginal-cost-intensive green products (MIGPs), which are representations of radical value addition or continuous value addition.

Key Learning

- The implementation of sustainability at any organization at the operations level demands its consideration during decision-making on product design, process and logistics design, and facility design.
- Design for environment, sustainable product design, eco-design, green design, and design for remanufacturing are the popular terms reflecting sustainable products, each with a different emphasis on one of the related aspects.
- The primary drivers for sustainable product design are state regulations and policies, and updated and renewed voices of customers demanding green products and clean environment.
- The designer of a green product needs to take into account greenness factors during the production of the product, during its use by customers, and even during its disposal. Various companies have successfully launched their green products, and the market showed signs of acceptance.
- The entry of electric vehicles into the automobile sector is a great example of acceptance of society, particularly the intense environmental consciousness of the market. Interestingly, history shows that electric vehicles were in existence decades ago.
- Sustainable design also takes care of the extension of the useful life of a product.
- The ten golden principles are supposed to support implementation of this design process.
- The role of a sustainable product as an enabler of competitiveness and the relevant cost factors for its production are to be considered when designing the product. In this pursuit, products may be characterized as development-intensive green products (DIGPs) and marginal-cost-intensive green products (MIGPs), depending on the enhancement of either fixed costs or marginal costs.

3.2 SUSTAINABLE PROCESS DESIGN AND PLANNING

Several studies and researches in last two decades or so, including that of Rao and Holt (2005), justify the fact that the greening of industrial processes not only improves the health and overall living conditions of the society but also results in substantial cost savings, sales growth, and exploitation of new market opportunities. However, traditional conflicts and trade-offs between cost and customer service in production and logistics usually underestimate the environmental effects. **This is quite prominent in some reports showing that efficient practices like centralization of inventory, just-in-time (JIT) strategy, multiple sourcing strategy, and the like are not environment-friendly.** However, industry practitioners and researchers are opting for various techniques, which meaningfully balance the need for cost savings, customer satisfaction, and environment-friendliness.

Green operations in the industry include the following areas:

- Green manufacturing
- Green sourcing and inventory

In this context, it needs to be mentioned that in this book there is no exclusive chapter on green logistics, although in the section about facility design in this chapter, reverse logistics in Chapter 4, and carbon footprint analysis/life cycle analysis in Chapter 5, it is discussed quite elaborately. Green logistics primarily addresses the following decision problems associated with sustainability issues.

- Carbon emission reduction concerning the engine of the vehicle and efficiency of fuel consumption
- Reduction of "empty load" movements
- Shared load movements to the maximum extent
- Conversion of traditional vehicles to electric vehicles
- Cost-benefit analysis on the use of FTL and large-sized vehicles by minimizing carbon emissions per tonne-kilometer
- Redesigning the distribution network for better sustainability

3.2.1 GREEN MANUFACTURING

Sustainable or green manufacturing (GM) means making the manufacturing process sustainable. An interesting conflict may arise in this context: whether it is "greening the manufacturing" or "manufacturing the greens." The former seems to be more appropriate, as later focuses on the product, not the process. Since 2008, researchers and experts went on attempting to capture the meaning, role, and concept of GM in a universally accepted definition. The most cited and popular definition came up little later, suggested by the US Department of Commerce: *"the creation of manufactured products that use processes that minimize negative environmental impacts, conserve energy and natural resources, are safe for employees, communities, and consumers, and are economically sound"* (Haapala et al., 2013). So sustainable or green manufacturing is a set of processes that considers environmental impacts (e.g., air or water pollution and solid wastes), conservation (natural resources, including water and energy, particularly non-renewable ones), and safety of human beings, along with economic aspects. There is another definition, which seems to be more detailed one. It describes sustainable manufacturing as *"a set of processes and systems demonstrating reduced negative environmental impact, offering improved energy and resource efficiency, generating minimum quantity of wastes, providing operational safety, and offering improved personnel health, while maintaining and/or improving the product and process quality with overall life cycle cost benefits"* (Badurdeen and Jawahir, 2017). The concerned areas of this set of processes are environmental degradation, resource consumption, health and safety, and the cost of business. The BCG and CII report (Bhattacharya et al., 2011) includes some relevant issues on GM.

Although "green" and "sustainable" are synonymous in terms of application, some experts use the term "sustainability" as the ultimate goal to reach. In this context, we may explain the creation of sustainability value by the transformation of traditional manufacturing strategy at three different levels: *lean, green, and sustainable manufacturing*. This explanation may be expressed by considering four life cycle stages: pre-manufacturing (product designing and procurement), manufacturing, use at the

TABLE 3.2
Three Levels of Transformation of Manufacturing for Inclusion of Sustainability Value

	Manufacturing	Use	Post-use
Lean manufacturing	Reduce	Reduce	
Green manufacturing	Reduce	Reduce	
		Reuse	
			Recycle
Sustainable manufacturing	Reduce	Reduce	
		Reuse	
			Recycle
			Recover
	Remanufacture		

customer's end, and post-use. *Using the popular R format, the primary goal of lean manufacturing is achieving 1R (reduce waste and resource consumption), whereas that of green manufacturing is implementing 3Rs (reduce, reuse, and recycle). Sustainable manufacturing emphasizes the implementation of 5Rs (reduce, reuse, recycle, recover, and remanufacture) or in some cases 6Rs (including redesign).* It means, if we do differentiate between green and sustainable manufacturing, then the latter may be considered as highest level of sustainability, with remanufacturing as a unique form of value-additive business. This highest level of manufacturing includes the product recovery process after primary use. Table 3.2 displays these three levels of transformation.

> *Sustainable manufacturing primarily aims to minimize the adverse environmental impacts, conserves natural resources, and considers human quality of life and health, along with economic aspects. Lean, green, and sustainable manufacturing are the three levels of achieving all possible Rs.*

Any manufacturing process uses energy and material inputs, both of which are the outputs of industrial processes and exploit natural resources in single or multiple stages of conversion processes. On the other hand, customers dispose of the products after their use, and the products once again enter into the nature through the natural decomposition. The environmental degradation takes place during conversion of natural resources to usable input materials, during the manufacturing process, and during the use of the products by the customers. Sustainable or green manufacturing is supposed to consider both the conservation of natural resources (non-renewable in particular) and reduction of environmental degradation during the process of gaining economic value from manufacturing. Figure 3.2 depicts the outline of both the natural ecosystem and man-made socio-economic system along with their linkages.

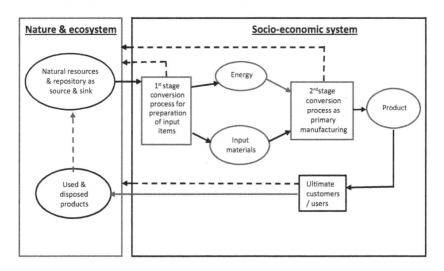

FIGURE 3.2   Interactions between the man-made socio-economic system and nature.

The figure shows some dashed lines representing the unwanted environment degrading elements like emission of pollutants (air, water, and solid wastes) during the two stages of conversion process (initial value addition for the preparation of input materials and the main manufacturing process) and during the use of product at the customer's end.

Green manufacturing involves transformation of industrial operations with due consideration to the design of green products (already discussed in previous subchapter), the use of green energy, and green business processes. *Green energy* means clean energy in terms of its generation. It further includes the replacement of energy from non-renewable resources with renewable ones and the efficient use of energy. *Green processes* strive to lower all types of resource consumption, waste generation, and emission of any form during any of the processes.

Now the question is, why is the industrial community be interested in adopting green manufacturing practices? Experts identified following driving forces, which are likely to play as motivators for adopting sustainable manufacturing in place of the traditional one.

- Scarcity of natural resources, including water
- Visible trend or future projection of increase in energy and other input costs
- Adherence to legislative restrictions and implementation of national policies related to environmental pollution, health hazards, and resource conservation
- Technological advances in clean technology and the use of alternative energy and processes
- Taxes and other penalties for environmental degradation
- Availability of rich knowledge and research outputs on new designs for sustainable products

- Strong customer demand for green products and a cleaner environment
- Inclusion of sustainability as a strategic goal for enhancing corporate value and image
- Production of sustainable product for differentiation in a competitive market

Design and planning for the management of green manufacturing is managing three types of green activities relevant to three items of industrial processes. Figure 3.3 depicts how these activities meet sustainability goals for gaining competitive advantage.

The following four sets of items clearly show how some green activities enable the three types industrial items achieve the strategic goals.

1. *Relevant items*: **Input materials, energy**, and **wastes** are the areas requiring primary attention for the efficient management of green manufacturing activities.
2. *Green activities*: By green activities here, we mean the specific activities meant for enhancing greenness in the manufacturing process. These include **cleaning** (pre-manufacturing activity), **utilization**, and **retreatment**. Cleaning is not a direct activity under the control of the manufacturer. It actually means green sourcing and procurement and the use of green materials and energy in the manufacturing process. Green input materials mean they have been produced sustainably. On the other hand, clean or green energy is generated in an environment-friendly manner and mostly from renewable natural resources. Utilization is planning and execution of appropriate methods with improved efficiency of materials and

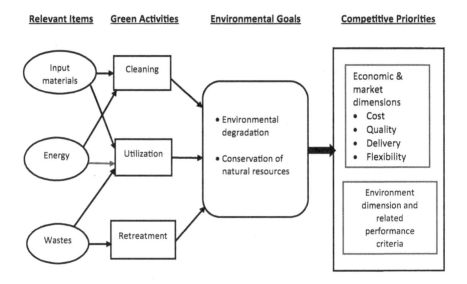

FIGURE 3.3 Basic components and simple activities required for implementing sustainability in industrial organizations.

energy use and reduction of waste generation. It does follow the principle of leanness producing "more using less." Retreatment is treating the wastes generated during manufacturing. It includes reusing and recycling wastes (gaseous, liquid, and solid) for value recovery and/or value addition instead of their disposal to nature. Recycling wastewater for some meaningful use and reusing wastes or byproducts, like fly ash, scrap metals, plastics, and packaging materials, are the typical industrial examples of retreatment.

3. *Environmental goals:* The execution of sustainable processes using three basic green activities helps achieve two primary environmental goals—prevent environmental degradation and promote resource conservation. The goals may be expressed as mandatory compliance with national policy or legislative restrictions. The compliance with Indian policy reflecting its INDC per the Paris Agreement led to fixing the target of reducing the GHG emission rate of 2005 by 35% by 2030. The environmental goals may also reflect voluntary corporate sustainability goals on the basis of the mission of a firm (better brand image). Walmart's sustainability goals include use of renewable or clean energy and zero-waste generation. Reducing carbon footprint, reducing water usage, using clean energy, reducing energy consumption, and using recycled materials are some of the prominent environmental goals considered in the industry. Of course, here some other societal goals are not included, like health, safety, and other social responsibilities.

4. *Competitive priorities*: The achievement of these goals leads to gaining competitive advantage of the manufacturer. Proper matching of traditional manufacturing strategies with sustainable manufacturing strategies results in the simultaneous fulfillment of the economic and market-focused criteria (cost, quality, delivery, and flexibility) and the environmental ones.

Thus, green or sustainable manufacturing processes are expected to meet the following responsibilities.

1. *Emission control*: The emission of pollutants may be direct (generated by the manufacturing process itself) or indirect (no direct responsibility of manufacturer). For direct emission control, the manufacturer adopts tools like electrostatic precipitators, dust collectors or filters, cyclones, and dust suppression systems.

2. *Green sourcing*: Suppliers who produce materials using recyclable or recycled items and non-hazardous or non-toxic elements are selected for supplying input materials.

3. *Efficient process*: Plant layout and production planning system should be as efficient as possible with the minimum use of materials and energy. This is also economically beneficial.

4. *Green energy use*: Energy generated from renewable natural resources should be preferably used.

5. *Waste reduction and recovery*: The process should be designed with appropriate tools for minimal generation of all types of wastes.

Retreatment and recycling plants should be installed in the factory for the recovery of useful wastes and byproducts. Economic value may be extracted from byproducts instead of disposing them. Even spent energy may be recovered using a kinetic energy recovery system (KERS). Five to 25% of the energy is expected to recover by this system. Delhi Metro could earn carbon credits by implementing a regenerative braking system.

6. *Monitoring and control of environmental degradation*: Processes may be continuously monitored by stations at suitable places for assessment of ecological, carbon, and water footprints, along with other sources of environmental factors, like dust, vibration, and noise (applicable in mining industry).

7. *Product recovery*: Value recovery from used products may be added as an additional chain of processes along with manufacturing process of OEM. BMW, Volkswagen, and Xerox have their remanufacturing plants for production of resalable remanufactured products.

3.2.2 LEAN MANUFACTURING

Lean is green. Lean manufacturing not only improves efficiency and contributes to quality of the process; it also reduces non-value-additive activities and the generation of wastes and thus makes the process greener.

Lean manufacturing meets the objectives of enhancing efficiency, reducing costs, reducing wastes, improving customer responsiveness, and hence improving profitability and corporate image, whereas green systems attempt to reduce the consumption of materials and energy and also waste generation. Although apparently these two sets of objectives seem to be parallel, there are the common goals of improving efficiency of input resources and reduction of waste generation. *Lean manufacturing focuses on value creation. Value stream mapping* helps separate out value-additive steps from wasteful steps. Value addition can be achieved by implementing pull-based processes. So waste reduction is an intrinsic component of lean manufacturing system. Industrial experts agree with the fact that the "lean is green" slogan reflects a win-win scenario.

A lean manufacturing system is developed primarily based on following principles, each of which contributes to greenness.

- Kanban or pull-based system: This reduces excess inventory of raw materials or work-in-process materials, which in turn reduces rate of consumption of materials, meeting the goal of conservation of natural resources.
- SMED, or single-minute exchange of dies: With the reduction of change-over durations, there is enough possibility of controlling waste generation (Bergmiller and McCright, 2010).

- 5S: It includes activities like sort, set, shine, standardize, and sustain. These, along with better housekeeping, lead to control on waste generation and economization of materials consumption.

> *Lean is green. Lean manufacturing not only improves efficiency and contributes to quality of the process; it also reduces non-value-additive activities and generation of wastes and thus makes the process greener. It primarily follows the principles of Kanban, SMED, and 5S.*

3.2.3 HOUSE OF SUSTAINABILITY (HOS) MODEL

Mukherjee (2011) proposed the house of sustainability (HOS) model as an innovative method of capturing all sustainability dimensions in formulating sustainable strategies. The HOS model is primarily based on the house of quality (HOQ) model, used for product design per customer needs. Mukherjee explained its application in coal mining industry, the outcome of which helped the corporate decision-maker assess the impacts of mine design parameters. Thus, the specific alternative mining strategy may be selected or existing strategy may be improved.

The dynamics of sustainability, as depicted in Figure 3.4, includes two actors (coal mine management and government) and two other stakeholders (society and

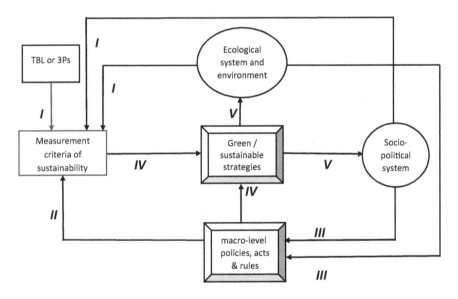

FIGURE 3.4   Basic dynamics of sustainable development.

(Source: Mukherjee, K. (2011). House of Sustainability (HOS): an innovative approach to achieve sustainability in the Indian coal sector. Handbook of Corporate Sustainability. Edited by M.A. Quaddus and M.A.B. Siddique, Edward Elger, Cheltenham, UK., 57–76.)

environment), which get affected and also react to impact of decisions taken by the actors. Please note that the concerned coal mining company is a public sector undertaking (PSU) with maximum share of ownership being kept by government of India. Figure 3.4 includes five sets of linkages connecting the actors and stakeholders, which, in totality, depict the dynamism of sustainability in a large organization like Indian coal mining company.

Set I linkages: Identification of sustainability criteria and measurement scale on the basis of the TBL principle along with the relevant parameters of the existing ecosystem and the surrounding socio-political system.

Set II linkages: The measurement criteria are refined by governmental policies and current priorities as instructed by governmental agencies.

Set III linkages: These linkages show how the current status of ecological and socio-political systems affect the macro-level policies or rules for further amendment.

Set IV linkages: These show the essential influences of sustainability criteria and macro-level policies and rules in formulating the sustainable strategies.

Set V linkages: These represent the impact on ecological and socio-political systems because of the implementation of green strategies of the mining organization.

Strategies on mine design are formulated on the basis of these sustainability factors and contemporary governmental directives, policies, or regulations (following the linkages of Set IV). Similarly, governmental agencies (e.g., pollution control boards at central and state levels) or policy-makers (e.g., ministry of environment and forests) formulate national policies or enact acts or rules on the basis of current condition of society and its level of awareness and demand for a cleaner planet (linkages under Set III). Once a mine design is implemented, resulting environmental status and societal opinion get updated (linkages under Set V). So the effectiveness of this decision-making process is determined by how correctly the sustainability factors have been captured and how they have been matched with technical factors of the mine design. Mine design parameters are classified under three groups: design of the mine (methods or techniques of mineral extraction in a mine, material-handling system design, location of mine entries, layout of the mine, etc.), design of the industrial complex (design of supporting systems: coal-handling plants, surface transport, workshops, mineral-processing plants, coal washeries on the surface, etc.), and design of the infrastructure for the employees and local community (construction of residences, roads, health centers, schools, etc.). For a detailed list of the parameters, readers may refer to the main source of this discussion: Mukherjee (2011).

Figure 3.5 is a simple and brief representation of planning framework for sustainable mining process. It may be noted that the mine design strategy involves the

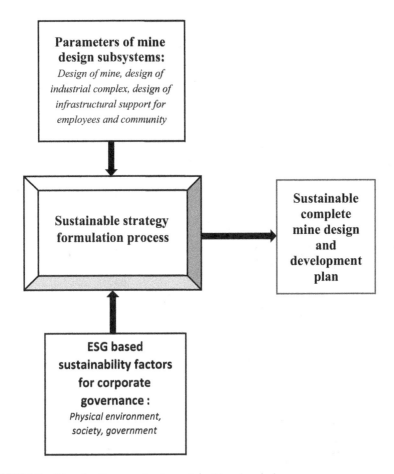

FIGURE 3.5   Planning framework of a sustainable mine design.

appropriate selection of several design parameters. For example, a mine production strategy includes the design and development of an underground mine with a specific mine layout and with a suitable level of technology. The design strategy of a mine also covers the decisions on a specific location for entries to the mine (location of shafts), the capacity of workshops, the need for a specific material-handling system, and so on. The impacts of all these mine design parameters relevant to mine production strategy are to be evaluated by the environmental sustainability factors, and accordingly, an alternative set of parameters is to be considered for mine production strategy design. Mukherjee (2011) developed the house of sustainability (HOS) model based on the principle of the well-accepted house of quality (HOQ) model, or the quality function deployment approach, applied to product design decisions.

Interestingly, mining operation is a well-known and highly profitable economic activity. Incidentally, in mining activities, we do not manufacture any physical

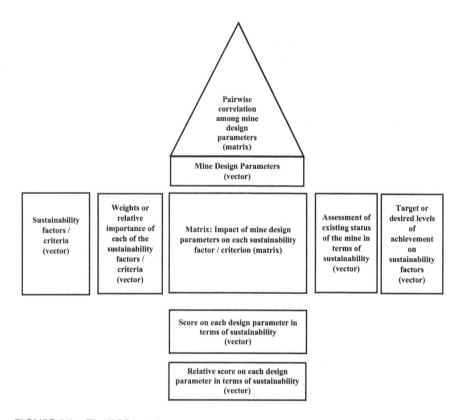

FIGURE 3.6   The HOS matrix.

(Source: Mukherjee, K. (2011). House of Sustainability (HOS): an innovative approach to achieve sustainability in the Indian coal sector. Handbook of Corporate Sustainability. Edited by M.A. Quaddus and M.A.B. Siddique, Edward Elger, Cheltenham, UK.)

product; rather, we dig the earth, dislodge or extract the coal or ore below the ground (most often at different levels of depth), and transport it to the surface to the right destination. We are to carry out this huge set of tasks by investing a lot and maintaining profitability both in operational strategy and planning.

The HOS model supports taking right decisions in selecting the mine design alternatives and process (extraction and logistics) design alternatives, keeping in mind both sustainability and profitability. Figure 3.6 demonstrates the interactions between the design parameters and sustainability factors/criteria through the model of HOS.

The impact of a design parameter on a sustainability factor may be assessed in different manner. We are considering here numerical assessment, like *strong (9)*, *medium (3)*, and *weak (1)*. These may be assessed by any other suitable and adaptable scale. The quantification of the impacts and the relative importance or weight estimated for each of the sustainability factors on the basis of opinions and directives of stakeholders help the quantitative evaluation of sustainability worthiness of all mine design parameters. This is an excellent decision support system for mine process design with a special focus on achieving sustainability.

The following steps are meant for evaluating the impacts of mine design parameters, which in turn help decision-makers formulate appropriate strategic process plan of a mine.

Step 1: Let s and p represent the indices for sustainability factors and mine design parameters respectively, then $F_s = d_s \sum_p R_{sp}$ for each s.

where F_s = Total impact score on s^{th} sustainability factor considering the proposed design parameters

d_s = Relative importance or weight of sth sustainability factor

R_{sp} = Impact of pth mine design parameter on s^{th} sustainability factor

The column vector comprising F_s of all sustainability factors show the current sustainability status of the mine.

Step 2: $H_p = \sum_s d_s R_{sp}$ for each p.

where H_p = Total impact score of pth parameter considering its impact on all sustainability factors

The row vector having values of all H_p at the bottom of the house in Figure 3.6 represents total impact scores of all mine design parameters.

Step 3: $I_p = H_p / \sum_p H_p$

where I_p = Relative impact score of pth design parameter among all parameters.

The bottom-most row vector HOS in Figure 3.6 shows the relative impact scores of all design parameters.

The important outcomes are shown in the values of F_s and I_p vector. The impact scores represented by the values of F_s are compared with values in column vector of the target or desired impacts. These may be the strategic targets for competitive advantage or the mandatory limits of environmental degradation as fixed by the Ministry of Environment, Forest, and Climate Change (MoEFCC), central or state pollution control boards, or any other similar agencies. It may also depict the strategic goal or target of the organization. For example, it may be the maximum limit of annual carbon footprint fixed at the national level. It may also be the minimum quantity of recycling the waste products or emitted water from washeries, coal-handling plants, and the like to be subsequently used for planting trees or creating landscapes, for example. Values of H_p vector represent total sustainability impact of all the mine design parameters, which are the essential inputs for formulating sustainable process design of a mine. The row vector with I_p values of all mine design parameters provide required prioritization or relative importance of the design parameters in mine process design in order to meet the sustainability targets.

This HOS model is applicable in sustainable process design in other sectors as well. However, sustainable process design is more important in the process-focused industry than product-focused ones. However, it may also be equally applicable in designing green products. HOS may be treated as an effective model for designing sustainable processes in chemical plants, refineries, cement factories, paper mills, and so on.

> *Similar to the house of quality (HOQ), the house of sustainability (HOS) is proposed by the author for designing of either green product or sustainable process. It was explained as a model for process-focused industries.*

3.2.4 HIERARCHICAL FRAMEWORK FOR SELECTING A SUSTAINABLE PROCESS DESIGN

Quaddus and Mukherjee (2013) developed the analytic hierarchy process (AHP) model for solving mine project selection problems, incorporating all possible sustainability factors. A real-life case was taken from the Indian mining industry for assessing the validation and suitability of the model. Corporate management of the Indian coal mining company intends to evaluate two coal mine projects, considering all techno-economic and sustainability factors, with the ultimate intention of selecting one of them. The mine projects differ in terms of geo-technical factors (e.g., mineable reserves and overburden volume, quality of coal, types of industrial clients, capacity of production equipment), economic factors, environmental impacts, and social impacts. A decision support system is to be designed for helping decision-makers make appropriate decisions.

In order to implement the AHP, the researchers (Quaddus and Mukherjee) first identified five levels of decision-making for creating the hierarchical framework of the whole decision problem (see Figure 3.7). This is a commonly used framework for applying the AHP model in selecting any investment project developed by Professor T. L. Saaty.

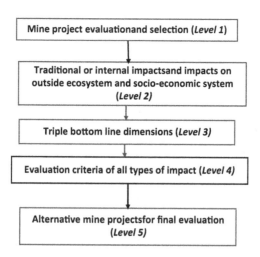

FIGURE 3.7   Basic levels popularly used in the AHP modeling framework for project evaluation.

Level 1 represents the problem itself, whereas level 2 includes the two types of impacts. Subsequently, we consider sustainable development dimensions (level 3), followed by evaluation criteria and finally the level (level 5) of alternative mine projects. In a mine project evaluation problem, the impacts are divided into internal impacts and external impacts. Here, internal impacts are assessed by traditional economic evaluation process of project appraisal, whereas external impacts are representations of contribution to sustainable development, and thus, these are further divided into sustainability dimensions in the next level. Level 4 shows the criteria of evaluation based on the dimensions identified in the previous level. Level 5, or the last level, includes alternative projects under each criterion. Interested readers may refer to Quaddus and Mukherjee (2013) for detailed description of the model and solution. However, Figure 3.8 gives a glimpse of the AHP structure.

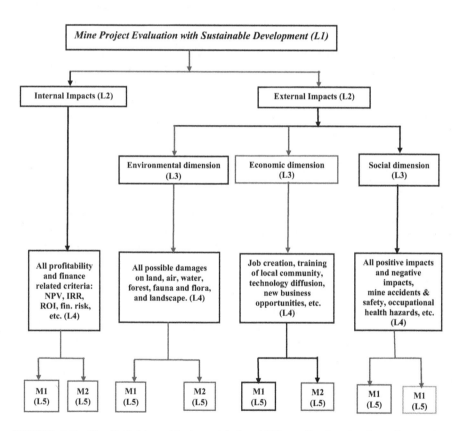

FIGURE 3.8 Detailed hierarchical model for AHP application in the mine project evaluation.

(Source: Quaddus, M.A. and K. Mukherjee, 2013. Hierarchical framework for evaluating mine projects for sustainability: a case study from India. Handbook of Corporate Sustainable Development Planning. Edited by M.A. Quaddus and M.A.B. Siddique, Edward Elger Publishing Ltd, Cheltenham, UK.)

The following are the parameters representing the impacts and dimensions or criteria in the AHP-based HOS model.

- Internal impacts: net present value, return on investment, required capital investment, payback period, financial risk involved in the investment, and the like.
- Environmental dimensions (external impact): amount of mineral deposits to be exposed, reclamation of the damaged or mined land, loss of farmland/productive land, deterioration of landscapes and aesthetics, loss of forest land, impact on flora and fauna, air deterioration/air pollution, water contamination/water pollution, acid mine drainage, and the like.
- Economic dimensions (external impact): generation of employment opportunities, training facilities to unskilled workers, sectoral improvement through technology diffusion, NBO (new business opportunities, such as transport, construction, spare parts manufacturing, and creation of new entrepreneurs, because of the project implementation).
- Social dimensions (external impact): POI (positive impacts like creation of public health centers, schools, etc.), NEI (negative impacts like increase in alcoholism, crime, etc.), SCD (social and cultural disruptions), AMS (accident and mine safety).

Problems represented in the AHP model can be solved by an algorithm developed by T.L. Saaty (1980) based on paired comparison of preferences at each level as input data followed by the solution technique based on matrix algebra.

The corporate management of the Indian coal mining company was interested in the critical assessment of two alternative mine projects (each with target production capacity of ten million tonne of coal per annum) on the basis of both financial (internal impact) and sustainability (external impact) factors. Input data on financial factors and some of environmental factors can be obtained from the detailed project report (DPR) prepared by the mine planning and design division of the company. Impacts on other factors are mostly assessed by experts or managers subjectively. For pairwise comparison of preferences at each level, two corporate executives have been consulted. Structured questionnaires were used to collect the data based on Saaty's scale. The widely known Expert Choice software was used for solving this decision problem. Subsequently, detailed sensitivity analysis further enriched the result of project selection problem. This model acts as a decision support system (DSS) tool for taking strategic decisions, combining both financial and sustainability goals. Interested readers may refer to the research report of Quaddus and Mukherjee (2013) for detailed analysis and ultimate conclusion.

> *The analytic hierarchy process (AHP) model may be applied for the selection of a production process or a project considering sustainable development criteria, along with other economic ones. Here we referred to an application case in selecting coal mine project, applying the AHP model.*

3.2.5 GREEN INVENTORY AND SOURCING

The management of inventory plays a crucial role in overall management of any industrial business unit or the whole supply chain. Whether it is input materials, WIP, spare parts, or finished goods, the management of this "necessary evil" needs special attention for maintaining the smooth flow of items and improvement of efficiency and/or responsiveness with due consideration to uncertain demand fluctuations of downstream processes or the ultimate end customers. Management literature is already quite rich with various models, methodologies, and techniques in this pursuit. After the successful intervention of the Toyota Management System, the advent of JIT, VMI, lean process, JIT-II, and the like came into practice. The primary decisions in inventory planning are on the quantity of procurements and the frequency of procurements. We may enlarge the scope of this decision domain by including decision areas like supplier selection, estimation of required supplier base, and determination of procurement-mix, which are often called sourcing decisions.

Now let us consider the dimension of sustainability in this decision domain in addition to the existing economic and market consideration. The following four areas may be addressed in this context.

1. Ordering and sourcing
2. Inventory or stock-keeping
3. Emergency procurements in case of shortage or backordering
4. Multiple ordering or quick response for seasonal or perishable goods

The quantity and timing of the order are decided, keeping in mind the achievement of two primary objectives—betterment of customer service (often by responsiveness) and minimization of relevant costs. With the volatility of market demand and competitive market condition, most of the businesses look for a responsive strategy as the prioritized goal. On the other hand, inventory management decisions are also made by making it more efficient (particularly in core sector with more stability of market demand), by controlling the relevant cost parameters. The relevant cost parameters in this decision area are the costs of placing an order/procurement, costs of inventory holding and costs of shortages, or backordering costs. The higher the order size, the lower the annual ordering costs will be, but the higher the inventory-holding costs will be. Further, because of the infrequent orders, there may be a chance of shortage in meeting customer demand. However, in such cases, shortages may be avoided by keeping extra safety stock, which results in higher inventory-holding/carrying costs. So either you should have full knowledge on future demand (which is almost impossible) or there should be balancing of all relevant cost parameters. Moreover, the right choice on the mode of transport (vehicles with higher travel speed), use of full truck load options, containerization, freight consolidations, and so on may also help in balancing costs and also responsiveness objectives. Contemporary distribution schemes like hub and spoke, last-mile delivery, and cross-docking may also be considered in this context. JIT-based inventory management is regarded as an excellent system as it helps simultaneously in inventory reduction and responsiveness enhancement.

Now let us add sustainability as the third objective in this decision situation, which results to some changes in overall outcome. If the order size increases (even by FTL, containerization, or consolidation mode of distribution), there will be more chances of environmental degradation because of higher emissions with large-sized vehicles, overstocking and longer inventory-keeping in warehouses, and more material-handling activities. On the other hand, frequent transport (with smaller order size) increases the transportation activities and thus emission. In a JIT-based system, this is quite prevalent. Simultaneously, a faster logistics mode (like air service) for better responsiveness may also increase emissions. Thus, although JIT contributes positively to efficiency and responsiveness, it may badly affect the sustainability goal. The ordering or procurement strategies are to be redefined or redeveloped with more comprehensive perspectives, if sustainable operation is the goal of the corporation. In other words, if we consider all the three goals simultaneously in an inventory management decision—efficiency, responsiveness, and sustainability—then the decision-making process will be little more complex because of multiple trade-offs.

The cost components for keeping or holding inventory primarily include interest on working capital, cost of owning and maintaining the storing facilities, in-house material-handling, obsolescence, and pilferage. The process of managing the inventory and the decision on the quantity of procurement obviously strive to reduce these costs, as mentioned earlier. In sustainable operations management, the factors like environmental deterioration are also to be considered along with these cost items. Incidentally, sustainable inventory management is essentially dependent on the items being stored and the in-house storage facility. There are certain environment deterioration factors associated with keeping the stock. Indirect emission in storing facilities relate to the energy use in maintenance of required environment inside the stores like freezing or heating the facility. This may also affect the ozone layer depletion (in case of refrigeration). Efficiency in electric energy consumption simultaneously reduces the cost of stock-keeping. Moreover, the movement of diesel-operated material-handling equipment emits pollutants. Walmart identified this issue in its warehouses and opted for more environment-friendly trucks to control its carbon footprint. There may also be some leakage of hazardous materials or pollutants in case of storing certain chemicals. For items with a short shelf life or high perishability (like food items, including dairy products and vegetables), overstocking may lead to generation of wastage and increase in disposals. Waste generation is also an unsustainable phenomenon for scarce natural resources, including water, considering the need to conserve these resources. Disposal, on the other hand, degrades the environment. This is also true for items with high obsolescence, like mobiles or similar electronic products (examples of infamous e-waste generation, particularly in developing countries like India). So the cost of holding inventory should suitably include these environmental impacts, or the inventory planning decision should simultaneously minimize both inventory-carrying costs and environmental costs, which indicates a multiple-objective decision. In the famous news boy (or news vendor or Christmas tree) model, for a single period with uncertain demand, the critical ratio (C.R.) is calculated as the probability that the demand will be satisfied with order quantity Q (i.e., $F(Q)$).

$$F(Q) = C.R. = \frac{c_u}{c_o + c_u},$$

where C_u and C_o are the underage and overage costs respectively.

As green inventory-keeping discourages inventory-holding, $C_o > C_u$ and thus $F(Q)$ should be lower. This is achieved by reassessment of overage cost (i.e., C_o). It is expected to add cost of disposal and environmental degradation (often as opportunity cost) to this overage cost.

In case of inventory-planning to meet the demands of multiple periods, we may apply the traditional EOQ model, opt for the computation of reorder point or level (for continuous review policy) or that of the order-up-to levels (for periodic review policy). In all these three techniques, we may simply add greenness-related costs to inventory-carrying cost per unit in the computation.

The third situation arises, if either the profit margin is too high or the manufacturer cannot afford to accept any occurrence of shortage for some highly important customers. As a preventive measure, the manufacturer will be inclined to keep extra stocks, which triggers more environmental degradation. Alternatively, a corrective or reactive action may be considered in case of any shortage. This is opting for quicker mode of transport for procuring additional units of product, which often leads to the selection of a mode with higher potential emission, like air service.

We may now consider fourth case of inventory-keeping and ordering strategy. In case of seasonal goods (with a short shelf life, like fashion goods), if the supply lead time is relatively long, the retail houses often opt for placing second or even third order once they acquire sufficient information on the trend of demand during the remaining period of the season. This is known as *quick response* or *multi-order procurement strategy*, which is quite common in the apparel industry. Additional orders result in additional inbound transport and thus higher emission of pollutants.

If we expand the context, we may also add sustainability in the sourcing decision or in the management of suppliers. It may be achieved by the following management actions.

1. Sustainability criteria are considered during supplier selection, along with cost, delivery time, and quality. These criteria may be expressed as carbon footprint generation, water consumption, use of non-biodegradable materials, use of recycled materials, and energy efficiency at the supplier's end.
2. The vendor scorecard should include environmental or emission issues as KPIs for keeping control on performance of existing suppliers.
3. The manufacturer may extend support or training facilities to introduce greenness among suppliers.
4. A penalty and reward system may be associated with decisions on the renewal of orders to suppliers.

While selecting suppliers, global sourcing is often considered to be more appropriate because of the widening the search space for cheaper inputs or for better quality inputs. But in terms of sustainability, the distance of travel from a faraway location adversely affects sustainability goal because of the higher emission in inbound logistics. This somewhat relates to classical "food mile" concept in the US. Perhaps the reverse movement of "global to local" may be better in this context. Some of the developed countries have started thinking in this direction. Corporations in those countries are encouraged to opt for training local people in place of gaining the benefit of global sourcing. Moreover, like quality certificates (i.e., ISO 9000 series), it is expected that suppliers should have ISO 14000 certificates. The maintenance of sustainable sourcing may also demand other initiatives of external standards by the prospective or existing suppliers, like Global Reporting Initiative, Dow Jones Sustainability Index, and FSC standards for wood items.

In this context, let us refer to the case of *Mathura Refinery of Indian Oil Corporation Limited (IOCL)* in India. This refinery attracted public criticism because of its close proximity to Taj Mahal, a famous monument of India, and to the Bharatpur Bird Sanctuary, an Indian ecological landmark that is a popular tourist hub. IOCL took various measures contributing to sustainable process design or modification. The increase of the height of stacks for all furnaces and chimneys of the power plant, maximum use of low-sulfur crude, installation of facilities for continuous measurement, and the control of pollutants are some examples of such measures. Subsequently, with the Auto Fuel Policy, new standards were to be followed for ultra-low-sulfur diesel. IOCL installed additional process units for sulfur recovery, which resulted in more emissions. IOCL again opted for process redesigning for the treatment of air and effluent emissions and management of oily sludge being emitted from the new facilities to reduce sulfur emission from the refinery.

Walmart, a global retail giant, extensively strives to meet its sustainability goals. It aspires to use items only from renewable resources, not to generate any wastage, and deal with only environment-friendly products. Its efficient strategies for maintaining "every day low price" simultaneously supports establishing green logistics system. Its trucks travel efficiently by reducing empty-mile and out-of-route drives and also by better packaging. The Sustainable Consortium (TSC) was formed to reflect sustainability indices of Walmart by some key performance indicators. It helps both retailers and suppliers to assess their performance through a sustainability scorecard. Various toolkits have been devised for the measurement of KPIs, like the intensity of water use, quality of emitted air, deforestation, chemical fertilizer use, GHG emissions, product take-backs, recycled contents, and so on, applicable for variety of products retailed by Walmart.

McDonald's, the global fast-food chain, keeps sustainability issue as one of its key missions and corporate goals. Mr. Francesca DeBiase, executive vice president and chief supply chain and sustainability officer, clearly commented that McDonald's had been continuously raising the bar of achievement as a responsible company committed to people and the planet. Its focus on planet and people is quite visible from its climate action (commitment to emission reduction), beef sustainability (working with farmers for improvement on beef farming and production practice), packaging and recycling (improved packaging, reduction in wastes, and increase in recycling), and commitment to the community. It is also in the process of elimination of using artificial preservatives for its classic lineup in the US. In 2006, McDonald's judiciously handled the Greenpeace protests against loss of rain forests of Brazil because of huge production of soybeans for feeding to chickens used in McDonald's kitchen.

This was a star case of considering the extended supply chain in managing supply chain processes and maintaining its sustainability. A similar case came to light by successful management of *Mattel* (a global leader in children's toys) in rejecting the products supplied by contract manufacturer that uses unsustainable paints (with lead content).

> *Sourcing decisions and inventory management are essentially affected by sustainability issues. Although the successful tools like JIT or VMI may show their effectiveness in efficiency and responsiveness, they require redefinition while we add sustainability criteria along with these two.*

Key Learning

- Designing and planning sustainable processes require implementation of green manufacturing, green sourcing, and inventory planning.
- Green manufacturing includes a set of processes that not only takes care of costs and quality factor but also considers environmental impacts, energy efficiency, waste management, resource conservation, and human health and safety issues.
- The key enablers of green manufacturing are the scarcity of natural resources, increased cost of energy and other inputs, technology advances, legislative restrictions, pressure from green-conscious customers, and strategic moves for differentiation.
- Sustainability-focused activities under green manufacturing are emission control, use of efficient or lean manufacturing processes, green energy use, waste reduction and recovery, monitoring of carbon footprint and the recovery of used products.
- House of sustainability (HOS) and analytic hierarchy process (AHP) may be applied for selecting the right set of sustainable processes.
- Sustainability issues are also to be considered in ordering and sourcing, inventory planning, emergency procurement in case of shortage, and multiple ordering decisions for seasonal products, in addition to implementation of green manufacturing processes.

3.3 SUSTAINABLE FACILITY DESIGN

The creation of a sustainable facility (factory, warehouse, distribution center, etc.) demands decision-making on two aspects. The first one is the selection of right location of the facility, and the second one is the design and in-house management of resources used for the facility.

3.3.1 SUSTAINABLE OR GREEN FACILITY LOCATION DECISION

Facility location is an important strategic decision in operations management or in supply chain management, considering downstream and upstream business partners of relevance. Various techniques may be applicable for this decision-making, which differ in terms of assumptions, conditions, and viewpoints of the developer. Many models, algorithms, and tools may be accessed in published reports or may be obtained as commercially available decision support systems.

These techniques may be classified under various schemes. One of such popular schemes groups the techniques into two classes—single-location models and multiple-location models. Single-location models represent binary options in decision-making and essentially attribute-focused. Here, the decision is selection of one location for a particular facility from a set of alternative locations. Each optional location is assessed by a set of attributes, such as investment costs, operational costs, local taxes and/or subsidies, availability of skilled manpower, availability of utilities, proximity to destinations/demand centers/primary resources, and so on. On the other hand, multiple-location techniques help decision-makers select a set of facility locations, interdependent in terms of capacity building, transportation linkages, and sharing of some resources. A decision is made by considering all factors and achieving some strategic objectives. The single-location problem may be treated as a multiple-attribute decision problem, or MADM problem. Here, the decision is dependent on both the assessed value of each alternative location in terms of criteria and the relative importance of each criterion as perceived by the decision-maker. In case of choosing a sustainable facility location, we are to consider criteria like the loss of forest or productive lands, water consumption, generation of GHGs, or even the number of families to be displaced along with other economic and business criteria. In reality, quantification of criteria and getting appropriate relative importance of each criterion seem to be quite difficult or in some cases impossible. Attempts have been made to capture this complexity through various methods like AHP, ANP, ELECTRE, and PROMITHEE, some of which really gave rise to satisfactory results in practice. On the other hand, multiple-location problems are more complex, even in the formulation stage. Traditionally, warehouse/distribution center locations are determined by analyzing judicious trade-offs among inventory-holding costs, transportation costs, and delivery time. It is also possible to simultaneously determine the optimal number of facilities and their locations in a supply chain network design. Mixed-integer programming or similar models in operations research are quite suitable in this strategic decision. Now, sustainability issues may be added to this decision-making by introducing the emission of CO_2. This emission takes place because of the number of trips, distance traveled, vehicle size, load being carried, road condition (mainly slope), and empty truck movements. Although the JIT strategy helps in inventory savings, it may lead to higher emission of pollutants because of multiple trips. It is further to be noted that if the distribution or outbound logistics is managed by the manufacturer, then this pollution is treated as Scope 1, whereas if it is outsourced to TPL provider, the emission will be considered as Scope 3 under the GHG protocol. However, in both these cases, this adds to carbon footprint.

For the assessment of carbon footprints in this decision endeavor, Martinez and Fransoo (2017) proposed use of the Network for Transportation and Environment methodology, named after the organization Network for Transport Measures (NTM).

Input data for such computation include fuel consumption (f), distance traveled (d), and weight per shipment (w). The formula may be expressed as following:

$$E = l\left[d\left(f^e + (f^f - f^e)\frac{w}{W} \right) \right]$$

where

E is Total emission in grams of CO_2 for a shipment

l is Constant emission factor (2921 g of CO_2/lit)

f^e and f^f are fuel consumption of the empty and fully loaded vehicle (lit/km), respectively

W is the truck capacity

The common parameters for evaluating alternative locations in this type of models are costs and GHG or CO_2 emissions. Both cost and CO_2 emission rates may be computed for various distance traveled and demand loads, considering alternative distribution designs. The situation may be more complex, if multimodal distribution network (including last-mile delivery) is considered, which is incidentally practiced in most of the supply chain distribution networking.

3.3.2 SUSTAINABLE FACILITY DESIGN DECISION

Facilities like factories and warehouses are major contributors to rise in GHG emissions, water consumption, energy consumption, and waste generation (particularly in industrial sectors like chemical plants, textile factories, paper mills, etc.). As indicated by Bartolini et al. (2019), warehouse activities contribute around 11% of the GHG emissions generated by logistics sector worldwide. Application of life cycle analysis (LCA) gave rise to the fact that 65–90% of energy consumption in a warehouse is due to heating, ventilation, and air conditioning (HVAC) (Gazeley, 2008). In case of food, pharmaceutical, or other chemical products, which require extensive maintenance of specific freezing environment, this energy consumption will be much more. With continuous increase of global manufacturing and related activities, the construction and maintenance of new facilities are expected to create unsustainability in the following ways:

- Extensive use of building materials of various forms, leading to the exhaustion of non-renewable natural resources
- Huge consumption of energy, exhausting the finite fossil fuel resources and resulting in unclean processes of acquiring of these mineral resources
- Consumption of energy during the construction phase and for lighting and HVAC when managing the facilities
- Extensive use of air conditioners, leading to ozone layer depletion, along with carbon dioxide emissions, global warming, and climate change from other unsustainable activities
- Enormous consumption of water and huge generation of solid and liquid wastes

The sustainability concerns and goals in designing and maintaining sustainable facilities are summarized in Table 3.3. Keeping in mind the primary scope of this book, we are limiting our discussion only to the environmental dimension (not the social one) of TBL.

TABLE 3.3

Summary of Sustainability Concerns and Goals for Design and Maintenance of Facilities

Sustainability/Greenness Concerns	Decision-Making Goals
1. Environmental protection	
• Environmental pollution (air and water)	• To minimize pollutant generation in the facilities
• Emission of hazardous and toxic materials	• To arrest the emission of pollutants to the atmosphere
• Maintenance and enlargement of sinks for absorbing unwanted pollutants	• To reduce deforestation
	• To reduce the use of productive farmlands for creating new facilities
• Maintenance of biodiversity	• To encourage afforestation around the facilities as sinks
• Protection of flora and fauna	• To create new and improved habitats
	• To protect the ecosystem in all activities in facility management
	• To minimize or, if possible, to eliminate the use of toxic or hazardous materials in facility design and operations therein
2. Conservation of natural resources	
• Efficient use of natural resources	• To improve production, transport, stock-keeping, and other value-additive processes for better efficiency involving the facilities concerned
• Least use of non-renewable resources	
• Research and development for creating substitutes of non-renewable resources	• To implement a lean manufacturing process and Six Sigma whenever possible
• Replacement of non-renewable resources by renewable ones	• To use recycled and recyclable materials in facility design and in various related processes
• Efficient use of water, the most valuable and scarce natural resource	• To minimize water consumption and recycle wastewater for reuse
	• To prioritize local suppliers as the "food mile" model in construction
3. Energy management	
• Efficient use of energy	• To economize and minimize energy consumption in all processes and in lighting and HVAC
• Replacement of non-renewable energy sources by renewable ones	• To maximize the use of solar or other cleaner source of energy, whenever practicable, particularly in designing the buildings or facilities
4. Waste management (both solid and liquid)	
• Reduction of waste generation and accumulation	• To minimize waste generation, implementing the Japanese method of waste elimination
	• To reuse waste products or byproducts for beneficial use
	• To design effective solid waste disposal and reuse facilities, particularly in urban environments

Sustainable facility design is planned on the basis of the principle, which is prevalent for any green building design and Akadiri et al. (2012) propose that its primary goals are resource conservation, cost-efficiency and design for human adaptation. However, control on GHG emissions may also be considered in this pursuit. Resource conservation focus covers strategies for energy, materials, water, and land conservation. As some of the resources are rare or scarce, it is better to replace them by non-scarce or renewable resources. Improvement of productivity or efficiency also reduces this resource consumption. Energy use in this context may be considered in two ways. Operational efficiency is required for maintaining the environment inside the facility (lighting and HVAC), whereas indirect energy is used in production of construction materials, construction of the building, and so on.

> *Sustainable facility design can be achieved in two stages—by choosing proper single or multiple locations and subsequently by considering the sustainability issues during the complete life cycle of the facility or facilities.*

3.3.2.1 Resource Conservation Initiatives in Sustainable Facility Design

The following decisions may be taken for achieving the objective of resource conservation.

I. Energy Conservation

The following are some mechanisms for achieving this.

1. **Choice of appropriate materials and construction methods**: This is for the reduction of unwanted heat loss or gain inside the facility and the use of materials, which consumed low energy during its creation. Aluminum, for example, consumes high energy during its production.
2. **Insulating building envelope**: It is the installation of an insulation against heat loss or air leakage by some coating.
3. **Design for deconstruction and recycling**: Like the design for remanufacturing in product design, buildings may be deconstructed or dismantled after its useful life, and the components or materials may be directly reused for new facility construction. The materials may be recycled before its use. This saves energy and natural resources. This calls for exploring the cheaper and simpler technology for dismantling or disentanglement of components or subsystems of a building.
4. **Use of energy-efficient technology**: Energy efficiency should be an inbuilt component of the objectives in a construction project and also during its operations (compact layout, least movements of material-handling equipment, choice of energy-efficient machineries, etc.).
5. **Use of passive energy design**: Facility design should prioritize natural modes for reduction of energy consumption levels (natural lighting by exposing to sunrays or ventilation by free flow of air; use of existing water [or recycled water] for evaporation, cooling, and landscaping; etc.).

II. Materials Conservation

The choice of materials is very important in this context. It is to be noted that mineral resources are nonrenewable, and mining activities do have significant contribution to environmental degradation. Further, the replacement of mineral resources by man-made products means an increase of embodied energy. So perhaps following decisions may help in maintaining sustainability in materials conservation.

1. **Design for waste minimization:** It may be achieved by the reduction and recovery of wastes and the reuse and recycling of wastes.
2. **Proper management of materials use**: Identification and use of durable materials, use of locally available products or materials, and avoidance of toxic or hazardous materials.

III. Water Conservation

These activities are meant for both the construction and operations of facilities. The following strategies may be adopted in water conservation.

- Use of water-efficient (low flow and/or low pressure) plumbing fixtures
- Use of recycled water or graywater in activities like toilet flushing, gardening, and machine cleaning
- Rainwater harvesting and graywater storage by the recirculation of wastewater
- Designing low-water-demand landscaping

IV. Land Conservation

This is primarily related to efficient layout design and material-handling system, which simultaneously economize the space (in all three dimensions) and cost.

Sustainable facility design also takes into account the cost-efficiency incorporating costs of conservation-related activities and reuse strategies. The total costs may be analyzed for the whole life cycle of a building and the operations and dismantling of the facilities as life cycle cost analysis. Three such primary segments of costs may be considered in this pursuit.

1. *Development or initial costs*: This calls for analysis of costs like acquisition cost of building or land, professional consultant's (including designer, planner, etc.) cost, cost of materials and construction, cost of commissioning, and so on. Cost reduction of this category requires supply from local markets, opting for modular or standard designs, maximum possible outsourcing of construction works, and the use of recycled or reclaimed materials or wastes.
2. *Operations costs*: These are the costs meant for running the facility. These cost items include labor cost, energy cost, cleaning, repair and maintenance cost, and also material-handling cost within the facility. Better operational efficiency and layout design significantly reduces these costs. However, costs of energy consumption, maintenance of environment within the facility, and material-handling are somewhat related to environmental degradation issues of green facility design.

3. *Cost of deconstruction*: Like any project life cycle, a facility has its limited economic life, beyond which it is to be dismantled and some investments may be recovered. Reuse and recycle of construction components and materials not only save costs but also contribute to energy savings, conservation of scarce materials, and overall reduction of generation of wastes and pollutants.

Sustainable facility design also takes care of human health and safety issues. It addresses the thermal, acoustic, and illumination comfort of human beings. Green facilities strive to maintain well-ventilated space and safer operational environment along with fire and natural hazard protection.

Appropriate decisions are taken for resource conservation while sustainably designing facilities, which include energy, materials, water, and land.

3.3.3 Green Building Rating System

Green buildings are expected to consume lesser resources and are more environment-friendly. In this pursuit, various rating systems of greenness in building construction are available internationally. Relatively popular rating systems are the Building Research Establishment's Environmental Assessment Method (BREEAM) in the United Kingdom, the Comprehensive Assessment System for Building Environmental Efficiency (CASBEE) in Japan, the Green Building Tools (GB Tools) software developed by the International Framework Committee under the Green Building Challenge process, the Hong Kong Building Environmental Assessment Method (HK-BEAM), and **Leadership in Energy and Environmental Design (LEED)** in the USA. Among them, LEED is the most widely used and accepted green building rating system in the world. The rating system was developed by the US Green Building Council (USGBC) (see www.usgbc.org) with around 1.85 million square feet of construction space being certified almost every day. LEED came up into practice in 1998. Since then, various versions evolved, which actually made the LEED certification more effective today. Projects registered after 2016, for example, are to follow the scheme of LEED v4. The rating system under LEED includes benefits to people, planet, and profit of TBL in assessing five main categories: *building design and construction, operations and maintenance, interior design and construction, homes,* and *neighborhood development*. The category "green building design and construction" is further divided into subcategories—*new construction, core and shell, schools, retail,* and *healthcare*. Facilities under various categories are evaluated across **six credit categories (or attributes)**: *sustainable site, water efficiency, energy and atmosphere, materials and resources, indoor environmental quality,* and *innovation in design*. Assessment is made out of 100 total points (per LEED 2009 version). Facilities can be classified under following four levels of certification.

- **Platinum: 80 points and above**
- **Gold: 60–79 points**
- **Silver: 50–59 points**
- **Certified: 40–49 points**

Sustainability goal in facility design may be achieved by applying LEED as a tool during implementing the following key strategies.

1. Site selection and commuting policies: reduction of both development and commuting (Scope 3 per GHG protocol) impacts
2. Consideration of facility and material life cycle in green facility designing: future adaptation, deconstruction or recovery of facility and materials, and use of renewable, durable, and recyclable materials
3. Balancing natural light (sun) with heating and cooling leads: site orientation and good insulation
4. Installing energy-efficient systems: for lighting, HVAC, and office equipment
5. Water efficiency: low-flow fixtures and use of native plants
6. Installation of recycling and reuse plants for water and other wastes or byproducts
7. Overall sustainable operations: proper housekeeping, regular maintenance of equipment, and green supplies

LEED certification is gaining popularity in India, and the Indian Green Building Council (IGBC) intends to enhance status of India in global arena by having 10 billion square feet of green building footprint. Almost 14 lakh houses in India are now green (https://realty.economictimes.indiatimes.com/news/industry by Ravi Kumar Diwakar, 27 November 2018). The first LEED Platinum–certified building in India is the CII—Sohrabji Godrej Green Business Centre, Hyderabad. The certificate was awarded in 2003 for the 20,000 square feet built-in space by 18% increase in cost, which was paid back in seven years. In Kolkata, the first LEED Gold certification was achieved by the Technopolis building in Salt lake (LEED—CS 1.0 Pilot) in 2006, covering 72,000 square feet (see Figure 3.9). One of the most successful green buildings of India in last decade is the Platinum-rated Suzlon One Earth, Pune, which receives 90% of illumination from natural daylight. Interestingly, it has been observed that the incremental cost requirements for achieving greenness show a downward trend over the time, which may be due to the continuous technology development.

ITC hotels became one of the largest chain of hotels with maximum LEED Platinum–certified properties. Welcomhotel Amritsar became Punjab's first LEED-certified hotel. It got the certificate in 2011 for India NC (New Construction) project within seven months of its launch. The green strategies adopted by Welcomhotel Amritsar during this new construction project are the following:

- Using 33% less energy
- Reducing water consumption by 46%
- Developing a rain-harvesting system

FIGURE 3.9   LEED-certified Technopolis building in Kolkata.

- Utilizing 75% of the roof area by covering it with solar reflective index tiles as the maximum capacity of the system
- Using more than 50% of wooden products, which are certified by the FSC (Forest Stewardship Council)

ITC may be credited for its green building endeavor, because of its LEED Platinum–certified corporate building and the biggest Platinum-certified green building in 2004. Moreover, ITC Gardenia is the first LEED Platinum–certified hotel in Asia Pacific in 2009. In the next year, ITC Maurya was declared as the first hotel with LEED Platinum certification in the world in the Existing Building category. Subsequently, ITC Grand Chola became the largest LEED-certified hotel in 2012. More than 18 properties of the ITC chain of hotels have been LEED certified, and most of them got Platinum certifications.

Keeping in mind its unique business ecosystem, an Indian rating system has been developed in 2007, named as **GRIHA (Green Rating for Integrated Habitat Assessment)** in India. It was initially developed by TERI taking into account the provisions of the National Building Code 2005 and Energy Conservation Building Code 2007 and subsequently endorsed by Ministry of New and Renewable Energy (MNRE) of Indian government.

GRIHA is a five-star rating system applicable for buildings of commercial, institutional, and residential nature. A National Advisory Council (NAC) has been constituted by MNRE to operationalize the National Rating System (NRS) per the GRIHA norm. It also trains, advises, and incentivizes in implementing the GRIHA scheme. GRIHA assesses the greenness of a building during its whole life cycle. The process of GRIHA certification starts with the registration by the GRIHA secretariat. There are five volumes of the GRIHA Manual. Volume 1 is the introduction with a briefing on green building rating system (see www.grihaindia.org for more information). In the GRIHA rating system, 34 criteria are considered for assessment, each of which has unique scale of point system. Some of these criteria are mandatory, and some are partly mandatory. Ultimate evaluation is made using a 100-point system, and a building is rated per the following points.

- **One star: 50–60 points**
- **Two stars: 61–70 points**
- **Three stars: 71–80 points**
- **Four stars: 81–90 points**
- **Five stars: 91–100 points**

It is quite understandable that green buildings do consume lower energy and lower resources (using recycled materials and materials from deconstruction of buildings), and they are essentially more environment-friendly (less emission of harmful wastes). GRIHA Manual Volume 1 in 2010 mentions the following specific benefits of a green building:

- It consumes *40 to 60% lesser electricity* compared to conventional buildings.
- It opts for *on-site electricity generation by renewable sources*, like a solar thermal system.
- It consumes *40 to 80% lesser water* compared to conventional buildings.
- It generates *lesser wastes*.
- It generates *lesser environmental pollutants*.
- It maintains *health, safety, and proper sanitation facilities*.
- It *restricts the use of ozone-depleting substances*.

There exist globally accepted rating systems for assessing the degree of greenness of a building. Among all such rating systems available, LEED seems to be the most popular one. It assesses the sustainability of a building with four levels of certification: Platinum, Gold, Silver, or simply Certified. Subsequently, India established its own system of green certification with GRIHA. It rates a building with one, two, three, four, or five stars.

Key Learning

- Decisions on sustainable facility design actually address two problems. The first one is the selection of right location of the facility, and the second one is the design and in-house management of resources of the facility.

- Facility location decisions may be made applying two schemes—single-location models and multiple-location models. Single-location decision is represented as a binary choice—whether to create the facility in a particular location or not. Relevant attributes representing sustainability of a facility location are to be considered in this decision-making, like loss of agricultural lands, deforestation, GHG emissions, and water consumption in water-scarce regions. Multiple-location decisions are more complex, as issues like interdependence, transport linkages, and sharing of some resources are to be considered along with emissions and other environmental issues.
- The main concerns in adding sustainability in designing facilities include four sets of concerns: environmental protection, conservation of natural resources, proper energy management, and waste management (both solid and liquid).
- Design should take care of the conservation of various resources and enhancing operations efficiency in their management.
- Energy conservation can be made by choosing appropriate materials and construction methods. Materials conservation requires waste minimization and proper sourcing and identification of materials. Water conservation demands the use of appropriate plumbing fixtures and water recycling. Efficient layout design and material-handling system lead to land conservation.
- Sustainable facility design also requires improvement in cost-efficiency through the life cycle of facility. The focus should be on three cost parameters: developmental cost, operational cost, and cost of deconstruction.
- Sustainability in facility management can be assessed and benchmarked by green building rating systems. Leadership in Energy and Environmental Design (LEED) seems to be the most popular global rating system in this context. LEED certification is valid for five main categories of facilities, each of which may be assessed by six types of attributes. Based on this assessment, a facility may be certified under one of the four levels of certifications: Platinum, Gold, Silver, or Certified. In 2007 TERI developed a rating system which exclusively captures Indian ecosystem, and it is named as GRIHA (Green Rating for Integrated Habitat Assessment). Under GRIHA each building is assessed by 34 criteria and is rated as having a one- to five-star certification based on the total points.

Prior to the discussion session, it is expected that student groups will be formed. Now each of these questions may be discussed among the group members. The objective of the discussion session is to encourage students to think threadbare and explore all related issues, not arriving at the answer or solution to the problem,

Discussion Questions

1. What should be the additional management concerns often encountered by corporations while designing a product sustainably?

2. How can you compare and contrast design for environment (DFE), design for disassembly (DFD), and design for remanufacturing (DFR)?
3. Throw some light on the current status and popularity of electric cars in India. What are the unfavorable issues that may raise some doubts on achieving overall sustainability by replacing fossil-fuel-driven cars with electric cars? What infrastructural facilities should be added or improved to achieve its success in the Indian economy?
4. How do the product design factors affect extending the two types of product life? Which product life is more influenced by sustainability in product design?
5. How do the criteria like, efficiency, responsiveness, and sustainability interact among themselves in sustainable design of a process?
6. Is lean always green? Then why are JIT or pull-based processes not necessarily sustainable?
7. Can you design a sustainable strategy involving one process or a set of processes, which eventually optimize efficiency, responsiveness, and sustainability simultaneously?
8. Develop an HOS model for the paper industry/oil refinery/thermal power plant, which ultimately gives rise to prioritization of process design parameters.
9. If we compare among the three critical managerial activities—product design, process design, and facility design—with the goal of making them sustainable, then which of the three will be the most difficult and problematic?

REFERENCES

Akadiri, P. O., Ezekiel, A. C., & Olomolaiye, P. O. (2012). Design of a sustainable building: A conceptual framework for implementing sustainability in the building sector. *Buildings, 2*, 126–152. doi:10.3390/buildings 2020126.
Badurdeen, F., & Jawahir, I. S. (2017). Strategies for value creation through sustainable manufacturing. *Procedia Manufacturing, 8*, 20–27.
Bartolini, M., Bottani, E., & Grosse, E. H. (2019). Green warehousing: Systematic literature review and bibliometric analysis. *Journal of Clean Production, 226*, 242–252.
Bergmiller, C. G., & McCright, P. R. (2010). Parallel models for lean and green operations. *Proceedings of Industrial Engineering Research Conference, 1*(1), 22–26.
Bhattacharya, A., Jain, R., & Choudhary, A. (2011). *Green Manufacturing—Energy, Products and Processes*. New Delhi: Report of BCG and CII.
Billatos, S. B., & Basaly, N. A. (1997). *Green Technology and Design for the Environment*. Washington, DC: Taylor & Francis.
DOE. (2014). The history of the electric car. http://energy.gov/articles/history-electric-car. Accessed on 8 April 2019.
Gazeley, (2008). *Sustainability Report 2008*. London: Gazeley UK Ltd.
Giudice, F. (2008). Product design for environment: The life cycle perspective and a methodological framework for the design process. In: Gupta, S. M., & (Fred) Lambert, A. J. D. (Eds.). *Environment Conscious Manufacturing*. Boca Raton, FL: Taylor & Francis and CRC Press, 33–89.

Haapala, K. R., Zhao, F., J. Camelio, Sutherland, J. W., Skerlos, S. J., Dornfeld, D. A., Jawahir, I. S., Zhang, H. C., & Clarens, A. F. (2013). A review of engineering research in sustainable manufacturing. *Transactions of the ASME, 135*(4), 1–16.

Luttropp, C. (2017). Principles of ecodesign in sustainable supply chain management. In: Bouchery, Y., Corbett, C. J., Fransoo, J. C., & Tan, T. (Eds.). *Sustainable Supply Chains.* Los Angeles, CA: Springer Series in Supply Chain Management, vol. 4, 303–316.

Luttropp, C., & Lagerstedt, J. (2006). Ecodesign and the ten golden rules: Generic advice for merging environmental aspects into product development. *Journal of Cleaner Production, 14*(15–16), 396–408.

Martinez, J. C. V., & Fransoo, J. C. (2017). Green facility location. In: Bouchery, Y., Corbett, C. J., Fransoo, J. C., & Tan, T. (Eds.). *Sustainable Supply Chains.* Los Angeles, CA: Springer Series in Supply Chain Management, vol. 4, 219–234.

Mukherjee, K. (2011). House of Sustainability (HOS): An innovative approach to achieve sustainability in the Indian coal sector. In: Quaddus, M. A., & Siddique, M. A. B. (Eds.). *Handbook of Corporate Sustainability.* Cheltenham, UK: Edward Elger, 57–76.

Qian, L. (2011). Product price and performance in one market or two separated markets under various cost structures and functions. *International Journal of Production Economics, 131*, 505–518.

Quaddus, M. A., & Mukherjee, K. (2013). Hierarchical framework for evaluating mine projects for sustainability: A case study from India. In: Quaddus, M. A., & Siddique, M. A. B. (Eds.). *Handbook of Corporate Sustainable Development Planning.* Cheltenham, UK: Edward Elger Publishing Ltd, 2nd edition, 161–177.

Rao, P., & Holt, D. (2005). Do green supply chains lead to competitiveness and economic performance? *International Journal of Operations and Production Management, 25*, 898–916.

Saaty, T. L. (1980). *The Analytical Hierarchy Process.* New York: McGraw-Hill.

Shahan, Z. (2015). Electric car evolution. https://cleantechnica.com/2015/04/26. Accessed on 7 April 2019.

Short, T., Lee-Mortimer, A., Luttropp, C., & Johansen, G. (2012). Manufacturing, sustainability, ecodesign and risk: Lessons learnt from a study of Swedish and English companies. *Journal of Cleaner Production, 37*, 342–352.

www.apartmenttherapy.com/ifea-sustainability-zero-waste-2030-recycle-products-259562. Accessed on 15 March 2019.

https://cdm.corporate.walmart.com. Accessed on 25 July 2019.

www.greenbiz.com/research/2002/6/12/green-product-design. Accessed on 12 March 2019.

www.perspectives.devalt.org/?p=2343. Accessed on 15 March 2019.

www.sustainabilityconsortium.org. Accessed on 12 March 2019.

www.unido.org/sites/default/files/2008-05/PR-6-Examples_0.pdf. Accessed on 15 March 2019.

4 Product Recovery Management—Cradle-to-Cradle Initiatives

4.1 PRODUCT RECOVERY MANAGEMENT— GLOBAL AND LOCAL PERSPECTIVES

In the recent past, the field of economics experienced enrichments and various innovative inclusions of new areas of interest in order to capture the changing thought process and expanding human knowledge base, along with the remarkably vibrating and turbulent socio-economic environment surrounding us. New concepts emerged addressing judicious exploitation, management, and use of both natural resources and man-made products.

In this pursuit, one of the most discussed terms among economists, environmental economic scientists, business scientists, and leaders of global forums on relevant themes of discussion is circular economy (CE).

4.1.1 Circular Economy—A New Concept of Industrial Economics

The emergence of circular economy can be simply explained as a change of viewpoints in the production and consumption process from the traditional linear chain of activities like take-make-dispose to a circular endless chain of restorative use of resources. Circularity is the key word, and it means that using raw materials does not end with discarded waste. In fact, CE was derived from a strong foundation that emphasizes significant losses in value chain in the traditional model of resource extraction, production, consumption, and disposal. The Ellen MacArthur Foundation (www.ellenmacarthurfoundation.org/circular-economy/concept, accessed on 5 February 2020) argues that CE attempts to redefine growth focusing on positive society wide benefits, and it is expected to build economic, natural, and social capital. CE is conceptualized on the basis of the following three principles.

1. **Design out waste and pollution**: Creation of wastes and pollution may be considered as design flaw.
2. **Keep products and materials in use**: Any product/component/material is never to be treated as waste at any stage. It may be reused, repaired, or remanufactured for further use.
3. **Regenerate natural systems**: Natural resources regenerate or return to the soil and other ecosystem of natural.

DOI: 10.1201/9780429195600-7

CE is actually a logical synthesis of several schools of thought or concepts, as shown in the following sections.

4.1.1.1 Cradle-to-Cradle

This concept was proposed and developed by the chemist Braungart and the architect McDough (McDough and Braungart, 2002) as an approach primarily on design philosophy involving industrial and commercial processes. Cradle-to-cradle (C2C) design is inspired by the "biological metabolism" of material flows in industrial systems. Products should be designed for increased efficiency in minimizing the negative effects and for circular recovery or reusability. C2C does not encourage any waste generation in industrial processes. Minimization of waste generation is a result of maximization of value extraction. It also prioritizes use of renewable materials and energy and maintenance of diversity in natural and social system.

4.1.1.2 Performance Economy

Walter Stahel, founder director of the Product Life Institute, Geneva, developed the concept of performance economy as an outcome of his works on the "functional service economy" (Stahel, 1994). It represents an efficiency-focused service economy through resource utilization and product-life extension, adding environmental benefits. It aims for circularity and the maximum use of the value of the products with substantial reduction of inputs and energy used for a service.

4.1.1.3 Biomimicry

Biomimicry is an approach imitating the designs and processes existing in natural systems for the development of man-made products and processes (Benyus, 1997). Solar cells are designed based on the functioning of leaves of a tree. Benyus thinks that biomimicry is primarily relying on consideration of natural products and processes as the model for measuring, mentoring, and replicating during the development of a product or process. Similar concepts are being used in the healthcare sector for designing artificial limbs and organs and also in developing algorithms for solving complex problems, such as genetic algorithm and ant colony algorithm.

4.1.1.4 Natural Capitalism

Natural capital refers to air, water, soil, and all living organisms. Natural capitalism represents the understanding and implementation of a system of interdependence in nature, production processes, and man-made activities through an induced synergy between environmental parameters and business processes. Hawken et al. (1999) perceive this natural capitalism as the enabler for "next industrial revolution." It may be manifested by the following four principles:

- Increasing the productivity of natural capital by extending usable life of products
- Implementing biologically inspired production models with least waste, like closed-loop systems (similar to biomimicry)

- Using the "service and flow" model by providing value to customers as a flow of services
- Reinvesting in natural capital for restoration and regeneration of natural resources

4.1.1.5 Industrial Ecology

Industrial ecology originally emerged as a concept developed by Frosch and Gallopoulos (1989), and soon it became globally recognized as a popular field of study and research. *Journal of Industrial Ecology* is an internationally renowned journal of scientific reports. Industrial ecology is the study of the flow of materials and energy in all operations in the industrial system, pollution and waste generation, and environmental impacts of byproducts. Its main objective is implementation of an industrial ecosystem that integrates all industrial operations and environment in a closed-loop cycle to attain zero waste. Industrial ecology operationalizes three-directional approaches: analytic, procedural, and proactive.

4.1.1.6 Blue Economy

Blue here refers to the color of sky and ocean: the primary components of the natural environment. Pauli (2010) developed this concept on the basis of several real-life case studies. For attaining sustainability, blue economy suggests understanding of local environment and efficient use of resources available locally.

4.1.1.7 Regenerative Design

This concept of regenerative design for products is based on the assumption that energy and materials used in product design can be renewed and revitalized. This concept is actually a replication of intrinsic mechanism of nature's process of regeneration as depicted in biomimicry. Rodale (1983) first proposed the concept of regenerative agriculture. However, this concept experienced the real maturity after a decade by Lyle (1996), which emphasized creation of a framework, which can function with locally available (see blue economy) renewable resources minimizing the unnecessary transportation efforts.

> *Circular economy (CE) replaces the traditional linear process of take-make-dispose with a circular endless chain of restorative use of resources. The foundation of CE lies on the established concepts like cradle-to-cradle, performance economy, biomimicry, natural capitalism, industrial ecology, blue economy, and regenerative design.*

During the first decade of 2000, the concepts related to (or flesh and blood of) circular economy gave rise to operational strategies or business models like reverse logistics and closed-loop supply chain, which will be discussed in subsequent chapters of this book. Although these concepts are somewhat similar because of their common goal of circularity for better sustainability, they are also different in terms of the following characteristics:

- Primary focus on environmental, social, or economic dimension or any of their combinations
- Economic sector(s) primarily addressed by the concept
- Extent or degree of consideration of reusing waste generated by any process
- Role of the concept in creating a completely new business model or in influencing some operations in a supply chain
- Its influence in policy formulation at the macro level (national policy-making/framing guidelines) or its contribution in offering a new outlook to the society or a community
- Its impact on a specific stage of the product life cycle

Circular economy thus enables us to raise doubts on the traditional view of industrial economics primarily explaining the socio-economic dynamics of the make-use-dispose model. Product recovery management (PRM) represents a business model, which operationalizes the circulation of physical resources by extending the value-additive activities. Let us first look at the possible definition of PRM.

Product recovery management (PRM) is the management of all used and discarded products, components, and materials to recover as much of economic and ecological value as possible, thereby reducing the quantity of discarded waste (Thierry et al., 1995).

PRM may also be defined as the **management of all activities required for the recovery or extraction of all possible values from used or discarded products, components, or parts and subsequent value addition (if necessary) by converting them to any usable items (as products, components, or parts) for original or some other users.**

So the key points in this definition include the following.

- The inputs to PRM's industrial process are discarded items, either the whole product or part thereof discarded by the original or primary customer of the new product.
- The process attempts to extract or recover as much economic value as possible from the discarded items. This economic value includes both market (or demand) value and use (or reuse) value of the product. The recovered value, along with some added value, results in the creation of a product, component, or part, which will further be sold to the market. Thus, the whole set of activities give rise to economic benefit to the owner of the PRM process.
- As one of the goals of circular economy is reduction of discarded waste to the maximum extent, PRM attempts to extract ecological value so that discarded waste does not deteriorate the environment.
- PRM simultaneously satisfies the goals of economically benefiting an organization and controlling environmental degradation. So PRM may be treated as one of the powerful tools for achieving sustainability.

- PRM may be considered as a new business model by the original equipment manufacturer (OEM) that is manufacturing a recoverable product. A new entrepreneur may invest in creating a facility for PRM as a new business opportunity.

In short, PRM enables extension of industrial value chain, not linearly but circularly. So it means PRM converts the traditional industrial linear model of *take-make-use-dispose* to an endless industrial circular model of *take-make-use-takeback-recover-make-use*. However, "endlessness" is relative and may not continue after some repetitions, as after some repetitions of the circular cycle (comprising recovery of old parts and addition of new ones), it would be as same as manufacturing of new products.

In this context, let me reiterate the meaning of two closely similar established concepts representing state of a product over a timeline.

Product life cycle is progression of a physical item or product through the four stages in its lifetime depicted by its existence in the market. In other words, it is a set of four stages (introduction, growth, maturity, and decline) perceived from the viewpoint of market demand. It is a popular term used particularly in some areas of management, like production management, marketing management, and R&D management. Product life cycle is also explained as three phases of life—beginning of life (BoL), middle of life (MoL), and end of life (EoL)—considering the activities of the product's owners. This timeline is not made on the basis of market demand but on the basis of activities or processes under various ownerships. BoL covers activities like research on NPD (new product development) and product design, production, and distribution to customers under the broad ownership of manufacturers and distributors. MoL is controlled by the customers, and it includes activities like the use of the product and its service and maintenance by the user, and it begins from the moment the user owns the product. EoL starts from the end of the product's useful life of the product as perceived by the customer/user (i.e., end of its ownership). EoL extends till the final disposal of the product for landfilling or incineration including all possible product recovery processes. When it is decided that no further recovery is possible/feasible or economical, the EoL phase ends. PRM, in general, takes care of managing the collection of used products from end customers during EoL, inspection, sorting/selecting recoverable ones, managing the product recovery process, discarding the non-recoverable waste materials/parts, and redistributing remade products (output from product recovery processes) to an appropriate market. Here, the source for collection of used products is the market of *primary customers* and the destination for selling the remade products (recovered value) may be termed as the market of *secondary customers*. The two markets may be same or different. Here, it is further to be noted that there is another term which seems to be quite popular in this context. This is extension of useful life (EUL). Actually, this is the outcome of product recovery process, which starts at the beginning of EoL of a product. In other words, we mean that successful EUL reduces the product's age (makes it younger) and the recovered product again returns to users/customers (same or different), and thus, MoL starts second time.

> *The product recovery process, being a business application of CE, extracts value from used products or components collected from the primary market and then, with some additional value, converts the used product or component to near-new ones. These are sold in the secondary market.*

It should also be noted that *PRM, like other industrial businesses, is management of an industrial organization having ultimate interest of profit-making. It is unique in the sense that the key raw material to the production process is a used product and the output is the same product with recovered modules or parts.*

PRM is the management of primarily two processes—product recovery process (PRP) and reverse logistics (RL).

PRP involves process options that intend to extract the value from the used products, recover them, and if necessary, add new value to the used products (as new parts) so as to extend the useful life of the concerned product.

Details of RL will be discussed in subsequent sections. However, it may be briefly understood as the additional focus on reverse distribution of used products from the customer end to recovery points and subsequent forward distribution to market of the remanufactured products or customer locations of recovered items along with necessary inventory management.

The following is an outline of possible options of PRP based on various propositions by management scientists, including the most famous one by Thierry et al. (1995).

Reducing: Here *the value of a product is conserved at the user's end* for the extension of its usable period through better use, distribution design (for vehicles), lean manufacturing (for machines), and layout (for material-handling equipment). The application of industrial engineering techniques or better maintenance management often help achieve this goal. This PRP option is treated as an overall initiative for achieving both efficiency and sustainability. This is not technically acknowledged as a full-fledged product recovery process.

Reusing or directly reusing: This is also ideally not a product recovery process but is essentially an option of extending the usable life of a product. The reuse option is *reselling the used product to the secondary customer* by the primary customer. The primary customer takes up this option once he/she thinks that the usable life of the product has ended because of various reasons, including change in global trends. There exists a thriving unorganized market of used cars (known as secondhand cars) and other used products. Maruti Suzuki True Value is the platform for buying secondhand Maruti cars and selling used cars. Cars may be purchased even with warranties like the new ones. Moreover, new platforms are also available online for selling and buying used vehicles as third-party service providers. Some of such examples in the Indian market are Cardekho, Carwale, and Droom. For other products, there exist the platforms like Secondhand Mall and Second-hand Bazaar. Here the process is primarily a change of ownership, although there may also be some inspection and cleaning activities before selling.

Repairing: Once a product loses one or more of its functional values, through repair the owner of the product *gets back the lost functional value(s)*. The working order of the used product is returned. Repair includes some services on the product, overhauling for improvement of some functional quality (e.g., efficiency, fuel consumption, emission rate) or replacement of some part(s). It may involve limited disassembly and reassembly of the used product. Repair is considered as the *functional recovery process of a product*. It may happen any time during MoL.

Cannibalization: This is a type of *limited product recovery process*, as the recovery takes place for a small portion of the product. Actually, only a few parts of the whole product are reused after disassembly, and the remaining portion of the product may be discarded. These parts may be *reused as spare parts* during maintenance service, as replacements during the warranty period, or as components in manufacturing new products. Thierry et al. (1995) used the term "cannibalization" to mean collecting (or extracting) some parts from a product for their use in another. It is quite popular in aircraft industry, which deals with very expensive set of spare parts. The IC chips from a used computer may be reused as components during manufacturing of electronic toys. Same may be true for reusing other small electronic gadgets, sensors, wires, cables, and so on.

Refurbishing: It is typically a product recovery process, and the output of the process is the same product of better quality and usability than the used product. The used product is *disassembled till the module level*. Critical modules as inspected and, if necessary, replaced. The accepted modules are subsequently reassembled. Sometimes modules are replaced by upgraded versions. Most of the aircrafts are refurbished after completion of some flying hours. Of course, the refurbished product is not comparable with the new product in terms of quality and reliability.

Remanufacturing: Remanufacturing is treated as *the highest level of product recovery process*. Here an attempt is made to convert a used product to a new one, in terms of all quality and reliability. Remanufacturing entails *extensive disassembly to the parts level*, vigorous inspection and quality checking, and reassembly of recovered parts along with some new parts, if necessary. Fleischmann et al. (1997) claim that a remanufactured product may be treated like a new one in terms of quality standards. The warranty offered to the new product may be equally applicable for the remanufactured product. A remanufactured product may be introduced along with the new one with a cheaper price. Remanufacturing may be combined with technological upgrade, like the refurbishing process. Automobile products, locomotives, and photocopiers are remanufactured in most countries.

Recycling: Being the last possible recovery option, its main goal is *recovery of materials or parts still retained in good condition* in the used product. It is also the oldest form of product recovery (paper recycling, for example).

In some cases, the product is melted and valuable content is recovered, even applying some chemical processes. Unlike other recovery options, recycling does not intend to restore the old product. So logically, the last activity like reassembly or any other similar value-additive activity is not carried out under the recycling process. *The used product loses both its identity and functionality.* However, any recovered material or part may be reused either as input material in manufacturing of a new product or as a spare part during the warranty period, if its quality is found to be acceptable. In this context, it may further be mentioned that *if the purpose of remanufacturing is the recovery of a product's value to the fullest extent, the purpose of recycling may be described as recovery of materials from the used product.* Paper, glass, and metal recovered from used cars or even the valuable debris generated from demolished old civil constructions are excellent examples of recycling and meaningful use of the recycled materials.

The list of product recovery options clearly indicates that all are practically not product recovery activities. Reducing is an operational strategy for improving resource utilization and rationalizing the use of the product in order to extend MoL so that EoL starts later. Reusing is also a method of extending useful life by changing the ownership of the product to a new user or second user. The product might have lost its usable value at the first user's or customer's end, but the second customer is ready to use it for some more years. This is a very common business practice in developing countries, known as the secondhand product market. The remaining five options are technically the product recovery processes. These processes start with the acquisition of used products. In PRM the used products are often known as *cores*. Core is the raw material and the backbone of any remanufacturing organization. So onward let us interchangeably use both the terms (core or used product).

> *Indirect product recovery processes are reducing and reusing, whereas direct product recovery or value recovery processes are repairing, cannibalization, refurbishing, remanufacturing, and recycling.*

Incidentally, other than reducing (primary objective of efficiency improvement), the remaining six processes essentially extend the useful life of the product or recover some value.

Table 4.1 displays a comparison of the recovery options based on certain characteristics. It clearly shows that in case of cannibalization and recycling, product identity cannot be retained, as we are not getting back the same product. On the other hand, the quality of used products or cores improves significantly by the remanufacturing process. Some experts opine that the quality of a remanufactured product may even surpass the quality of a new product because of repetitive checking, testing, and repairs.

Interestingly, the recovered items of the core may be used at different stages of the product life cycle. In this context, it is further to be admitted that the value

TABLE 4.1

Comparison of Product Recovery Options

Product Recovery Process	Disassembly Level	Quality Improvement	Replacement of Parts	Extension of Useful Life	Technology Upgrade	Warranty Period	Existence of Product Identity
Reusing	Nil	Nil	Nil	Not applicable	Nil	Remains unchanged	Retained
Repairing	Modules/components	Low	Moderate	Low	Normally very low	May be partwise	Retained
Cannibalization	Components/parts	Nil	Nil	Not applicable	Nil	Not applicable	Loss
Refurbishing	Modules	Low	Low at module level	Moderate	Yes	Moderate	Retained
Remanufacturing	Parts	Very High	High/moderate	High	Yes	New warranty as-good-as-new products	Retained
Recycling	Parts	Nil	Not applicable	Not applicable	Not applicable	Not known	Loss

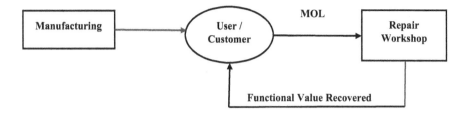

Functional Value Recovery (REPAIR)

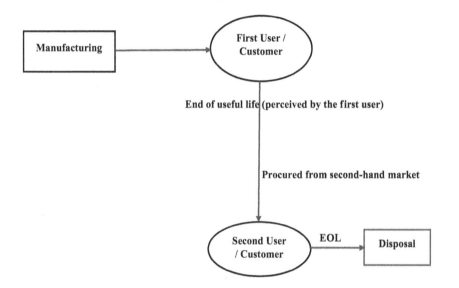

Use Value Recovery (REUSE)

FIGURE 4.1 Role of product recovery options (repairing and reusing) in the three phases of the product life cycle.

recovery activities in repairing and reusing are somewhat less intensive than those in the remaining four product recovery options. Figure 4.1 and Figure 4.2 show the contribution of the product recovery processes during the product life cycle (i.e., in BoL, MoL, and EoL) as the source of materials for supply and use. Figure 4.1 depicts impact of the repair and reuse in phases of product life cycle. The impacts of remanufacturing, refurbishing, cannibalization, and recycling are shown in Figure 4.2.

Figure 4.2 shows that the outputs from remanufacturing and refurbishing are contributing to the BoL and MoL stages of product life. Moreover, the cannibalized outputs are often different products and fulfill the demands of other markets. Among all product recovery options, remanufacturing and recycling are considered to be the recovery options of highest level, both as contributors to sustainability and as a popular business model. Often, we treat the remanufactured products "as good as

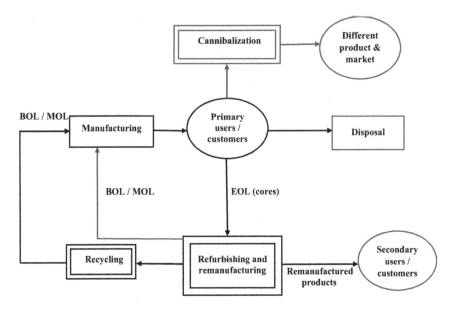

FIGURE 4.2 Role of product recovery options (remanufacturing, refurbishing, cannibalization, and recycling) in the three phases of the product life cycle.

new" products. Here this "as good as new" phrase connotes that the remanufactured product is same as the new product in terms of its use, functionality, efficiency, and even reliability. In other ways, we may say that the remanufacturing organization is offering to customers almost same product as the new one, but with a price significantly lower than the new product available in the market. The remanufactured products are sold with the same (or near same) warranty condition offered to a new product.

In some viewpoints, we may say that there are three types of recycling—*primary*, *secondary*, and *tertiary*. *Primary* recycling means that the recyclable material or product is recovered and reused without any change and often reused for *the same purpose*. Actually, a "direct reuse" product recovery may be considered equivalent to this type of recycling. Using materials or components of the used product (after its dismantling) for *some other purpose* without reprocessing it is termed as *secondary recycling*. Some simple activities are quite visible in unorganized sectors or at home, like cutting the upper half of a plastic bottle to use it as a pot, cutting and reshaping hard paper packets to make art, using empty paint containers as buckets (quite popular in the Indian construction industry), and using battery casings as water containers, which are typical examples of secondary recycling. *Tertiary recycling* refers to a process that involves chemically altering used products to reuse some recoverable items or materials. Actually, this type of recycling is popularly known as simply recycling, the last option of product recovery. It requires facilities for reprocessing, chemical processing, heat treatment, and so on. Extracting hazardous but reusable materials by melting or other processes (e.g., lead from batteries), paper recycling, glass recycling, and plastic recycling are the typical examples of tertiary recycling.

In fact, use of IC chips from recycled laptops for some other electronic product is often classified as cannibalization or tertiary recycling.

Let us explain the total dynamics of product recovery activities with reference to the product life cycle by a set of simple equations considering a particular type of recoverable product.

Total products available at a time = Total BoL products under manufacturers + Total MoL products owned by the customers + Total EoL products generated.

Total EoL products generated = Total EoL products used for recovery + Total EoL products disposed of (A).

Total EoL products used for recovery = Recoverable total EoL products + Unrecoverable total EoL products (B).

Recoverable total EoL products = EoL products recovered by all PRPs (i.e., Reused + Repaired + Cannibalized + Refurbished + Remanufactured + Recycled) + Rejected or Disposed EoL products/their contents after treatments (C).

A + B + C = Incinerated products + Landfilled products.

A: Total EoL products disposed of at the customers' end as junk products.

B: Total EoL products disposed of at the end of the owners of the product recovery process, because they are not judged by initial testing to be economically worthwhile for carrying out recovery operations.

C: Total EoL products or their components/parts disposed of during product recovery process.

All these product recovery options, characteristically being different, may be related to some extent. For example, repair is not simply an independent product recovery activity. During refurbishing, each module or subassembly may be repaired, if necessary, before reassembly. Remanufacturing may need repairing of parts or components. A repairable part is repaired instead of being replaced because of economic benefit. However, reliability factor is given due consideration in this decision-making, as a warranty contract is normally offered in marketing the remanufactured products. A used part from a core may be reused as an input part to a different product in the cannibalization process. Recycling, of course, rarely requires any repair activity. Moreover, while testing each module during refurbishing, we may carry out further disassembly of the module for functionality testing. Subsequently, there may be replacement at the part level. This near-to-complete disassembly, complete testing, and reassembly may transform the whole process to remanufacturing only. Further, rejections may occur at the product, module, or part level, which generates inventories. These rejections are the outcomes of testing or checking for technical feasibility (functionality, quality, and reliability). However, the activities like the disassembly, repair, and reassembly of items from one further level may make the refurbishing business economically infeasible. So the rejection of items at module level may be considered economically more viable option, and thus, this adds to the generation of the inventory of rejected items. Moreover, a functionally healthy module or part with lower reliability is rejected by a remanufacturer, although the same may be included in the reassembly by another remanufacturer. This is actually because of the difference in expectation of their customers of remanufactured products. In one case, the remanufactured products are sold with higher price and longer warranty period compared to those in other case. So items with lower reliability may not be accepted in former case. A rejected module, component, or part in a remanufacturing or refurbishing process may be used for cannibalization. Besides, the rejected modules or parts, which are in good condition can be reused as spare parts or sometimes, even in manufacturing of new products or modules. Users of these recovered items may be the OEMs (original equipment manufacturers), workshops (managing maintenance and repairs), or suppliers of the OEMs. Rejected items during remanufacturing, refurbishing, or cannibalization may also be sent for recycling. Of course, some non-recoverable parts are ultimately disposed of.

The recycling business of end-of-life vehicles (ELVs) generates usable spare parts for running vehicles (MoL vehicles). After removing the usable spare parts, some portions of internal upholstery of the vehicle are burnt down, and the bulk of outside metallic portion is crushed and sold as scrap metal. The steel industry uses the steel parts for recycling, and similarly, other metallic parts, like copper, lead, and aluminum, are recycled for their subsequent use. In this context, the need for shredding machines became evident. The shredder scrap is normally divided into ferrous metals, non-ferrous metals, and non-metals. In terms of weight, an average passenger vehicle is expected to contain 65–70% ferrous metals, 9–10% plastics, and 7–8% non-ferrous metals scrap.

Let me draw the attention of readers to a fact pertaining to traditions and culture.

Traditionally, the Oriental thinking is not simply discarding a product after it ages, although its performance might have deteriorated. We continue to use it till it reaches the stage when no further repair or part replacement is going to improve its usability. Traditionally, we avoid the habit of frequently replacing existing products for better performance or better technology, like the practice being followed by Western world. That's why, perhaps, traditional Indian families are normally large in terms of family members, showing happy coexistence of couples with their parents, who are senior citizens. So traditional Oriental practice is more sustainability-friendly, as it supports frequent repair and part or component replacement (refurbishment) for delaying the disposal of a product. However, globalization and modernization have slowly diminished this distinct difference of Oriental and Occidental lifestyles and perceptions of life. Heterogeneity in consumer behavior among countries is reducing day by day. Now, most of the customers look for newly launched models of a product and do not hesitate to discard an existing model and replace it with a better model. So the justification for PRM is its endeavor to mix two different cultures.

4.1.2 Primary Motivators or Drivers

PRM seems to be a meaningful business activity that significantly contributes to sustainability and also pays back to the owner of the process in the form of profit. This also results in the extension of useful life of the product. So apparently, it is an attractive business option. But the reality is not exactly matching with this logic. Product recovery business is surely not a very popular business option in the business community, both globally and nationally. Is it practically worthwhile for a manufacturer to opt for remanufacturing instead of manufacturing? Why will OEMs invest in remanufacturing its used products instead of funding new product development and the expansion of product lines? What will attract the business community towards funding businesses only meant for the recovery of the products already used by customers or any services related to PRM activities?

Various experts and researchers studied this topic and tried to identify the factors, which are expected to drive the business community to take up the product recovery activities. Diversified viewpoints have been proposed. One group classified the drivers into environmental, economic, and social dimensions, like the 3Ps of sustainability concept. Another viewpoint is considering the drivers as corporate strategies of the manufacturers:

- A product recovery strategy may be treated as an operations strategy option (combining product recovery strategy with conventional manufacturing strategy).
- Capital maintenance (e.g., refurbishing during overhauling) may be considered as a product recovery strategy in place of replacing the old machine with a new machine.
- A product recovery process may be opted as an environment-friendly or sustainable manufacturing strategy for enhancing corporate brand values.

The following may be considered as the list of motivators that may drive the corporations to select product recovery business options. Broadly, let us first group them under external and internal factors.

4.1.2.1   External Factors as Drivers

4.1.2.1.1   Laws and Legislative Directives

During the last three to four decades, the entire world had shown its serious concern for the untamed progression of global warming and the depletion of natural wealth, which is becoming too harmful for current generation and dangerous for the generations ahead. The primary areas of serious attention in this context are carbon emissions, depletion of the ozone layer, conservation of natural resources, water crisis, elimination of green cover, shortage of landfilling spaces, and danger on human health because of hazardous wastes. So various norms, guidelines, and directives have been formulated and communicated by global institutions like UNEP, UNFCCC, and IPCC. Moreover, annual COPs also frame targets and plans for the global communities. Some of these targets are mandatory for countries that are the signatories in the treaties and conventions, whereas these may lead to voluntary agreements for some other countries. Each country makes its policies accordingly so as to achieve the targets. Some of these policies are nothing but guidelines or directives, whereas others, being critical ones, are enacted and converted to acts, laws, or regulations. Chapter 2 of this book includes elaborative descriptions on this topic. National laws and legislations thus became operative, some of which address management of end-of-life (EoL) activities of the used products because of the worsening landfilling and waste disposal problems. European countries are forerunners in acknowledging the "extended producer responsibility" and "polluter pays" principles, and they first converted them to nationwide laws and practices.

In the automotive sector, end-of-life vehicles had drawn enough attention among environmental scientists. It has been apprehended that automobile ownership perhaps has been increasing at a rate higher than the population (Sakai et al., 2014). These vehicles are likely to be converted to wastes after the end of their useful life. It is estimated that the automotive sector is going to generate 5% of global industrial waste (Simic, 2013). So environmental problems came into prominence in society because of this waste generation and management of these disposed vehicles. The rapid development of sustainability awareness seems to trigger this consciousness in the society. The need for reclaiming disposed vehicles also arose because of the scarcity of space for landfilling. Activities like recycling started some time in 1960s, although per some other views, its origin is during WWII. The main objective of recycling is getting the most out of a used and discarded item.

Per the data from Eurostat (https://ec.europa.eu/eurostat/statistics-explained/index.php?title=File:End-of   life_vehicles,_2008%E2%80%932017_(number).png# file, accessed on 28 April 2020), around five million vehicles' useful life expire annually. They surely create huge waste annually.

In 1972, the Organizations for Economic Cooperation and Development (OECD) recommended for the first time the "polluters pay" principle in framing environmental policies in a country. It actually means that the owners of the processes that pollute the environment and the users of scarce natural resources are solely responsible for environmental pollution and shortage of resources. They are thus supposed to pay for any environmental costs and social costs arising out of these problems. Another form of this principle is extended producer responsibility (EPR), first formally introduced by Thomas Lindhqvist of Sweden in 1990 in a ministerial report on environmental issues. EPR was proposed as a policy under waste management. Its main objective is the reduction of environmental impact due to generation of waste out of disposed products after their use. As the manufacturers are the primary economic gainer from the production and use of the products, they are to bear the additional responsibility of managing the environmental and social issues related to the handling the used products. Thus, the manufacturers are to be responsible for collecting, recycling, and finally disposing used products. Practically both the "polluters pay" principle and EPR are closely related because the organizations responsible for creation of pollution should also be responsible for payment of environmental and social costs associated with activities during and after product use (i.e., disposal-related issues). This drives the manufacturer to plan proactively during BoL. Normally these principles are implemented by three possible approaches—mandatory, negotiated, and voluntary. Here by negotiated approach we mean incentivization like subsidies or carbon trading.

Some of the widely known directives and legislations in this context are the European Commission's (EC) directive on waste electrical and electronic equipment (WEEE); EC directives on ELV and reusing, recycling, and recovering of motor vehicles; EC directives on packaging and packaging waste; Japan's Home Appliance

Recycling Law (HARL); and Turkey's regulations on waste oils, EoL vehicles, spent batteries, and accumulators. EC's Directive 2000/53/EC is quite popular globally in policy-making on the recycling of ELVs. Subsequent directives like Directive 2005/64/EC are also meant to enforce reusability, recyclability, and recoverability of motor vehicle designs.

In the US, there is still no specific ELV directive or legislation that is so effective and popular like that in the EU. One of the primary reasons may be that the US is yet to suffer from shortage in waste disposal sites. However, the US Environmental Protection Agency (EPA) has been trying hard to promote the recycling business of used vehicles. In this context, we may also mention the well-established Universal Waste Rule by the EPA in 1995 of North America, which makes the manufacturers responsible for collecting the hazardous waste like batteries at their own cost for disposal or recycling. Massachusetts prohibits landfilling or incineration of cathode ray tubes (Guide and Wassenhove, 2001). Japan, like Europe, is prioritizing recycling ELVs after the implementation of Japan's recycling laws and initiatives. Japan, being a small country in terms of geographical area, faces scarcity of landfilling sites. So it could turn this problem to a revenue-earning opportunity through usable metals and other materials obtained from the shredded cars. Japan particularly enacted EPR and product recovery for electrical appliances like TV sets, refrigerators, ACs, and washing machines. Australia has its policy of recycling ELVs as well.

These legislative pressures and policies at the country level do have significant impact in activating product recovery businesses. Take-back legislative measures are also influencing these endeavors. These are equally effective in managing e-waste or WEEE. Even in the US, which is supposed to be little slower in prioritizing these environment-focused policies compared to Germany, the Netherlands, and France, almost 25 states have created e-waste take-back programs. WEEE includes EoL wastes of computers, TV sets, refrigerators, and cell phones, and it is one of the fastest growing waste streams in the world. It is anticipated that e-waste generation in the EU will reach 12 million tonnes by 2020. This problem was addressed by WEEE directives and RoHS directives on restrictions on the disposal of hazardous items. The first Directive 2002/96/EC was enforced in 2003, which was subsequently updated by the second Directive 2012/19/EU. This, along with ELV directives, extensively pushed the remanufacturing and recycling activities in the EU.

In India WEEE is named as e-waste. Interested readers may go through the document "E-Waste in India," by the Rajya Sabha Secretariat of the Government of India, in June 2011, for detailed discussions on the composition of e-waste and management of e-waste. Reports by the UN predicted that by 2020, e-waste from old computers is expected to jump by 500% in India, with reference to the level of 2007. Per the Central Pollution Control Board (CPCB) report, 6.2 million tonnes of hazardous waste is generated every year, of which the recyclable hazardous waste is around 3.08 million tonnes. Of course, the Hazardous Waste (Management and Handling) Rules 2003 is applicable for managing hazardous portion of e-waste. On the other hand, the Guidelines for Environmentally Sound Management of E-Waste was published by the CPCB in March 2008, which emphasizes use of EPR, and the State Pollution Control Board has been empowered to implement it for managing e-waste. Subsequently, India worked out on framing an exclusive directive or regulation for managing e-waste. This led to creation of the E-Waste (Management and Handling) Rules 2010, which was amended later on in 2018 for proper implementation. EPR thus became applicable for managing e-wastes. Similarly, with enormous growth rate of Indian car sales during last decade or so, there is huge risk of generating post-useful-life cars ready for disposal. Thus, ELV management issues are quite prominent in this country, which is further aggravated because of the non-existence of any stringent laws on ELVs and absence of any effective initiative in formalization of take-back policies. Most of the car-recycling activities in India are managed informally by unorganized industrial units. However, there is already a guideline on declaring ELV status of vehicles. After completing 15 years from the year of registration for petrol vehicles and 10 years for diesel vehicles, the vehicles are to be declared ELVs, provided they are not discarded earlier due to some serious damage or accident. However, because of huge volume, the period has been relaxed to 15 years for all the vehicles. Delhi is the first city that implemented this ELV declaration. Unfortunately, there exists a huge mismatch between world-class manufacturing facilities and recycling infrastructure in India. Unlike in the Western world, the Indian scrapping and shredding system is very weak in capacity and in technology, compared to the vehicle population. In India, the existence of formal recycling is either negligible or primitive in use of technology.

This shows that various laws, directives, and global and national initiatives play the role of being the primary driving force for carrying out product recovery activities during EoL. Mostly, these are mandatory activities among manufacturers, or these create enough scope for taking it up as a business endeavor. In brief, we may say that these primary and most popular enablers for product recovery represent the serious concerns at the macro or governmental level. They essentially include mandatory obligations like legislative take-back obligations along with EPR in managing EoL, expensive disposal or disposal bans, restrictive landfilling or incineration, and similar external forces.

4.1.2.1.2 Green Consciousness of the Prospective Customers and Corporate Image Building

With easy access to almost all information and free flow of global data, society is becoming more aware of environmental issues, global warming, and scarcity of limited natural resources of this planet. People are conscious of waste disposal problems along with landfilling issues. Customers are thus looking for green products nowadays. Waste generation is slowly being avoided, and it is being perceived as an unacceptable activity during the production, use, and end of use of a product. The changing attitudes of customers subsequently changed the market expectations and created the new market for remanufactured products, which has motivated various industrial corporations to convert it to a new marketing strategy. Organizations are understanding the economic advantage of remanufacturing instead of creating a new product line. On the other hand, this strategy is also associated with creating a new corporate image in the market. Green corporate image not only helps with marketing but also increases the overall brand value of the organization. So many corporations opted for voluntary take-back and product recovery programs so as to build a green corporate image and thus enrich their corporate strategy. Examples of voluntary take-back programs for recycling and remanufacturing are quite prominent in the electronics sector and the camera industry, manufacturing the products like photocopiers, cameras, cell phones, and computers. Kodak had actually initiated the design for the environment in its cameras years ago, and it reduced the weight of its camera even by 86%. Kodak started a take-back program for disposable cameras in 1990, although some old cameras are still believed to end up in landfills. In addition to the reduction of water use and its treatment, Fujifilm considered 3Rs (including recycling) as its corporate strategy. In fact, Kodak's single-use cameras did draw enough criticism from the media because of their characteristics as disposables/throwaways. They were earmarked as the indication of wastefulness. These made Kodak and Fujifilm launch take-back programs and recycling. It has been found that voluntary take-back programs ultimately create an impression of environmental consciousness among the prospective customers, leading to increase of sales volume and thus the profit contribution. IBM (particularly IBM Europe), Digital Europe, and Xerox have improved their corporate image through voluntary PRM and thus could gain economic benefits. IBM Europe initiated its product recovery programs in 1990, initially in Switzerland, Germany, the UK, and the Netherlands. It was reported that in 1992 IBM Germany recycled about 4,000 tonnes of used mainframe and personal computers. IBM offers its global asset recovery services to business customers. It was further reported that in 2018, it could process more than 25,000 metric tons (i.e. tonnes) of EoL products. Most of the waste products are recycled, reused, or resold. Hewlett-Packard has received positive media coverage for investing in a recycling infrastructure during EoL of computing products. Samsung and Nokia also have voluntary take-back and recycling programs. Xerox's initiative on its asset recycle management program is recognized as a success story for asset recovery and product stewardship. This voluntary product recovery is also popular in automotive sector. BMW and Volkswagen in Germany run take-back and recovery services for their used automobiles. Even in the apparel, textile, or shoe manufacturing industries, this strategy is often adopted voluntarily. For example, in 2013 Puma launched its program for collecting biodegradable or recyclable sneakers, jackets, shirts, and other items to implement cradle-to-cradle initiatives.

So one of the well-established attractions among corporations for taking up product recovery activities is, of course, the creation of green image in the market and uplifting the corporate brand value, which is expected to substantially contribute to higher profitability and growth potentiality.

4.1.2.1.3 Recovery of Valuable and Scarce Materials or Resources

One of the important motivations for initiating product recovery is recovery of some valuable and scarce materials, parts, or components, which may be useful as resources in any industrial operation. Organizations, particularly OEMs, get the benefit of not procuring the same resource or input for production or other essential activities. This benefit is really striking in cases where the procurement of these inputs is expensive or difficult, like imported items or scarce ones with limited availability. Actually, this motive is also associated with recycling and separating materials that are non-biodegradable, and so landfilling does not seem to be an appropriate disposal option. These materials can be directly reused or reprocessed for their use in industrial operations. Thus, the following simple situations show the three motivators for PRM.

1. Water recycling by treatment of the rejected or emitted water as *scarce resource*
2. Extraction of *non-biodegradable, valuable,* or *hazardous* parts or materials appropriate for reuse
3. Remanufacturing of spare parts for their use in the post-product-life-cycle period because of the *non-availability of new spare parts*

Some examples of recovery of materials for waste or used products are shown here:

- Stripping table lamps, computer cords, mobile charger cords, or wires/cords of other electrical appliances for recovery of electric wires
- Extracting metals, such as copper or even gold, silver, and platinum, from smartphones, old computers, or other electronic gadgets
- Recovering materials like nitrogen, phosphorous, and potassium from sewage for their use as fertilizers
- Recycling wastewater and treating and converting it to clean and potable water
- Recovering heat from wastewater from the outlets of industrial processes
- Recycling and reusing plastic bags, packets, and the like
- Creating compost from wastepaper, plant matter, food scraps, and the like
- Recycling ferrous and non-ferrous metals from used products for their recovery and reuse

Among valuable metals, aluminum, tin, cobalt, and rare earth metals are extracted from electronic wastes as important resources in manufacturing processes. Aluminum is used for structural, electrical, and thermoelectrical functions in electronics industry, particularly in air conditioners, personal computers, and commercial and personal printers. Personal computers, air conditioners, and refrigerators use large quantities of solder, which is the main source of tin in e-waste. Cobalt is largely considered as scarce and expensive resource. It is an important component of lithium batteries. Rare earth elements are also extremely scarce and expensive items representing 15 metals. Their importance in electronics is for their use in strong and permanent magnets, as well as in lamp phosphors in display screens. Of course, its recovery quantity is usually very low. Their presence is primarily in TVs (as lamp phosphors) and computers in permanent magnets used in hard disks.

The third type of recovered important items is spare parts, once it is not available in the market. This situation arises when the associated model is outdated and the manufacturer (OEM) of the model stopped its production. So the scarcity of these spare parts arises in post-product-life-cycle period. As the product is no longer manufactured, the supplier of a part also discontinues its production. But users of the model may still exist in the market. If the part fails and requires replacement by a new one, the manufacturer or its service center will not be able to supply the spare part. This leads to dissatisfaction among the users of the product or model of the product and thus may affect the image of the OEM. So this calls for creation of a separate storage of this type of spare parts by remanufacturing of the returned ones or used spare parts. This is the case of recovery of spare parts for their use at the end of the product life cycle (in the marketing or demand perspective) of the parent product. As a typical example of need for this type of recovery in India, we may cite the case of spare parts of Maruti 800 model, which is no longer manufactured now. In this context, we may refer to the research paper authored by the author of this book and Karl Inderfurth (Inderfurth and Mukherjee, 2008).

Thus, the industrial organizations may also be interested in product recovery business so as to make available some scarce and valuable items, which can replace the expensive and difficult procurement processes or by which any negative impact due to non-availability of an important item may be avoided.

4.1.2.1.4 Direct Contribution to the Improvement of the Environment

This may be perceived as voluntary contribution. But the implication of this factor is not mere image building but more as an intentional service to the society because of the awareness of current environmental degradation. In fact, this is the primary objective of any product recovery process. With high rate of obsolescence of products and fast development of new products, customers are quickly discarding old models and switching to new ones. Consequently, producers are intensifying product portfolio expansion and creating facilities for new product lines. Consequently, there is significant increase in industrial emissions, higher consumption of natural resources, higher use of space for landfilling or additional pollution due to incineration, and generation of non-biodegradable wastes along with toxic and harmful materials due to disposals.

As the market will continuously look for change of existing products or models, and business community will also support it because of higher profitability. The business goal of mass customization will remain forever. The product recovery process is the only win-win solution in this situation. Product recovery reduces the negative effects of unsustainability. Some of the customers are fully satisfied with remanufactured or refurbished products in place of highly priced new products. Moreover, this results to delayed disposals, and recovered hazardous materials may be suitably treated. So once corporations are conscious of this multidimensional effect of sustainable operations, like product recovery, they will be tempted or motivated to implement this new business process. This also enables national policy-makers frame, implement, and enact new policies, guidelines, circulars, laws, and acts on remanufacturing, recycling, take-backs, disposals, and so on.

4.1.2.1.5 Some Other External and Beneficial Factors

In supplier management or management of sourcing, one of the crucial decisions is the size of supplier base for each item. Years ago, there was hot debate among experts on the suitability between two extreme options, larger base versus smaller base of suppliers, or in other words, American viewpoints versus Japanese viewpoints. Ultimately, most of the experts agreed that due to high direct and indirect transaction costs, it is better to opt for a smaller number of suppliers in the panel for the procurement of each item. The common practice is thus to earmark two to three suppliers for each item. One potential drawback of such a strategy is an increased dependence on these suppliers. Manufacturers are to tolerate delays in deliveries and accept it with some loss. *In such a case, recovered components (output of PRP) may act as an extra source* whenever available suppliers fail to fulfill demand in a timely manner because of some reasons at their end. It may also be a *protection against the risk of delays or disruptions in inbound logistics.* It may be considered as *a viable option in case of sudden uncontrollable price rise at the supplier's end.*

Some organizations, like Mercedes-Benz, offer their customers the possibility to have the engine in their old cars replaced by a recovered engine of the same or other type with relatively lower price and an acceptable warranty.

A recovered product may be made available at the less time than producing a new product. This may save delivery time and storage costs or may be quite helpful in case of urgent demand from an important client.

Sometimes, *some local authorities ban the disposal of a specific product or part thereof for a fixed period.* This may compel the customers to sell or return the used products to the producer, who can initiate product recovery operations for getting some economic benefit out of them.

External enablers include the factors like laws and legislative directives, green consciousness of customers, corporate brand image, recovery of valuable and scarce materials, direct contributions to the environment, and other motivators for operational benefits.

4.1.2.2 Internal Factors as Drivers

4.1.2.2.1 New Strategic Initiatives

Product recovery is actually a new component of corporate strategy for large organizations. This is known as the *sustainable corporate strategy*, which was non-existent earlier. Product recovery has been accepted as a new achievable goal in sustainable strategy. Industrial organizations can now enrich their green image by adopting green procurement, green supply chain collaboration, and closed-loop supply chain strategy. *Remanufacturing becomes a parallel production strategy along with its original manufacturing, as prevalent now in industrial sector producing photocopiers, automobiles, printers, and so on—that is, remanufacturable products.*

This practically leads to creation of new business sectors in global economy, like recyclers, third-party remanufacturers, and certified (e.g., Green Dot in European EPR rules) PROs (producer responsibility organizations) as third-party organizations facilitating the take-back responsibility of producers and reverse logistics service providers.

It is also a new strategic initiative in marketing for an OEM, which is planning to introduce remanufacturing in order to capture certain market segment. A class of price-sensitive customers may be satisfied by the remanufactured products with much lower price for the as-good-as-new products. This is actually a typical market for secondhand products, primarily controlled by dis-organized or informal sector of economy. But secondhand products are normally poor in quality and reliability and also have no warranty. Normally the price of the remanufactured products is 20–30% lower than the new ones available in the market. For other customers, it is a case of enhanced choice base between the remanufactured and the new ones, depending on the need.

So the popularity of product recovery activities is significantly strengthened in global economy because of these new strategic initiatives taken internally by the corporation.

4.1.2.2.2   Scope for Cost Reduction

It may be estimated that product recovery *can save 40–60% cost* relative to manufacturing a completely new product. Actually, we are getting back most of the components, which have already been manufactured, having no need to procure materials for production of these components or outsourcing from suppliers. Moreover, sometimes there may be enormous *savings during the production operations, amounting to around 80% of original requirements* (Guide et al., 1997). In fact, by remanufacturing, we may avoid substantial amount of processing and manufacturing time, energy, or input costs, other than what we require for assembling (or more precisely, reassembling). The cost of expensive products like buses, locomotive engines, and aircraft may be substantially reduced by using remanufactured products instead of new ones. For example, instead of a new bus costing $2,20,000, a remanufactured bus may be used for urban transport, which costs only $70,000 (Amezquita et al., 1995). Xerox estimated cost savings of around $76 million in 1999 through this product recovery programs (Guide et al., 2003). Using recovered products as a cheaper option is quite visible everywhere. The simplest example is frequent use of retreaded tires by the truck drivers as a replacement. Most of the heavy machinery items like oil rigs or products like aircraft and locomotive engines are either refurbished or remanufactured. *The remanufactured Cummins engine is sold at a much lower price than its new version, but with same warranty like the new one.* In fact, some industry experts claim that the remanufactured products are expected to be *better than the new ones in terms of quality and reliability* because of repetitive checks and inspections. In China, it is further estimated that in comparison to new product manufacturing, remanufacturing can *save the energy consumption by 60%, lower materials consumption by 70%, and reduce emissions by 80%.* The quality and performance of remanufactured products may exceed the new ones in several cases, with their costs being only 50% of the new products (Deng et al., 2017). Recovery of acceptable components, parts, or materials facilitates their direct use in subassembly or final assembly process. Naturally, this means avoidance of upstream processes of manufacturing the components and creating the subassemblies, which saves the energy consumption and other resource consumption a lot. Moreover, it *saves costs of sourcing and ordering of inputs.* As the market is already aware of strengths, weaknesses, and the functionality of the product, there is no need for any marketing expenditure or sales promotion of the remanufactured products. However, reference to remanufacturing is to be made in marketing of new product itself, while the manufacturer narrates the product design (DFR or DFE). *Further, as price of remanufactured products are much lower than the new ones, it is expected that sales volume increases. So the increase in sales along with lower costs culminates in improvement of profitability.*

The recovered parts are also used for replacement during warranty period as a cheaper option. The spare parts suppliers may procure these recovered parts from third-party remanufacturers for subsequent supply to OEMs.

Thus, these economic benefits seem to motivate the business community to invest for product recovery business.

4.1.3 PRM and Waste Recovery in Practice

According to Global E-Waste Monitor 2017, *India generates about two million tonnes of e-waste annually.* Based on general ranking, the US is the highest producer of e-waste, which is followed by China, Japan, and Germany, and the fifth rank in this order goes to India. But unfortunately, India rarely maintains enough of organized facilities to treat this enormous e-waste. About *90% of e-waste is recycled in informal sector in a crude manner.* Of course, now the scenario has changed significantly. The e-waste (management) rule was enacted in 2017, and more than 21 products (Schedule I) were included under this rule. PRM, along with e-waste treatment, is slowly gaining popularity among Indian factories. Top bearing manufacturing companies like Timken India Pvt Ltd offers remanufacturing as a repair service option to customers. Incidentally, most of the cartridge making firms of India are engaged in refilling, rather than remanufacturing. Indian railways started Diesel Loco Modernization Works (DMW) at Patiala, India, in 1981. It added remanufacturing as operational activities in 1989. Several spare parts, like microprocessor control systems, AC-DC power transmission, fuel-efficient engine kits, roller-bearing suspension systems, traction motors, crankshafts, revolving chairs, fans, and large-sized sliding windows are remanufactured every year, resulting in huge savings in energy, materials, and capital costs.

Globally remanufacturing became an effective and popular business option initially in Europe, which slowly spread its acceptability in other economies as well, like the US, Japan, China, Taiwan, and so on. Even in 1994, paper recycling in Europe amounted to 27.7 million tonnes in an annual growth rate of 7% and a recovery rate of 43%. The glass recycling in the same year went up to little more than 7 million tonnes with recovery rate of 60% (Fleischmann et al., 1997). Big corporations like Union Carbide, Xerox, Deare & Company, IBM Europe, Delphi, Dupont, and General Motors had shown evidence of fruitful introduction of product recovery operations almost three decades ago. Interested readers may refer to the popular research article of Thierry et al. (1995) for detailed case discussions on the practices of BMW, IBM, and Copy Magic years ago. The seminal report of Geraldo Ferrer was published as a working paper during his PhD research in INSEAD (Ferrer, 1996). This included brief case studies on product recovery business of some corporate giants. Ferrer touched upon remanufacturing and recycling activities involving automobiles (BMW, Volkswagen, PSA, Daimler-Benz, Chrysler, Ford, General Motors, Mazda, etc.), photocopiers (Rank Xerox), personal computers, disposable cameras, and machine tools.

Normally, there are three different business cases.

Case I: This is applicable primarily for remanufacturing or refurbishing. The OEM is managing all the activities related to PRM. Xerox and European automakers follow this business model in remanufacturing their used products.

Case II: This is once again appropriate for remanufacturing and refurbishing. An OEM restricts its PRM-related activities only to remanufacturing, and other related activities are outsourced to external agencies (third-party service providers). It means, there are separate businesses, which are involved in collection and reverse logistics for supplying the cores to remanufacturers (OEMs).

Case III: This is a business situation, where OEMs are not involved in product recovery activities. OEMs, in fact, decide to restrict themselves to manufacturing of new products as the core business process. There exist separate businesses for product recovery like remanufacturers, recyclers, and so on. These remanufacturers may carry out the additional related activities or may outsource them to collectors of cores, third-party logistics providers, which are also separate business units.

In practice there exist three types of business models for product recovery. Case I: OEM owns all activities of PRM. Case II: OEM does the remanufacturing, and other activities are managed by other companies or 3PLPs. Case III: OEM is not involved in PRM; it only manufactures.

4.1.4 EXCLUSIVE CHALLENGES IN PRM

4.1.4.1 Acquisition Issues

It is quite apparent that sourcing of raw materials in product recovery process is quite unique and different from that of any traditional production process. In any production process raw materials are procured based on the requirements of the master production schedule, inventory levels, and more precisely, MRP. This originates from the demand of finished goods by the customers. But in PRM, particularly in remanufacturing management, the procurement of raw materials (used products/returns/cores) is significantly dependent on willingness of product users or primary customers.

So, the challenges in PRM are (a) selection of mode of collection of cores, and (b) uncontrollable uncertainties.

Possible modes of collecting cores are *take-back, buyback, off-lease, auction, seed-stock, exchange offers, and warranty returns.* The decision of collection management is primarily on the quantity and age of used products ready for collection. The decision-making is also affected by collection costs (including the cost of inbound logistics), expected cost of remanufacturing (depending on the age of the cores), and willingness of the user to disown the product (influencing the cost of collection).

During the acquisition, the remanufacturer is further supposed to take care of two types of uncertainty.

First, if the owner of product recovery process is not OEM, it is difficult to know the exact design, materials, fasteners, and other parts inside the product. If the used product was not manufactured based on standard design practices, a recycler faces the challenges in planning for treatments of its hazardous content. However, if the OEM is remanufacturing, this problem rarely arises.

Second, the *timing of availability* of used products, the *quantity of cores* or returns and *their conditions* are beyond the control of the remanufacturer, as they are exclusively user-dependent. Although the remanufacturer may use various incentives like buyback price or exchange offers, the user may or may not get rid of the used product at a particular time. This phenomenon is also associated with some emotional attachment, which makes the user reluctant to disown it, although the incentive offered by the remanufacturer may be financially attractive. On the other hand, for the same reason, it is difficult for remanufacturer to forecast the quantity of acquisition. Further, as the work environment and past data of failures at the user's end are also not known by the remanufacturer, the remanufacturer faces difficulty in assessing the condition of the used product. The owner of the product recovery process can only know about its condition after disassembly and inspection of the cores. Thus, there exists a series of challenges in managing the acquisition of used products in any product recovery process.

4.1.4.2 Balancing Cores with Demand

Acquisition of cores or returns should match with the demand for remanufactured products or recovered items. Otherwise, either there would be unnecessary inventory of returns or poor service level (shortages). It is another area of challenges for remanufacturers. This balancing becomes more difficult as the remanufacturers or recyclers do not know the exact recovery rates beforehand. It is found that most of the remanufacturers keep excess inventory of recovered items or parts as stock for future use, which may be used for maintenance purpose (Guide, 2000). As mentioned earlier, there may be three business models for PRM. Further, the remanufacturers may avoid keeping excess stock of cores by acquiring cores from core-brokers or third-party logistics providers instead of directly from users, in case of excess market demand for recovered or remanufactured products. Thus, PRM should take into account this unique characteristic when devising the acquisition process and production planning.

4.1.4.3 Inventory Planning

In the remanufacturing process, it is required to manage *three types of inventories— inventory of cores or returns, inventory of recoverable products, and inventory of recovered items*. Inventory of recoverable products demands some special attention in this context. Recoverable inventory includes inventory of used products, which are recovery worthy or *remanufacturable*. So long as the returns are not disassembled or checked critically, it is difficult to estimate the quantity of recoverable products and those to be disposed. So planning and control of this inventory is significantly difficult because of *uncertainty of recovery rate*.

4.1.4.4 Disassembly Process

Once the remanufacturer or recycler receives the returns, after initial cleaning, it is first disassembled. Disassembly apparently seems to be just reverse of assembly process. But much uncertainty is involved in this process because of lack of complete knowledge on the design and condition of the used product. These cores are recovered from different users of the product with varied work environment. Further, if the cores are collected from third-party core suppliers (not the customers of OEM), the remanufacturer may not be fully aware of the design. Moreover, the remanufacturer also does not have full knowledge on the failure history of the core. This will be more difficult for third-party remanufacturers, which collect cores of different makes with varied design. Disassembly scheduling and facility or tools requirements are thus quite challenging for remanufacturers.

4.1.4.5 Uncertainty on Processing Time

Total processing time of product recovery operation is primarily a combination of various times for inspection or checking, disassembling, replacement of worn-out parts by new ones, and reassembling. All these are directly affected by quantity, age, design, and condition of the cores. So the related processing times are difficult to predict during planning and scheduling of remanufacturing operations.

4.1.4.6 Parts-Matching Problems

This challenge is expected to be encountered at the operational level, particularly during reassembling operation. Normally, an inventory of recovered parts is created after the disassembly of cores. Cores may be from different sources and thus may be of various models and even makes, specifically in case of independent remanufacturing organizations. These remanufacturers often face difficulty in matching parts from various origins while reassembling for production of remanufactured products.

4.1.4.7 Reverse Logistics Network

Unlike the traditional (forward) logistics networks, which are applicable for distribution of finished goods to customers' end, the PRM requires reverse distribution or logistics activities from the core owners (i.e., customers) or core suppliers to the remanufacturing plants. This is an additional challenge to the remanufacturers. This is unique in two ways. First, forward logistics is a network of movements from few factories to many customer centers, whereas reverse logistics is a set of movements from many users to few remanufacturing facilities. Often it is outsourced to 3PL providers. Second, its complexity is due to the uncertainty in the sources of cores, quantity of cores, and timing of acquisition. It would be more difficult for the Case III business model, or independent remanufacturers, which collect cores of different makes and various shapes and sizes.

4.1.4.8 Uncertainty in Demand Forecasting

Forecasting of demand for remanufactured products or recovered items is more difficult than that of new products, as the later follows certain pattern, trend, seasonality, and product awareness. Demand for remanufactured products is affected by the attitude and perception of customers toward a used product. It entirely depends on the trust and faith of customers toward the quality assurance of the remanufactured products. Perhaps in countries like India, this attitude has made product recovery an unpopular proposition. Moreover, there is market resistance and competition with secondhand products, which is a thriving business in India, particularly operated by informal and unorganized sector. In most cases, the remanufacturers identify a specific market segment for selling the recovered products. Thus, the traditional forecasting methods may be hardly applicable for demand forecasting of recovered products.

The main challenges encountered in PRM are acquisition issues, balancing cores with demand, inventory planning, uncertainly of processing time, demand forecasting, disassembly, parts-matching issues, and reverse logistics.

Key Learning

- Circular economy (CE) is the foundation of product recovery management (PRM), which focuses on restorative use of resources.
- CE is a meaningful synthesis of seven schools of thought—cradle-to-cradle, performance economy, biomimicry, natural capitalism, industrial ecology, blue economy, and regenerative design.
- PRM is management of processing of used and discarded products or components for maximum possible recovery of their remaining value for further use in place of their disposals.
- Any product life may be divided into three phases—beginning of life (BoL), middle of life (MoL), and end of life (EoL).
- In practice there exist various product recovery processes—reducing, reusing, repairing, cannibalization, refurbishing, remanufacturing, and recycling.
- Primary drivers for product recovery may be classified into external and internal factors. External factors include legislative compulsions, green consciousness of customers and corporate image building, resource recovery, contribution to the environment, and other factors. Internal factors are new strategic initiative and scope for cost reduction.
- Recovery of waste products, particularly e-waste and wastes from heavy industry, is a popular practice across the globe. Remanufacturing has become another popular business proposition in automotive sector, particularly in Europe.
- Certain unique challenges are often associated with PRM. These relate to acquisition issues, balancing supply with demand, inventory planning, disassembly process, unpredictability of processing time, parts-matching problems, reverse logistics, and uncertainty in demand forecasting.

4.2   REMANUFACTURING—PROCESS AND MANAGERIAL PERSPECTIVES

Remanufacturing seems to be the most effective product recovery business option with the intention of complete value recovery from cores. Perhaps it is also second most popular product recovery process after recycling. The concept and definition of remanufacturing have evolved over the last couple of decades. The most acceptable and comprehensive definition seems to be the one proposed by the Remanufacturing Industries Council (RIC). The council defines remanufacturing as *"a comprehensive and rigorous industrial process, by which the previously sold, worn or non-functional product or component is returned to a like-new or better-than-new condition and warranted in performance level and quality"* (www.remancouncil.org, accessed on 10 June 2020). Per RIC, the most commonly remanufactured products are categorized as aircraft components, automotive parts, office furniture, electrical and electronic equipment, engines and components, medical equipment, printing equipment, and restaurant and food supply equipment. Of course, the list may be expanded with automotive products, locomotives, and so on. It shows that based on the 2012 report of US International Trade Commission, the total value of remanufactured product amounts to $ 43 billion by 2011.

The remanufacturing business has the following direct contributions to economy, society, and the earth.

1. Reduction of input materials—*conservation of scarce natural resources and cost*
2. Reduction of energy consumption by avoiding production of new products or components—*conservation of energy or coal in case of thermal energy sources and less pollution*
3. Reduction of production activities (only the reassembly of recovered modules or components)—*less pollution and cost*
4. Reduction of waste generation associated with disposal due to reuse of used modules—*less pollution*
5. Availability of products with lower prices with comparable quality—*less requirement of manufacturing of new products*

Experts propose that a product should have following seven characteristics to make the remanufacturing feasible or remanufacturable.

- The product should be durable.
- The product may fail functionally.
- The product should not be a customized one; rather, it should be a standardized one to the maximum extent and parts of a set of same products should be interchangeable as far as practicable.
- The remaining value of the used product should be sufficiently high.
- The cost of getting the used product should be lower than its remaining value.
- The product technology is by and large stable.
- There should be awareness among users of the product that remanufactured version of the product is available in the market.

A significant number of reports have been published during last 20–25 years showing the popularity of remanufacturing in global economy. Discussions have also been made at different forums to drive remanufacturing as a viable business option which results in gaining various benefits. The following is the summarized account of the benefits:

- Materials cost savings may be expected to be between 40% and 65%, leading to both economic benefit and conservation of resources.
- Remanufacturing may require around 15–20% of energy consumed during the manufacturing of new products.
- Remanufacturing is primarily more labor-intensive compared to manufacturing process, and it may require relatively more trained and skilled workers. So it may be an appropriate business in a country where cheaper labor force is abundant like India.
- Remanufactured products may be sold at a lower price because of lower production costs. The difference of price varies from product to product, but by and large the price of a remanufactured product is expected to be within 30–40% of the new product. Thus, if the remanufacturer identifies right type of market, the profit may be quite high because of higher expected sales. It is estimated that remanufactured automotive products may raise a profit within the range of 27–33%.

The Remanufacturing Industries Council (RIC) defines remanufacturing as a comprehensive and rigorous industrial process by which the previously sold, worn, or non-functional product or component is returned to a like-new or better-than-new condition and has the same performance level and quality. The benefits of this business option include cost savings, savings in energy and resource consumption, lowering of sales price, substantial contribution to sustainability, enhancing brand image, and higher profitability.

4.2.1 STATUS OF THE REMANUFACTURING SECTOR IN INDIA—AN EMPIRICAL STUDY

The detailed discussion on remanufacturing is being initiated by research on Indian remanufacturing sector only after some initial discussion on essential characteristics and its definition by the RIC. It seems to be somewhat unusual. *Actually, the author intends to ignite the inquisitiveness in the reader's mind at the very beginning on why the Indian remanufacturing sector is not flourishing like Western ones.*

This study investigates the remanufacturing status in Indian economy. A research project was initiated by the author at IIT (ISM) Dhanbad in early 2002. Subsequently, a doctoral research project on this topic was awarded in 2006 (Mondal, 2006), and it was published in 2006 (Mondal and Mukherjee, 2006a).

Initially, the attempt was made to know about the existing status of India in terms of remanufacturing business. After much survey, only six companies could be identified as the remanufacturers out of 1,000 companies engaged in manufacturing of remanufacturable products at that time. Those are Xerox India Ltd, United van der Horst Ltd,

Soft-AID Computers Pvt Ltd, Kores Printer Technology Pvt Ltd, Transdot Electronic Pvt Ltd, and Timkin India Ltd. But at that time, remanufacturing was quite well known in Europe and other western countries. The business community and market of those countries had already accepted it as an established business sector.

So the obvious question was why Indian business community was so reluctant in taking up this product recovery business. Is it because of market infeasibility or operational infeasibility of remanufacturing business in India? Initial study and search in this direction gave rise to the following facts.

- Indian customers are highly price-sensitive, and India has a thriving market for secondhand products or reused products.

  Comment: Indian customers are more attracted to the low-priced version of any product. They are also ready to purchase old and used products in the thriving secondhand product market.

- It has been estimated that around one million customers may be created if the price is reduced by Rs 75,000 or, in other words, a 25–30% price cut is achieved for small and medium passenger cars.

  Comment: Demand in Indian market clearly follows the inverse proportionality to price. So low-priced product models are expected to draw customers significantly.

- Automakers are getting worried about input cost escalation and resultant hike in car price.

  Comment: Escalation of input costs and inflation may result in the rise of market price, at least in car market and thus reduction of sales volume in a price-sensitive market.

- Machine tools manufacturing requires expensive raw materials with 30% import duty.

  Comment: Machine tools (remanufacturable ones) manufacturing business uses expensive and imported raw materials.

- Personal computers are becoming obsolete in two to three years, causing disposals and resultant environmental degradation.

  Comment: Huge availability of used PCs may be expected in the next two to three years.

- Indian economy is growing and showing upward trend of sales of remanufacturable products.

  Comment: A significant growth of demand for remanufacturable products is envisaged in India in near future.

- High demand for products like cars and mobile phones is causing entry of new models, which leads to making older models obsolete in faster pace.

  Comment: Huge availability of used cars and mobile phones means significant supply of cores or returns (inputs) in remanufacturing business.

- Indian labor force is relatively cheap.

  Comment: Any business option demanding relatively more manpower than technology intervention seems to be a feasible proposition in Indian economy.

So it may be inferred that there is growth in demand for cheaper products, used products are being available due to higher rate of obsolescence, and there is availability of cheaper labor force in India. *Thus, logically remanufacturing should have been a suitable and appropriate business activity in Indian economy.* But fact showed that India was yet to reach its level of maturity in product recovery business. This motivated the researchers to search for the causal factors, which gave rise to the following research questions.

1. *Why are Indian manufacturers not opting for remanufacturing their products?*
   This research question aims at identifying the factors that demotivate Indian business community to invest for remanufacturing.
2. *Is there any industry-wide commonality among these demotivating factors?*
   The researchers attempt to segregate the factors sector-wise.

Here we had limited our study only to Case I type business model (see Chapter 4.1)— that is, remanufacturing is an extension of business activities by the OEM in addition to its existing manufacturing.

The following fourteen issues were identified on the basis of extensive literature review, which are the possible barriers in remanufacturing. These are treated as the variables for constructing the survey instrument in this study.

1. Nonexistence of specific market for remanufactured products
2. Low demand for remanufactured products
3. Active unorganized or informal sector
4. Thriving secondhand market
5. Low confidence of customers on quality of remanufactured products
6. Mindset of customers not similar to that from western countries
7. Not profitable business
8. Technically not viable
9. Expertise not available
10. No environmental compulsion
11. Uncertainty in timing, quantity, and quality of acquisition of cores
12. Difficulty in acquisition
13. Difficulty in reverse distribution
14. Expensive logistics activities

The next step is conducting a survey and subsequently an empirical study so as to respond to those research questions. The primary data for this study is obtained through questionnaires constructed on the basis of the fourteen variables. Here the population of the survey is the set of manufacturers from Center for Monitoring Indian Economy (CMIE) database. From CME, 972 companies are chosen, which are manufacturing the remanufacturable products.

Remanufacturability is judged by satisfying the following conditions as proposed by Lund (1998):

* The product technology is quite stable.
* Most failures experienced by the product are functional failures.

- The product design by and large is standardized and thus parts may be interchanged among products.
- There exists high scope of value additivity.

A questionnaire was developed for getting the responses from the manufacturers of remanufacturable products on the importance of each of the 14 variables as a demotivating factor in a 5-point Likert scale. Sample size was estimated statistically (keeping it 110 with some safety factor) and questionnaire was sent to 110 companies randomly chosen from 972 manufacturers of CMIE database. Finally, 41 valid responses were obtained, which were considered as input data for the research study.

The respondent companies were classified into four industry types—automobile (18), electronics and computers (9), consumer durables (2), and industrial machinery (12), the numbers within the parentheses being the individual share from 41 valid responses. Analysis of the responses showed high correlations among some of the variables. So the researchers presumed existence of some underlying factors, which are binding these variables. Thus, the subsequent step may be conducting factor analysis for identification of these factors or constructs by grouping the variables. Factor analysis on the responses resulted in identification of following six factors, which critically influence the decision on whether to invest for remanufacturing or not.

1. Acquisition of cores and reverse logistics issues
2. Technology related to technological feasibility and product design
3. Market-related factor
4. Attitude of customers, reflecting the general mindset of Indian customers
5. Profitability of remanufacturing business as perceived by the manufacturers
6. Laws in India showing stringency or some restrictions and mandatory activities

This provides significant insights and some understanding on the barriers or challenges in taking up remanufacturing activities in India and the underlying demotivating factors for the OEMs corresponding to the first research objective.

As a response to the second research question, another research investigation was carried out on the "natural" grouping of the respondent organizations. Application of cluster analysis and development of dendogram resulted to identification of the critical factors under three clusters representing three different industry types. Table 4.2 displays the dominant factors of the three clusters.

These three clusters are distinctly the computers and electronics sector, the industrial machinery sector, and the automotive sector. Logically, as the computers and electronic industry is highly technology-focused, the technology-based design issues are the critical ones for taking any decision on whether or not to remanufacture. Similarly, the industrial machinery sector requires special transportation mode for effective logistics because of the large size and various shapes of cores and resultant remanufactured/refurbished items. In case of automobile industry, what critically matters are the acceptance of customers to buy the remanufactured cars. Moreover, the legislative pressure on treatment of the end of product life of cars also matters a lot in motivating or demotivating manufacturers toward the remanufacturing business.

TABLE 4.2

Dominant Factors Influencing the Three Industry Clusters

Clusters	Industry Types	Critical Demotivating Factors
I	Computers and electronics	Technology (design focus), customer attitude, market, and profitability
II	Industrial machinery	Cores acquisition and technology (reverse logistics focus)
III	Automotive	Market and legislation

(Source: Mondal and Mukherjee, 2006a)

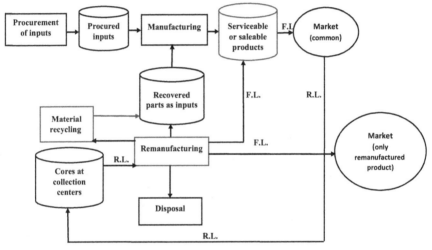

F.L. stands for Forward Logistics
R.L. stands for Reverse Logistics

FIGURE 4.3 A comprehensive framework of managerial activities for the OEM involved in both manufacturing and remanufacturing operations depicting the complexity of inventory management (four types) and logistics management.

4.2.2 REMANUFACTURING OPERATIONS AND MANAGEMENT

Remanufacturing management may be explained as a management of a set of operations keeping due consideration to some external and internal managerial factors. Figure 4.3 shows a general network of activities, if the OEM initiates remanufacturing of its own products along with its existing manufacturing business. Its existing market supplies the cores or returns for remanufacturing. The remanufactured products are once again sold to the market. Of course, the market for remanufactured products may be different from that of manufactured ones in some sectors. During remanufacturing operations, some items are earmarked as disposable ones, and some others are sold to material recyclers. Readers might have noticed that the framework in Figure 4.3 includes four facilities for maintaining stocks. In case of manufacturing, input materials are kept in stores after their procurement. A separate storage facility is maintained for keeping stock of recovered healthy parts or remanufactured spare parts. Some acceptable parts (immediately after disassembly) may be directly used in manufacturing operations as input materials. Both the manufactured and remanufactured products are kept as the finished goods inventory in the downstream storage facility. These are often known as serviceable inventory containing products for a common market. If there is an exclusive market for remanufactured products or it is remanufacture-to-order type of business strategy, then the remanufactured products are directly sent to the customers. Examples may be the remanufacturing or refurbishing services to big and heavy machineries, oil rigs, and so on. The fourth inventory facility represents the storage of cores or returns as the inputs to remanufacturing operations.

It may further be noted that some of the processes under remanufacturing operations are outsourced to external service providers, depending on the core competence and other relevant managerial and techno-economic factors. Various processes and related decisions under remanufacturing business are discussed further here.

Core acquisition is the first operation, and core availability actually determines the feasibility of any remanufacturing business. Cores or used products are under possession of the customers of original product or disposers' market. If the policy of extended producer responsibility is prevailing, the OEM may have already established take-back strategies for core collection. Otherwise, some compensation mechanism is to be implemented, like exchange offer, discounts and direct purchase from customers. In case of the latter option, remanufacturers will have sufficient control on the quantity and quality of cores by varying price offers and compensation amounts. Core brokers and retailers may also be the sources of core collection. Acquisition and collection may be treated as two separate activities, or both may be carried out by the same business unit.

The second crucial process in remanufacturing is the transfer and transportation of cores from collection centers to remanufacturing facility or *reverse logistics*. Reverse logistics demands special attention in designing and operating reverse distribution network and also in planning for other logistics-related decisions, like warehousing, inventory management, and material-handling. Forward logistics networks are applicable for distributing large quantity of limited types of products or models from a few factories to various demand centers, whereas reverse logistics networks are meant for distributing a small amount of various type of models from various sources to a few remanufacturing plants. The latter is more complex because of uncertainty in quantity and timing of acquisition. Moreover, variation in types of cores and availability of only few cores makes the reverse logistics relatively less cost-effective. The management of reverse logistics becomes more difficult also because of packaging issues and loss of economy of scale, particularly for machineries and other engineering products.

As the next process at remanufacturing plants, the cores are *inspected and sorted*. Although a preliminary checking might have taken place at collection center, cores are thoroughly inspected at the remanufacturing sites so as to ascertain the remanufacturability of the cores. This step is economically relevant before proceeding further in remanufacturing operations, as costs of disassembly and reassembly carry maximum share of total costs of remanufacturing. Cores of poor quality are either disposed (even recycled) or disassembled to extract the parts of good quality, which may be further used for manufacturing of some other product (like cannibalization). Under this process of inspection and sorting, cores of different models or variants are sorted accordingly. The variation in design often occurs, as the manufacturer goes on updating the design for extending the product life cycle of the product, particularly if the product is in the decline stage of market demand in its life. The sorted cores are stored and kept as inventory.

The subsequent step in remanufacturing is *disassembly of cores*. Disassembly is usually a manual process, demanding labor hours and skill in disassembly. Workers should be skilled enough in handling hazardous materials safely, which are quite risky in terms of human health. However, as remanufacturing is gaining its popularity worldwide, manufacturers are opting for design for remanufacturing (DFR) or design for disassembly (DFD) as disassembly-friendly design for new products. This makes disassembly easier and less time-consuming. The information and knowledge gained during disassembly may be subsequently used during reassembly and quality improvement of the remanufactured products. The disassembly process continues till the desired stage of releasing subassemblies, components, or elemental parts.

These released parts are then *cleaned, inspected, sorted,* and *stored*. The cleaning may be done simply using water or by some specific chemicals. The inspection at this stage is more critical than the previous one. The parts may undergo *repair* or *rework* if required. This repair will help to get back the usability or acceptability of the parts. As the cores are collected from various sources, which represent different work environments, ages of cores, and processes of handling by the users, the time and resource requirement for disassembly also vary from core to core. After disassembly, the released parts are now sorted and stored separately, usually earmarked by standardized codes.

The last stage of remanufacturing is *reassembling*, followed by final *testing*. Like the process of assembly line in manufacturing, reassembly may be carried out with appropriate layout design involving multiple work stations. Here, parts mismatching may arise, as parts have been generated from cores with varied designs. Cores may represent dissimilar models of the same product. Two additional activities may also be included during reassembly. *First, some newly procured parts may be added in place of the old parts. Second, technology updating may be carried out by replacement of old components and by improving the quality or enriching specifications of the product.* This may help in marketing of the remanufactured products. Finally, the reassembled products are to undergo a series of tests and quality checks so as to meet the as-good-as-new standards. These products are to qualify for the required warranty standards (often same as that of the new products).

Although this list represents a generic list of processes in remanufacturing operation, there may be slight changes depending on the type of products and the business contract with the customers. For example, in case of big machineries, oil rigs, or locomotive engines, usually the customer asks for refurbishing or remanufacturing of a used product and same is true for the tire-retreading business. So the duration, complexity, and required resources will vary with the type of product under consideration and the requirements based on the type of contract from the

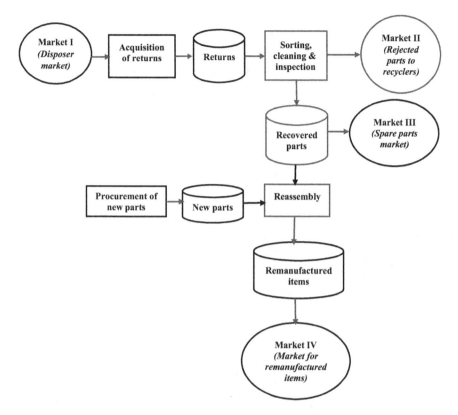

FIGURE 4.4 Activities involving collection centers and remanufacturing plants connecting four markets.

customer. It may be noted that inspection and sorting are carried out both on cores and parts. The second inspection is of course more crucial, as parts are now ready for final assembly. Figure 4.3 depicts the activities under collection center and remanufacturing plant. In some cases, the role of collection center ends at acquisition of returns, and then the returns are handed over to the remanufacturing plant. It is now the responsibility of the remanufacturing center to carry out the remaining activities of remanufacturing. In some other case, the responsibility of the collection center extends to the recovery of usable parts, and the remanufacturing plant does the reassembly and final marketing of remanufactured products. These two cases clearly show the variation of activities under two facilities—collection center and remanufacturing center. Figure 4.4 also depicts how four different types of markets are involved in the remanufacturing process. Among these markets, the *disposers market* is the source of returns or inputs to remanufacturing, and the remanufacturer produces usable spare parts and items for recycling along with the remanufactured products.

The detailed flow of activities under remanufacturing operations is shown in the flow diagram in Figure 4.5.

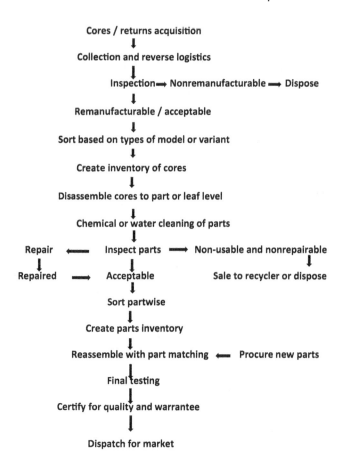

FIGURE 4.5   Detailed flow of activities in remanufacturing operations.

> *Remanufacturing operation comprises processes like core acquisition; reverse logistics; inspection and sorting of cores; disassembly of cores; cleaning, inspection, sorting, and storing; reassembly; and testing, all to be carried out sequentially.*

Let us now explore various decision areas addressing the following questions in remanufacturing operations.

4.2.2.1   What Remanufacturing Strategy Is to Be Adopted by the Remanufacturer?

This strategic decision practically influences all other decisions in remanufacturing operations. Like strategic manufacturing options, there exists two remanufacturing strategies—remanufacturing-to-stock (RTS) and remanufacturing-to-order

(RTO). RTS is meant for a situation with longer remanufacturing lead time, and relatively stable and known demand for remanufactured products. The second and most popular option is RTO or demand-driven remanufacturing. RTO is more applicable in remanufacturing business, as it is meant for uncertain demands. RTS is more proactive, and RTO, on the other hand, is a reactive production activity of remanufactured products. The selection of the strategy is also dependent on the ownership of remanufacturing operations, like whether it is OEM or an independent remanufacturer. If OEM remanufactures its own product, it must have studied the market beforehand and must have earmarked a specific stable market for this product. It may opt for RTS for further supply to distributers or retailers. RTS may also be applicable for suppliers of remanufactured or refurbished spare parts. On the other hand, RTO clearly means carrying out the remanufacturing operations when it is demanded or when order is received. Independent remanufacturers usually adopt this strategy. However, delivery lead time, inventory costs, and logistics issues are some other factors, which are to be considered during this strategic decision-making. In RTS the stock of remanufactured products is kept ready either at remanufacturer's facility or at the distributor's (or retailer's) end. The customers buy the readily available products. In most of the cases, a hybrid strategy is adopted. This is like postponement strategy or delayed differentiation in manufacturing, which combines make-to-stock (MTS) for production of components or subassemblies and make-to-order (MTO) for final assembly based on actual orders. Figure 4.4 indicates that there exist two types of inventories—inventory of cores and that of parts. There may also be a third type of inventory and that is the stock of remanufactured products after final testing for quality. These products are ready for shipping to the customers.

Now let us discuss on the *cost-benefit analysis* required for deciding on the right type of remanufacturing strategy in a business situation. For this purpose, let us assume that the criteria for this decision are *speed of delivery* (lead time between receiving the order and making it ready for dispatch or order fulfillment duration) and *inventory costs*. Remanufacturer intends to reduce delivery lead time (more customer satisfaction) and also the associated inventory costs. In other words, these criteria are *responsiveness* (delivery speed) and *efficiency* (inventory costs). The following simple example is considered for explanation. Some assumptions are also to be included in this example along with relevant data and symbols for associated parameters.

4.2.2.1.1 Assumptions, Data, and Parameters

- Analysis is for single unit of remanufactured product, which may be appropriately expanded with multiple units.
- The time unit considered for the example is day. This may be generalized to hour, depending on the actual case.
- The core is acquired and kept in stock on the 1st day of the month, and the order for the remanufactured product is received on 20th day of the same month.
- After product disassembly, α number of acceptable parts or leaves are released. We also assume that no repair or rework is necessary.

- C_1, C_2, and C_3 are unit costs of the core, part, and remanufactured product, respectively.
- w is the holding cost as percentage of unit cost of an item kept in stock for one day. It may be estimated keeping in mind the interest on working capital and other associated costs for storage.
- The estimated durations of disassembly and reassembly are assumed to be three days and two days, respectively.

4.2.2.1.2 Analysis

Here we are analyzing three possible strategies—Strategy X (RTS), Strategy Y (RTO with parts kept in stock), Strategy Z (RTO with cores kept in stock). Let us compute the impact of the strategies.

Strategy X

- Once cores are acquired, they are directly disassembled and then reassembled after relevant cleaning and inspection processes. Subsequently, remanufactured products are kept as inventory after final testing. So no stock of cores and parts are maintained. There is only stock of remanufactured products.
- Once the order is received on the 20th day, the remanufactured products are shipped directly from its stock. It is a case of quickest delivery. On 6th day of the month, the remanufactured product is ready and sent for storage.
- Inventory costs = $(20 - 6) * w * C_3 = 14 * w * C_3$

Strategy Y

- Once cores are acquired, they are disassembled to parts, and parts are kept in stock after the usual cleaning and inspection. No stock of cores is maintained.
- Once the order is received, the parts from the stock are reassembled and then shipped. There is no stock of remanufactured products. So on the 23rd day of the month, the remanufactured products are ready for final testing and subsequent shipping.
- Inventory costs = $(20 - 4) * w * C_2 = 16 * w * C_2$

Strategy Z

- Once cores are acquired, they are kept in stock after usual inspection for remanufacturability.
- Once the order is received, the cores from the stock are taken for disassembly, and subsequently the parts are reassembled. No stocks of parts or remanufactured products are maintained. So on the 26th day, the remanufactured products are ready for final testing and subsequent shipping to customers.
- Inventory costs = $(20 - 1) * w * C_1 = 19 * w * C_1$

Decision

- RTS keeps the stock of remanufactured products, so directly they may be shipped without any delay.
- However, cost of a remanufactured product is much more than that of parts or cores. In other words, $\alpha * C_2 = C_1 +$ disassembly costs and $C_3 = \alpha * C_2 +$ reassembly costs.
- The remanufacturer should judiciously take into account the conflicting parameters like inventory-holding duration and cost of item for selecting the right type of strategy.
- The critical factor in this strategy selection is uncertainty of demand. Any error in demand forecasting may make RTS very expensive as $C_3 > \alpha * C_2 > C_1$.

Similar to manufacturing strategies, the remanufacturer is to decide on the most suitable among the three strategies—remanufacture-to-stock (RTS), remanufacture-to-order (RTO), and RTS-cum-RTO. The decision is taken analyzing the impact of two conflicting criteria like responsiveness and cost.

4.2.2.2 What Are the Sources of Cores Acquisition? What Should Be the Modes of Core Acquisition?

These are the decisions related to acquisition planning. The following decisions are made, applying appropriate techno-economic analysis.

- Mode of acquisition (discount, exchange offer, take-back policy, warranty, etc.)
- Sources of cores (customers, retailers, core brokers, etc.)
- Timing of acquisition (age of cores)
- Acquisition price (influenced by the mode and timing of acquisition)
- Quantity (small or large quantity of acquisition)

In most of these decisions, it is suggested to take into account some or all of the following relevant parameters.

- Original purchase price of the product at the user's end
- Running cost of the product being paid by the user, including operating and maintenance or repair costs
- Current price of the product available in the market
- Estimated cost of remanufacturing, on the basis of some quality level of used product
- Failure history of the product
- Overall impact of acquisition cost and cost of reverse logistics on profitability expected from remanufactured products

4.2.2.3   How to Manage Inventory in Remanufacturing Operations

Practically, remanufacturing requires maintenance of four types of inventories. Type A, type B, type C, and type D represent inventory of cores, recovered parts, newly procured parts, and remanufactured products, respectively. Presence and absence of a particular type of inventory depend on the remanufacturing strategy adopted by the company—that is, whether it is RTS or RTO. RTS (Strategy X, as explained earlier) is a push-based strategy, like MTS. Remanufacturing to stock is applicable for known demand situation and with established market. If the OEM includes remanufacturing along with manufacturing facilities, the strategy may be RTS. A market of remanufactured items may be created by middle-class customers, who are price-sensitive and often look for secondhand products. Small retail service stores run their business on photocopying services in small towns of India. They may be the regular customers of remanufactured photocopiers. Small schools of Indian villages may be another group of purchasers of this product. The same is applicable for cheaper remanufactured refrigerators for small grocery stores and sweet shops in India. In the RTS strategy, inventory management of type D stock (stock of remanufactured products) is essential, as the cores are first disassembled and subsequently reassembled after carrying out necessary cleaning and inspection without any rest. However, type C inventory may also be maintained, as some new parts are to be added during reassembly in case some recovered parts are rejected. On the other hand, the RTO strategy has been developed on the basis of pull-based process, similar to MTO. It is a reactive strategy, and once the order for remanufactured products is received, the activities start at remanufacturing facility. As mentioned earlier, two strategies may be derived from RTO, like Strategy Y and Strategy Z. In Strategy Y, the remanufacturer disassembles the cores to parts and the inventory of parts is maintained as type B stock, whereas in Strategy Z, the inventory of cores is kept as type A stock. Strategy Y is more applicable in a case, where the order fulfillment duration is less and the parts are only reassembled for order delivery. Cores are to be disassembled, and subsequently the parts are reassembled for final delivery of remanufactured products in case of Strategy Z. Of course, the inventory of new parts is to be maintained as type C stock in both the remanufacturing strategies. Inventory-related issues encountered in case of the independent remanufacturer are much less. These independent remanufacturers are engaged in remanufacturing or refurbishing of old products or machineries from any customer. These are not simply replacement of some unacceptable parts by new parts (from type C inventory). A tire-retreading business includes similar set of activities. Usually remanufacturing-to-order strategy does not require maintenance of stock of remanufactured products (inventory type D). The uncertainty factor and supply-demand mismatch lead to estimating required safety stock in inventory planning, as the remanufacturer is to satisfy the customers (maintenance of availability or service level) and also to reduce inventory costs. In RTO, the release of items from stocks of upstream inventory types may be controlled by a pull-based or Kanban-based mechanism so as to reduce inventory costs. In other words, we may say that the reassembly will take place, once there is demand for remanufactured products, and disassembly of cores will start to replenish the stock of parts. Management of type C inventory is another issue of complexity. The newly procured parts in type C are used as replacement, once a part is found to be unacceptable for reassembly. Moreover, most of the remanufacturers add new parts or components (around 10–20%) in the remanufactured product as a mandatory policy of remanufacturing. This improves the reliability of the product and this enables better marketing, as perception of prospective customers becomes better. But remanufacturer faces difficulty in predicting the condition of parts beforehand and also identifying the specific new part to be procured and kept ready for reassembly for shipping without delay. Just-in-time may be applicable disassembling and reassembling activities for the replenishment of downstream inventory types. JIT may be equally applicable for the replenishment of type C inventory (new parts) by formalizing suitable contracts with suppliers. Further, another difficult task in inventory management is estimation of inventory cost for each of the four types of inventories. It is also difficult to estimate per unit cost of each part. How do we estimate the opportunity cost of non-availability of a part or a core? It is an excellent food for thought. However, we may assess all cost components associated with maintenance of the stock and consider the interest rate of working capital or ROI as an estimate for inventory carrying cost.

4.2.2.4   How to Manage Disassembly Operations

Disassembly is product explosion or separation operation. It is carried out stepwise, like product to subassemblies, subassemblies to components, and components to parts. The steps may vary depending on the type of product and its design. Stages or levels of disassembly are also expressed as parent (product) to intermediaries (subassemblies) to leaves (parts). Disassembly is often carried out per the planned depth of disassembly. This depth indicates the limit of continuing the disassembly operations. As disassembly requires extensive labor-hours and skill of operation, it is an expensive task. So, if necessary, disassembly may be stopped at the intermediate level or at the stage of subassemblies or components, depending on the quality and condition. The subassemblies may be directly reassembled, and this is the most common practice in refurbishing. Sometimes cleaning and critical inspection also guide in determining the required depth of disassembly. If the original product is designed for disassembly or designed for remanufacturing, the disassembly operation is expected to take less time and require less cost of disassembly. For example, using modules and clips instead of fasteners makes the design more like DFD or DFR and so quite suitable for quick disassembly. The parts generated from disassembly are sorted, codified, and stored.

In addition to many others, the most critical decision-making areas of remanufacturing management are formulation of remanufacturing strategy, sources and modes of core acquisition, inventory management, and management of disassembly operations.

4.2.3 Management of Disassembly Operations

The following simple guidelines are to be considered during scheduling disassembly operations.

- Used products are to be sequenced for disassembly. The expected release of parts is to be estimated from this disassembly operations. This is to be matched with the demand for remanufactured products by the customers, particularly in RTO.
- Disassembly line is to be established by each family of cores on the basis of the product configuration and its design or model.
- Similar to the layouts used in manufacturing operation, disassembly may be carried out in a specific layout (e.g., serial, parallel, U-shaped), if necessary.
- There may be various degrees of uncertainty on the outcome of disassembly depending on the quality of used products based on the prior knowledge on product design, work environment at the user's facility, and mode of handling and maintenance by the user.
- Parts may have physical defects (weight, specification, or dimension change) or functional defects. The parts with physical defects may be removed and replaced by newly procured parts. This enhances the reliability of remanufactured products. Sometimes, these parts might have been badly damaged and got stuck into the body or with other parts. This may lead to loss of more than one part repairing of the damages. The functionally defective parts may be sent for repairs or reworks.

4.2.3.1 Elemental Tasks in Disassembly

Disassembly is interestingly a unique process, as its objective is how to efficiently separate a part out of the main body of the product, unlike the assembly process of a manufacturing operation. Reassembly, on the other hand, is more like an assembly operation in manufacturing.

The completeness of disassembly is determined by the depth of disassembly. It may be either full disassembly or partial disassembly. Full disassembly is the complete separation of cores to the last level—that is, the parts level. It is more expensive, demanding more resources and capacity of disassembly. On the other hand, partial disassembly may be limiting disassembly to subassembly or component levels.

There may be three categories of elemental operations in disassembly or separation—*destructive*, *semi-destructive*, and *nondestructive operations*. Irreversible fasteners (e.g., weld) are removed using destructive tools (grinders, cutters, etc.) with possibility of *full destruction* of fasteners along with significant damage on parts or components. In *semi-destructive operations*, an attempt is made to limit the damage and destructions. For example, cutting off screw heads or drilling rivets minimizes the damage on the connected components or parts. *Nondestructive operations* can be carried out when the original product is fitted with suitable types of fasteners. Clips or undamaged screws surely help achieve nondestructive operations. Its main aim is to separate components without any damage. Extensive cleaning and removal of dusts or other undesired materials is required prior to any nondestructive operation. A modular design and design for disassembly (DFD) enable conducting nondestructive operations.

The following tasks are very common in disassembly.

1. Standard tasks

- *Separating* parts after the removal of fasteners or connectors
- *Rotating* a part like a bayonet coupling using pliers or spanners
- *Pulling or pushing* even with hammers, like separating bearings from shafts
- *Unscrewing* for disconnecting threaded connections for nuts or bolts
- *Emptying* a tank for the removal of fluid
- *Changing the form or shape* of a part for removal but not damaging it completely

2. Destructive tasks

- *Making a cut* for separating parts or finding access
- *Bending* a part completely for opening or getting access
- *Drilling* for separation or destroying a fastener
- *Milling* for separation and removal of parts
- *Crushing or shredding* for disposal of a part leading to its destruction

4.2.3.2   Design for Disassembly (DFD)

Using the DFD approach for a remanufacturable product during its manufacturing enables quick and easy disassembly, using the least amount of resources during the remanufacturing after its end-of-life. DFD also enhances overall effectiveness and efficiency of disassembly and thus remanufacturing operations. The product designer plans for implementing the DFD in the production process by suitably combining disassembly methodology, product/component/parts technology, and associated hazard/risk/human health factors. Most of the DFD principles are oriented toward ease and speed in removing the fasteners or connectors, preferably without destruction. DFD guidelines are as follows:

- Minimize the number of fasteners, as the removal of fasteners takes a lot of time.
- Use standardized and commonly used fasteners to reduce the variety of disassembly tools, which cause delay in disassembly because of the change-over times
- Use fasteners or connectors, which can be removed easily and quickly without destruction like clips.
- Build easy access to all fasteners and connectors between components or parts.
- Minimize the number of parts or components.
- Use parts that are standardized in dimensions and easily available.
- Minimize the use of hazardous and unsafe materials, which may require special devices for handling and difficult to be recycled or disposed.

4.2.3.3 Automated Disassembly

In developed countries, corporations often express their reluctance in remanufacturing business, because of expensive use of manual labor in disassembly operations. Unlike in developing countries like India, labor costs in most of the developed countries are very high and so the remanufacturing business lacks cost-effectiveness. Automated systems are rarely used in disassembly, as the cores collected from users or core-brokers are mostly found to be of different quality, size, or shape or even different makes or models, and automated systems may not be efficiently applicable for non-standardized cores. Prior information is also not sufficiently available about the model, variant, and design of the cores expected to be acquired from users of products. Moreover, the physical condition of the used products is unknown before the collection. The uncertainty because of these factors may be another reason for the non-usage of the automated disassembly system.

However, there is a recent development in this pursuit. Intelligent systems and smart tools have been created using *artificial intelligence (AI)*, which can be suitably used for disassembling cores that were manufactured with DFD. Smart robots are used in this automated system to carry out disassembly operations in place of manual labor. *Cognitive robots* may be used in this intelligent disassembly system (Vongbunyong and Chen, 2015). These robots emulate human behavior using the knowledge base (KB) developed through learning (may be *deep machine learning). The cognitive robots are composed of three modules—cognitive robotic module (CRM), vision system module (VSM), and disassembly operation module (DOM).* CRM is the brain of the robot. It is an AI-based system using KB and cognitive functions replicating the human mind, and it controls all other activities of the robot. VSM includes sensors, cameras, and hyper-spectral imaging devices. It attempts to identify fasteners and connectors, as well as defects among the components or parts. Its functions include recognition and localization of components of different types, specifications, and conditions. DOM is equipped with suitable tools for disassembly. The smartness of the cognitive robots is reflected by human-like cognitive functions involving KB, learning, and revision or updating the KB. This automated system, of course, makes the disassembly process more efficient with more precise outcomes.

Disassembly operations are normally carried out by some destructive and non-destructive tasks. If the products are originally designed by DFD or DFR, destructive tasks may be avoided to the maximum extent. Although disassembly operations are mostly carried out manually by skilled manpower, AI-based cognitive and intelligent robots are nowadays developed so as to make the disassembly operations more efficient and effective.

4.2.3.4 Analytical Modeling in Disassembly Scheduling

Disassembly scheduling is a complex decision problem, and various researchers proposed analytical models and their corresponding solution methods including optimization and heuristic-based approaches.

Inderfurth and Langella (2008) proposed four suitable models in this context. One of the important technical parameters to be used in this scheduling decision is estimated yield rate, or the number of released items (subassemblies or parts), by disassembling the cores. Let us formulate the disassembly scheduling problem as a mathematical model similar to the models proposed by Inderfurth and Langella (2008). The following are some characteristics of the decision problem and assumptions considered in this model.

- The remanufacturer is to decide on (1) *number of cores to be acquired*, (2) *number of cores to be disassembled* either fully (to part level) or partially (to subassembly level), and (3) *number of new parts to be procured* from outside.
- The remanufacturer takes the decision with the intention of *minimizing the total costs*—that is, the cost of acquisition, disassembly, and procuring new parts.
- Although disposal is a crucial activity in remanufacturing, it is relaxed in this model as this depends on estimation of non-usability of a part or subassembly, which is uncertain beforehand. Moreover, in a country like India, disposal is still not very expensive or problematic at present. So the disposal cost is not added to total costs.
- Let us consider a time horizon, say one year, which is divided into planning periods. The decision is taken for each planning period in a time horizon or annually.
- The requirement of parts for each period is known (deterministic) and extra release of parts is stored for its use in next period. Dynamicity of the situation is reflected by inventory keeping throughout the whole planning horizon.
- Disassembly costs vary depending on the depth of disassembly. On the other hand, disassembly at any level is associated with some yield, which is also supposed to be deterministic for the simplification of the model formulation.
- Any disassembly operation requires resources and facilities available in the existing capacity of the disassembly of the plant. The disassembly decisions are to be made keeping in view the available capacity in a planning period.
- As there is capacity restriction in disassembly operations and costs also vary with the depth of disassembly, the remanufacturer decides on the depth of disassembly to be undertaken in a period, depending on the available cores in that period. However, this decision is further influenced by inventory-holding costs of cores, subassemblies, and parts and the demand for parts generation in a period.
- By depth of disassembly, here we consider whether it is a separation of cores to parts in a single-stage process or cores to subassemblies and sub-subassemblies to parts as a two-stage process.

Model

i = index for cores
j = index for subassemblies

k = index for parts
t = index for planning period

Decision Variables

$X_{i,t}$ = number of ith core acquired at t^{th} period
$X_{k,t}$ = number of kth part procured from outside at t^{th} period
$X_{i,t}^{j}$ = number of ith core to be disassembled to subassembly level at t^{th} period
$X_{i,t}^{k}$ = number of ith core to be disassembled to parts level at t^{th} period
$X_{j,t}^{k}$ = number of jth subassembly to be disassembled to parts level at t^{th} period

Cost Parameters

C_{i} = unit cost of acquiring i^{th} core
C_{i}^{j} = unit cost of disassembly for each ith core to subassemblies
C_{j}^{k} = unit cost of disassembly for each jth subassembly to parts
C_{i}^{k} = unit cost of disassembly for each ith core to parts
C_{k} = unit cost of procuring for each kth part from outside
h_{i} = unit inventory-holding cost per time period for i^{th} core
h_{j} = unit inventory-holding cost per time period for j^{th} subassembly
h_{k} = unit inventory-holding cost per time period for k^{th} part

Technical Parameters

α_{ij} = yield rate of ith cores for disassembly to subassemblies
α_{jk} = yield rate of jth subassembly for disassembly to parts
α_{ik} = yield rate of ith cores for disassembly to parts
β_{ij} = resources used for each ith core for disassembly to subassemblies by using the existing capacity of disassembly
β_{jk} = resources used for each jth subassembly for disassembly to parts by using the existing capacity of disassembly
β_{ik} = resources used for each ith core for disassembly to parts by using the existing capacity of disassembly

Inventory-Keeping Parameters

$Y_{i,t}$ = inventory of ith core in t^{th} period
$Y_{j,t}$ = inventory of jth subassembly in t^{th} period
$Y_{k,t}$ = inventory of kth part in t^{th} period
A_{t} = total available resources or capacity for disassembly in t^{th} period
$U_{i,t}$ = used products or cores available for acquisition in t^{th} period
$R_{k,t}$ = total parts required to meet the demand for remanufactured products in t^{th} period

Objective Function

The disassembly scheduling aims to minimize total costs—that is, the combined costs of core acquisition, disassembly, new part procurement from outside, and inventory holding.

To minimize the total cost = $\sum_t \left[\sum_i \left(C_i X_{i,t} + C_i^k X_{i,t}^k + C_i^j X_{i,t}^j + h_i Y_{i,t} \right) + \right.$

$\left. \sum_j \left(C_j^k X_{j,t}^k + h_j Y_{j,t} \right) + \sum_k \left(C_k X_{k,t} + h_k Y_{k,t} \right) \right]$

Subject to the Following Constraints

1. Two options on depth of disassembly:
   $X_{i,t} = X_{i,t}^k + X_{i,t}^j$ for each ith core and tth period
2. Total cores available for acquisition:
   $X_{i,t} \leq U_{i,t}$ for each ith core and tth period
3. Limit on capacity for disassembly:

   $\sum_i \beta_{ij} X_{i,t}^j + \sum_j \beta_{jk} X_{j,t}^k + \sum_i \beta_{ik} X_{i,t}^k \leq A_t$ for each tth period
4. Inventory balance equation of cores:
   $Y_{i,t} = Y_{i,t-1} + X_{i,t} - X_{i,t}^j - X_{i,t}^k$ for each ith core at tth period
5. Inventory balance equation of subassemblies:

   $Y_{j,t} = Y_{j,t-1} + \sum_i \alpha_{ij} X_{i,t}^j - X_{j,t}^k$ for each jth subassembly at tth period
6. Inventory balance equation of parts:
   $Y_{k,t} = Y_{k,t-1} + X_{k,t} + \sum_i \alpha_{ik} X_{i,t}^k + \sum_j \alpha_{jk} X_{j,t}^k - R_{k,t}$ for each kth part at tth period
   Here all the decision variables are non-negative in nature.

4.2.4 PRODUCTION PLANNING AND SCHEDULING IN REMANUFACTURING MANAGEMENT

Like in manufacturing management, the production manager of remanufacturing plant takes appropriate decisions in relevant areas of operations management, like forecasting, layout design, capacity planning, lot sizing in production plan, materials requirements planning (MRP), outsourcing decision, and even the master production scheduling (MPS). MPS seems to be crucial if OEM gets involved in remanufacturing along with existing manufacturing operations for its own products. Many corporations, like Air France, Lufthansa, BMW, Volkswagen, Caterpillar, Nokia, Xerox, and Philips, maintain and manage their large disassembly plants. As discussed earlier, disassembly, in fact, is not reverse assembly, as the whole process is affected by uncertainty on various process activities and uncontrollable factors. The disassembly issues are not relevant to the cases like containers, which are directly reused after cleaning and repairs (as well as some rework, if necessary). In case of refurbishing, the limited depth of disassembly eases the production planning activities.

Demand management is the first step in overall operations management of remanufacturing. If the market for remanufactured products is explicitly separate, like that of photocopiers of Xerox, the pricing and other related decisions are not supposed to pose much difficulty. These organizations strategically identify separate market for low-priced recovered products. Xerox produces remanufactured products with product code, including R, as the distinguishing mark, and no issue of cannibalization arises. However, the pricing decision is quite important in case of overlapping market. Customers may prefer remanufactured products to the new one, provided the price difference between the remanufactured and new product is significant and acceptable by them. If the market is price-sensitive like that of India, customers will accept remanufactured products sold with a lower price. The demand for remanufactured products is influenced by the acceptable lowering of the price relative to that of the new ones. Moreover, often OEM offers price discounts for its new products to attract customers. In this case its new products offered with discounted price may cannibalize its own remanufactured products. Further the products available in the secondhand product market and the unorganized sector usually compete with the remanufactured products. Thus, the OEM should be very careful in taking pricing decisions if it produces both the types of products. Demand management is also to tackle the problem of balancing the demand with the inflow of returns.

The remanufacturing production planning is influenced by *Forecasting of cores availability, core acquisition decision,* and *forecasting of demand for remanufactured products,* which is further affected by the pricing decision.

Materials requirement planning (MRP) for remanufacturing is somewhat different from what is practiced in manufacturing plants. The availability of materials per bill of materials (BOM) in reassembly process is determined by the uncertain release of parts or components by disassembly, loss of parts by destructive disassembly process, and quality of recovered parts. Uncertain lead time of disassembly process is also to be incorporated in MRP (or modified MRP) for remanufacturing. Order release for adding new parts either as mandatory corporate policy or as replacement is to be considered in MRP and MPS (primarily for reassembly planning) for remanufacturing. *A suitable set of work stations are created as facilities in the remanufacturing plant for stagewise disassembly, checking, reprocessing, reassembly, and final inspection.* If the OEM is engaged in both manufacturing and remanufacturing, there should be appropriate linkages, particularly in MRP and in some common activities like inspection and painting. Recovered parts with good quality may be used in the assembly process for new products, in lieu of new parts. Plant layout design needs to take into account all these issues.

As remanufacturing is treated as process-focused sector, the job shop process and layouts are most common in remanufacturing plants. It is further reported that till the recent past, the remanufacturing-based production rarely uses CNC machines, and the production process is primarily carried out by a group of general-purpose machines, each being used for a specific operation. Operations management activities in remanufacturing are mostly affected by processing times (due to wide range of variants of a product from different sources) and uncertainty in core acquisition. Because of the uncertainty and variation in processing time, the remanufacturer is to critically plan for the inventory of different types in the plant, like cores, new parts, serviceable parts, spare parts, WIP, and finished goods (i.e., remanufactured products).

The survey report of Guide (2000) indicates that majority of remanufacturers do opt for mixed strategies involving both RTS and RTO. Like the hybrid of MTS and MTO in manufacturing, popularly known as postponement strategy, the mixed strategy in remanufacturing is also a combination of strategies, including the generation of recovered parts (by RTS) and their reassembly (by RTO). Once the cores are acquired and received by the remanufacturers, they are disassembled following the push-based RTS and kept as the stock of recovered parts. When the customers place orders for remanufactured products, the parts are reassembled and made ready for final inspection, painting, and dispatch. Around 20% of the remanufacturers surveyed implement pure remanufacturing strategies—that is, either RTS or RTO. For production planning and scheduling, several remanufacturers implement modified MRP, just-in-time (like Kanban), theory of constraints (like drum-buffer-rope), and classical inventory control techniques (like EOQ, periodic review system and continuous review system).

Almost half of the remanufacturers under survey reported that they try to balance the returns with demands, either forecasted or actual ones. Companies following pure RTO strategy try to limit core acquisition to the actual demand for remanufactured products. Production planning is more complex if the company adopts pure RTS. Decisions on core acquisition and disassembly are made in anticipation of future demand, which of course may lead to overstocking or understocking costs in case of any mismatch between supply and demand. However, the remanufacturer may make use of economies of scale by bulk disassembly and reassembly and reduce the cost of production. Excess parts are often sold outside as spare parts, may be used as input parts inhouse in manufacturing, or may be utilized as spare parts in response to the returns of new products during warranty period. Any excess cores may be sold to other remanufacturers as well. There may be few cases where these cores or parts may be sold to scrap dealers.

As mentioned earlier, there may be uncertainty in the lead time of disassembly operations. This, along with unknown material recovery rate or yield, poses difficulty in planning for disassembly sequence. Sometimes, remanufacturers practice reverse engineering to generate disassembly sequence. Products designed with DFD or DFR are normally associated with more predictable materials recovery and lower disassembly times and are expected to generate less waste. Parts generated during disassembly are less damaged in this case, and there is less cost of replacing parts or components.

Parts recovery is usually an uncertain phenomenon, as it is associated with the quality of cores, age, and product design. But the remanufacturers are to predict the recovery rate, as this determines the acquisition lot size or quantity of cores to be collected at a time, purchase lot size of new parts for replacement, and overall remanufacturing lot size. The estimated recovery rate or yield plays an important role in MRP of remanufacturing planning. In general, the purchase of new parts for replacement accounts for 10–35% of the total parts used in reassembly. Most often remanufacturers predict the recovery rate based on historical data, either by simple averaging or by applying any suitable forecasting tool. So forecasting of materials recovery rate or yield is another unique managerial activity in remanufacturing management, other than cores acquisition, inventory management, and forecasting of demand for remanufactured products.

The use of MRP in remanufacturing production planning is well accepted, although in a modified form. Remanufacturers usually implement customized MRP system for planning of procurements, remanufacturing order release, maintaining BOM, and inventory planning.

Most of the companies following RTS apply MRP, as it is the technique meant for push-based production planning. In this case, *if the OEM remanufactures, the master production schedule and overall production planning should be matched and partially combined because of three reasons. First*, the recovered and accepted components or parts may be reused in manufacturing along with its use in reassembly for remanufacturing. *Second*, the lot sizes of both manufacturing and remanufacturing are to be determined considering overall optimization of the financial performance identified for the OEM. The optimal mix of lot sizes will be the baseline for production planning of both the business activities. *Third*, the common activities like quality control, painting, and final testing may be carried out as services from centralized facilities located conveniently in the factory. This may not always be true, as some OEMs create completely separate production facility for remanufacturing as may be evident in organizations like Xerox India or some automobile companies in Europe.

RTO-based remanufacturing may have two options of production activities. First, once the order for remanufactured products is received, the remanufacturer acquires the returns or used products, and then the remanufacturing activities start. But it may not be a practical option, as availability of returns from the customers, quantity and timing, and the remanufacturability or recovery rate (pertaining to quality issues) are uncertain. So the modified version is more appropriate. It starts with returns acquisition proactively, whenever the returns are available continuously from its retail houses, independent vendors, or the OEM's own customers. The OEM or independent remanufacturer may opt for an economically viable or cost-effective mode of returns acquisition. Once the returns of right quantity are received, they are disassembled and tested. The recovered and accepted components or parts are now kept in stock as inventory. Whenever the order for remanufacturing is received, these parts are reassembled and converted to remanufactured products. This is relatively more popular practice in remanufacturing sector. The second option is practiced for big machineries like oil rigs or big products like aircraft and also for the tire-retreading sector. The customers place the order for remanufacturing, and also they supply the used products to be remanufactured or refurbished. The same used product of the customer is remanufactured. As in RTO, the delivery time is important, and the remanufacturer makes the resource planning and facility planning beforehand so that the lead time for remanufacturing reduces.

4.2.5 SOME REPORTED REMANUFACTURING PRACTICES

Some remanufacturing practices around the globe are discussed here so that readers get some exposure to existing activities in the business world.

1. **Machine parts and component remanufacturing in Brazil** (Muris and Filho, 2015)

    Four Brazilian cases are outlined here as reported in the cited article.

    The first case is a multinational manufacturer of automobile parts and components engaged in remanufacturing of 10% of its own manufactured products. The strategic goals of this Brazilian remanufacturer were primarily the sustainability and profitability. Initially it restricted the product recovery operations only to steering systems; however, later on, it expanded its activity base. It follows the RTO strategy and carries out the remanufacturing operations with around 50 employees. Its main problem is the existing poor perception of the customers about the remanufactured products. Twenty-five percent of the cores are acquired from its own customers, and the remaining 75% are collected from third-party core suppliers. Because of uncertain recovery rate, the company disassembles extra cores than what is demanded and, if necessary, keeps the recovered parts as inventory for future use. All the plastic and rubber parts are disposed. As automation level is low, variation of processing time often poses difficulty in maintaining the delivery time.

The second case is again an MNC in Brazil and OEM of clutch systems. Its remanufacturing facilities are meant for remanufacturing of 40% parts of clutch kits for better profitability and sustainability. The large remanufacturing plant has employed around 190 employees and has implemented RTO, MRP, and Kanban-based approaches in production planning. The company unfortunately cannot use its own name on remanufactured products because of some Brazilian legislative restriction, and this creates a chance of damaging the image by unauthorized companies. The company collects the returns from distributors, and the distributors also place the orders for remanufacturing, and the company attempts to match the lot size of returns with the orders. Most of the parts are reused for reassembly, if not discarded, only after cleaning and rarely machining or repairing.

The third case is a large manufacturer of parts for rail and road transport. Its remanufacturing includes four different types of products—compressors, valves, disk brakes, and air-handling systems accounting for approximately 1.5% of total remanufacturing products of Brazil. RTO is the remanufacturing strategy. The main sources of returns are 40% from automakers and 60% from stores of spare parts. Its strategy is somewhat similar to that of the tire-retreading business, which means that the customers of remanufactured items are also the suppliers of returns. The returns are categorized on the basis of possibility of reuse (e.g., 100% of reuse, 60–70% of reuse, and around 20% of reuse), and the price of remanufactured products is fixed accordingly.

The last Brazilian case is a remanufacturer of diesel engines of different types of vehicles, heavy machineries, and marine and mining equipment. It remanufactures either the complete engines or its parts, like crankshafts and pistons. The company is engaged in core collection, cleaning, and reassembly, whereas disassembly and some repair works, including machining, are outsourced. The returns are collected from distributors, and the customers are given the choice of replacing manufactured parts by remanufactured ones as an exchange offer.

2. **Photocopier machines in India** (Mukherjee and Mondal, 2009)

This photocopier-manufacturing company is actually a globally famous multinational organization, which established its business in India in 1982 and soon became popular Indian company having maximum market share in photocopier manufacturing. The parent global company started its product recovery business some time in 1991 by remanufacturing its own end-of-life photocopiers. By 1997 it could enjoy savings of over $80 million in Europe. The Indian business unit has also initiated its remanufacturing in 1998, establishing the Asset Management Business Center. The remanufacturing plant was set up approximately two kilometers away from its manufacturing plant. Although its primary aim was environmental cleanliness and profitability, this business initiative was largely driven by its after-sales service policy offered to its customers. Its policy provides free servicing to all new machines sold till the end-of-life (approximately ten years), although the cost of consumables for servicing and maintenance are to be borne by the customers. By adopting this policy, the company not only maintains a good relationship with the customers but also keeps all technical information of each model sold to the market during its useful life. It also acts as a sales promotional strategy. However, the company is to bear the additional after-sales service costs, which increases over time as the machines age. The management thus decided to implement the scheme of taking back the older machines by buying them back. Then the company initiated the remanufacturing of returned machines in place of their disposal for greater economic benefit. The company sells the remanufactured machines to a different market under new brand names (prefixed by "R" as a representation of a remanufactured model). In 2003 its production capacity was 60–70 units per month, with approximate sales turnover of INR 30 million. In fact, the company enjoys the existence of a monopolistic market other than some competition with the unorganized sector. The price of a remanufactured machine is almost 30% of that of a new one, and these are sold to small photocopy shops and other small business centers. Thus, there is no risk of cannibalizing its own new products. The remanufacturing process starts from the acquisition of its own old products from customers. These are collected by the company's 12 regional and several local service centers all over the country. Once there are sufficient stocks at the centers, they are transported to the remanufacturing plant in batches as economic lots. This transportation operation (reverse logistics) is outsourced, but other relevant operations are carried out inhouse by the company itself. The returns are first stored at the remanufacturing plant. Then they are disassembled, cleaned, inspected, repaired (if necessary), and stored as inventory of recovered and acceptable parts/components. Some parts are not reusable by nature, like plastic parts, bushings, adhesives, and fasteners, which are disposed. Some other unacceptable parts are either sold to recyclers or disposed. Per the policy there should be at least 20% new parts, which is mandatory. These are to be procured. Some other new parts are also procured as replacement of unacceptable recovered parts. The reassembly is the last operation in remanufacturing before final testing. Figure 4.6 shows the workflow of remanufacturing activities. The remanufactured machines are kept in store, and thus it is actually an RTS strategy. When the demand rises, the remanufactured photocopiers are sold. The company uses both permanent and contractual workforce, although it feels that the use of temporary workforce seems to be more cost-effective but less reliable.

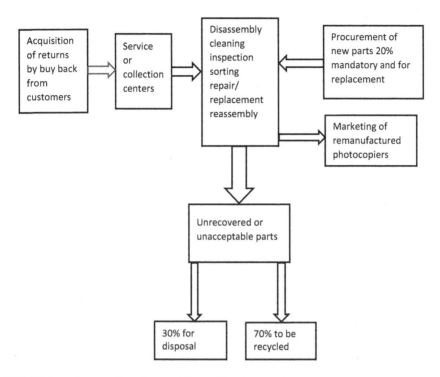

FIGURE 4.6   Remanufacturing workflow of the photocopier company.

(Source: Mukherjee and Mondal, 2009)

3. **Automotive engine remanufacturing** (Seitz and Peattie, 2004)

It is a case of an established European remanufacturer engaged in reman-ufacturing its own passenger cars and light commercial vehicle engines since 1986 in its dedicated remanufacturing plant. It meets the demand of spare parts from the customers and replacement requirements of failed engines or parts during the warranty period with cost savings of almost 40%. Once the used engines are received, they are first manually disas-sembled, then the parts are cleaned (primarily by deep chemical cleaning), inspected, and sorted. The parts are subsequently repaired, machined, and abraded for their reconditioning, and then stored. In response to the market demands for remanufactured products, the parts are reassembled to create an engine. The recovered and accepted parts are also sold as spare parts. Worn and rejected parts are replaced by new ones for remanufacturing of engines, which amount to almost 15% of total engine parts.

The company has been facing three types of challenges. First, it has a large and complex reverse logistics network of channels, which demand critical strategies for channel management. The second challenge is acqui-sition of sufficient cores for product recovery, and the third challenge is maintenance of effective relations with the large network of stakeholders.

Further, the cost analysis of remanufacturing operations reveals that 35% of prime costs is the labor cost only. So the company availed the opportunity of moving toward countries of lower wage rate (like some East European ones) for better profitability. On the other hand, parts variation occurs because of the frequent entry of new models of car in the market. This results in piling up more inventory of parts in the plant due to parts-matching issues. If high-quality and durable new parts are to be procured for the replacement of old ones, it may also cause delay in procuring and becomes expensive. The delivery times of remanufactured products range between 20 and 40 days. The company only remanufactures 30% of the parts, and the remaining 70% are outsourced, among them 15% are directly purchased from its own suppliers and 55% are externally sourced, most of which are remanufactured ones. Because of uncertainty in recovery from various types of returns and supply lead times, the company is to keep sufficient inventory so as to meet the delivery times to customers.

4. **Tire-retreading business in India** (Mondal and Mukherjee, 2012)

Tire retreading is a remanufacturing business where the customer offers the used tires to be retreaded (or remanufactured). Usually almost 25% of used tires wear out, which needs replacement, whereas 75% of tires remain reusable. These reusable tires are retreaded instead of replacement by new tires resulting in economic value addition and environment-friendliness. Using retreaded tires substantially saves cost and energy. Incineration of used tires emits toxic gases like polyaromatic hydrocarbons, CO, SO_x, NO_x, HCL, and smoke. Landfilling also causes land contamination. However, there are some beneficial uses of rejected tires as well, like road construction. So tire retreading leads to reduction of pollution and delays the disposal. In fact, by applying a new tread to an old tire, a new life (almost 80% of a brand-new tire) is offered to the old tire at a cost of 50% of the market price of the new tire. The transport sector is economically benefitted by availing the tire-retreading option. It is reported that the aircraft industry saves almost $80 million per annum, whereas the benefit of the trucking industry amounts to around $2 billion per annum by using retreaded tires in place of new tires. Moreover, there is enough savings of crude oil, which is scarce and mostly imported in India. For example, being a petro-chemical product, one new truck tire requires 22 gallons of crude oil, whereas the crude oil consumption of its retreaded version may be only 7 gallons. In case of passenger cars this reduction of crude oil consumption is from 7 gallon to almost 2.5 gallons. All these factors are the drivers for growing retreading sector. Almost 50–60% of demand for tires are met by retreaded tires in India. However, in recent past, with the introduction of radial and better-quality tires, particularly for trucks, buses, and light commercial vehicles (LCV), the growth of demand for tire replacement has been affected significantly due to the higher durability of new tires. Another reason for the reduction of this demand is improved road condition in India. But on the other hand, intensive growth in industrial and logistics activities contributed to the increase in demand for replacing tires, and so the retreading sector grew. Incidentally, the tire-retreading business is highly labor-intensive, and it operates in small-scale production units. So it is difficult to improve its profitability by technology enhancement or by economy of scale with higher capacity of production. The production unit for tire retreading thus plans for better utilization of capacity, efficient layout, and resource distribution by maintaining uninterrupted workflow.

Here let us discuss a case of tire retreading in India, managed by the TSC (Tire Service Centre) located at Dhanbad, India. It was opened in 1987, as a franchisee of Elgitread (India) Ltd. TSC is involved in retreading of three types of tires—cars, LCVs, and HCVs (heavy commercial vehicles). The used tires are collected from its own collection centers, roadside repair shops and other mechanic shops and the retreaded ones are delivered after some period. Any tire is a combination of different parts—inner liner, chafer, belt, carcass, apex, bead bundle, cap ply, shoulder, groove, and so on—along with tread at the outside of tire. The retreading process requires special care on these parts.

Six operations are carried out in retreading process. The first three are the common and compulsory ones, and another three depends on the condition of the tires. The first three operations are cleaning, buffing, and inspection. Buffing generates rubber powder, and after inspection, the tire may be rejected, if found unworthy for retreading. Both the rubber powder and rejected tires are sold to the market. The other three operations include case repair, retreading, and finishing. The operations of the tire-retreading process are displayed in Figure 4.7.

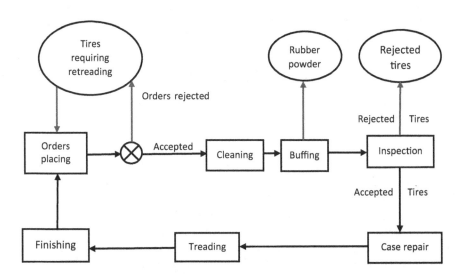

FIGURE 4.7   Tire-retreading process.

(Source: Mondal and Mukherjee, 2012)

TSC earns revenue from retreading jobs and also by sales of rubber powder and rejected tires.

5. **Some industrial initiatives for product recovery in the automotive sector**

It is estimated that by 2025, there may be generation of 21 million end-of-life cars in India, which means the creation of a huge dump yard for car wastes. In this context, the Indian environment ministry is advocating that the industrial sector focus on the formation of a rich sector for recycling of used cars. The Indian government is advising corporate sectors to establish recovery and recycling facilities with more than a 70% recycling target. But in reality, most of the recycling in India is carried out by the unorganized sector, primarily being small suppliers of spare parts.

Most of the European car manufacturing companies are already directly or indirectly involved in product recovery activities in some form or other. In the seminal work of Thierry et al. (1995), it is clearly mentioned that BMW has been engaged in recycling operations since 1990, and it was initiated in a plant at Landshut of Germany. Its strategy includes recycling low-value materials (e.g., plastics), reusing high-value parts, and remanufacturing high-value components. It reused the recycled plastics from used cars as a part of a new car (e.g., luggage compartment linings made up of recycled bumpers). BMW remanufactures high-value components, like engines and alternators, applying all essential activities like disassembly, inspection, and reassembly. Because of their high quality (as good as new), they are sold in place of new components for a new car. In most of the new product design, BMW follows the principles of design-for-remanufacturing and environment-friendliness. For a particular two-seater model, it could even claim that the complete disassembly of all-plastic body from the metal chassis would take exactly 20 minutes because of its special design. Over the time BMW could improve the efficiency of disassembly of its products by the disassembly-friendly design of cars and by use of specially designed equipment and tools for disassembly. It has established its disassembly and dismantling centers all over the world. Other car manufacturers, like Volkswagen (established first product recovery facility in Leer), Peugeot and Citroen (pilot plant for product recovery in Saint Pierre de Chandieu), Ford, and GM, followed suit and entered into the product recovery business of cars.

Caterpillar Inc. manufactures tractors, trucks, excavators, and loaders. It started remanufacturing diesel engines at the request of a corporate client as early as 1982 in Bettendorf, Iowa. Its first remanufacturing was the 1100 series engine as part of product support program. Slowly it established its remanufacturing business both for economic advantage and for contribution to circular economy. The company later on shifted the facility to Corinth. The quality standard of the remanufactured parts and components was exceptionally high. By bringing in synergy between manufacturing and remanufacturing, it actually implemented the extended producer responsibility policy. Its remanufacturing activities include both remanufacturing of complete smaller engines and repairs of larger engines using remanufactured parts. Remanufacturing is largely done following the RTO strategy. As the economics of remanufacturing business depend on the number of new parts to be added during reassembly, it tries to limit the new parts to 20% only. However, in most of the cases, depending on the condition of the component it increases, even to the level of 40% of the total number of parts. With its global remanufacturing activities, Caterpillar recycles more than 50,000 tons of product per annum. Caterpillar subsequently went on purchasing other remanufacturing companies and its profitability increased almost fourfold in last decade. It owns around 19 remanufacturing facilities across eight countries worldwide employing close to 4,000 persons. It remanufactured more than two million products only in 2010 (Jiang, 2006).

Key Learning

- Remanufacturing is an industrial process for recovering the available value from a used product and converting it to a product accepted by a customer in the market or a product as good as a new one.
- It helps in conserving natural resources, using less energy, and generating less pollutants and wastes, often making profits due to low production cost. It contributes directly to both economic and environmental dimensions of sustainability and indirectly to people by offering same service with affordable price.
- An empirical study on Indian business environment shows that OEMs are reluctant to opt for remanufacturing business because of six critical factors. Moreover, technology (design focus), cores acquisition, and market and logistics are the most critical factors in computers and electronics, industrial machinery, and automotive sectors, respectively.
- The remanufacturing business is actually the management of a serially linked business processes—core acquisition, reverse logistics, inspection and sorting, disassembly, cleaning, sorting, repair and storage of parts, reassembly and testing, and sales of remanufactured products, components, or parts. Thus, the key decision-making areas in this business include selection of most suitable remanufacturing strategy (that is, RTO, RTS, or their effective combination), effective mode of core acquisition, inventory management, disassembly scheduling, procurement of new parts, and marketing of remanufactured products without cannibalizing the OEM's own new products.
- If an OEM opts for remanufacturing, it is to incorporate design for disassembly or design for remanufacturing principles while planning for design of its new products.
- Five types of industrial cases (both Indian and global) have been discussed, which involve remanufacturing of products, components, or parts. The coverage of this discussion includes machine parts and components, photocopier machines, automobile engines, tire retreading, and some other initiatives in automotive sector.

4.3 REVERSE LOGISTICS AND CLOSED-LOOP SUPPLY CHAIN—PRINCIPLES AND MODELS

The inclusion of reverse logistics (RL) along with remanufacturing actually completes the activity list of product recovery process. RL may simply be understood as a "reverse" flow of goods. Unlike the "forward" flow of goods from sellers to buyers, as in forward logistics, RL takes care of the flow in reverse order—from buyers to sellers. RL has been existing in business process, which does not aim at achieving sustainable development. In that, RL prevails as a mode of extending after-sales service to customer and sales promotion. The return of unwanted or defective items by customers is quite common in industries, particularly in the retail and e-commerce sector. RL in that case indicates management of the collection, transportation, and stocking of those items. This return or reverse flow may also be applicable at various stages of a supply chain namely, from manufacturers to supplier, from distributors to manufacturer, from customers to retailer, and so on.

It is a serious management issue in online purchase as, unlike offline purchase, customers do not get the opportunity of physically checking or touching the products. For example, Amazon gets seriously involved in creating and managing the channels for RL to collect the returned items and sending the right product items, meeting the satisfaction of the customers within the stipulated time and following a standard procedure. However, RL for product recovery following sustainability is somewhat different.

Reverse logistics may be simply explained as management of reverse distribution (opposite to traditional forward distribution) along with the other usual components of logistics management like inventory planning and warehousing. Rogers and Tibben—Lambke (1999) described the process as *the process of planning, implementing, and controlling the efficient and cost-effective flow of raw materials, in-process inventory, finished goods, and related information from the point of consumption to the point of origin for the purpose of recapturing value or proper disposal.*

During 1997 to 2002, a European research group was created for carrying out a large research project named as REVLOG project (REVLOG, 1998) focusing on developing the theoretical foundation using quantitative modeling of reverse logistics process. Six universities of Europe were involved in the project. It was sponsored by the European Union and coordinated by Erasmus University of Rotterdam. *This working group defines the reverse logistics as the process of planning, implementing, and controlling the backward flows of raw materials, in-process inventory, packaging of finished goods from manufacturing, and distribution or use point to a point of recovery or point of proper disposal.* Any type of logistics management is management of flow of items, which may be materials, parts, products, wastes, information, or even cash. The direction of flow differentiates RL from traditional forward logistics, and that creates RL's distinguishing characteristics. Unlike forward logistics, the flow in RL originates from the point of consumption (mostly external entities) and terminates at the point of value recovery or reprocessing of the item(s).

RL was originally associated with waste management, which is an essential component of sustainable development. From the product recovery perspective, this waste includes disposable products, parts, byproducts, or undesirable pollutants, which are made ready for disposal at the user's end. RL in product recovery is meant for collecting and carrying this waste to a point, where there is possibility for some value extraction from wastes. *RL also differs from green logistics, which is a general concept primarily focusing on pollution reduction and energy efficiency with more applicability for forward logistics.*

A holistic view of supply chain, which combines both forward and reverse logistics, leads to the formation of a closed-loop supply chain (CLSC). In CLSC, the loop of flow is complete by reverse or return flow of goods to the point of value recovery of the waste products. The CLSC is actually meant for the remanufacturing business model, where the OEM itself is carrying out the value recovery operations. Almost same set of analytical approaches and models are equally applicable both for RL and CLSC. CLSC may be defined as the planning, design, control, and operation of a system to maximize value creation and value recovery by manufacturing a new product and remanufacturing the product after its use by customer with a combined objective of enhancing the efficiency, customer satisfaction, and environmental improvement from sustainability point of view. *CLSC is a set of processes for value addition (as applicable in manufacturing as an essential component of the traditional supply chain) and value recovery and creation by reverse logistics and remanufacturing by the same OEM.* If we consider the traditional SCM as management of cradle-to-grave, the CLSC management may be treated as a cradle-to-cradle or reincarnation. Along with the maximization of economic benefits, the CLSC management seeks to decrease consumption of resources and energy and to reduce emissions of pollutants. Readers may refer to Figure 4.2 for a simple framework of CLSC process. In this book, we have separately discussed the issues related to remanufacturing in section 4.2, although CLSC combines both remanufacturing and RL, along with the existing manufacturing and forward logistics processes.

Perhaps the two primary factors distinguish RL from forward logistics in terms of their management. First, forward logistics manages movements *from few to many*, whereas RL means management of movements *from many to few*. Second, RL is affected by the *uncertainty of sources and the quality, quantity, and timing* of acquiring wastes, returns, or used products.

Remanufacturing or other value-recovery processes are part and parcel of RL, and it is very difficult to identify the sources and the condition of the used product, parts, and components beforehand, which are to be distributed backwardly to the point of remanufacturing or value recovery or OEM facility (in case of CLSC). This makes the planning and management more complex than the forward logistics. Our discussion on RL will cover the following three areas of management.

- *Management of acquisition of returns or used products*
- *Inventory management of returns*
- *Management of reverse distribution*

Reverse logistics, as defined by the European experts in Revlog project, is the process of planning, implementing, and controlling backward flows of raw materials, in-process inventory, packaging of finished goods from manufacturing, and distribution or use point to a point of recovery or point of proper disposal. Closed-loop supply chain (CLSC) is the process created once the OEM gets involved in remanufacturing along with its existing manufacturing business. CLSC covers all the processes like manufacturing, remanufacturing, forward logistics, and reverse logistics. The complexity of managing CLSC arises because of the interactions among these processes.

4.3.1 MANAGEMENT OF RETURNS ACQUISITION

The most important activities in managing the returns are the identification of the sources of returns, returns forecasting, and selection of the mode of acquisition or collection.

The primary source of the cores is the customers or owners of the cores. It may also be the retailers, OEM's service centers, brokers, third-party vendors, independent repair shops, and showrooms. Returns may be collected from single source or multiple sources. If the product recovery is carried out by independent remanufacturer, it may collect the same product of different makes or different models, although in this case parts-matching would be somewhat difficult task during remanufacturing. If the OEM remanufactures, the collection of returns may be undertaken by the OEM itself through distributors or retailers, independent remanufacturers, third-party logistics providers (TPLPs), take-back collection centers, and NGOs on green activities. This requires an appropriate planning and coordination with reverse logistics and remanufacturing activities. There may be various modes of collection of returns, some of which are listed as follows:

- *Take-back*: It is the legislative requirement that the manufacturers are supposed to take back the used products after some period of its use or at its end-of-life because of a prevailing take-back policy (e.g., Volkswagen and Opel in Germany) or some contract already made known during the sale of new products to customers.
- *Buyback*: In buyback, the remanufacturers purchase the used products from the customers at a cost through its retailers or distributors or directly from the customers.
- *Off-lease or off-rent*: Lease and rental contracts specify that the manufacturers take back products after contract expiry. In this case, manufacturers can thus predict the quantities and timing of these return flows quite accurately and returns forecasting is easier.
- *Auction*: Many organizations sell off their scraps and used products (machines, spare parts, etc.) through auctions. Remanufacturers may collect by participating in this auction process. This is the usual practice in most

of the units managed by departments or organizations managed or largely funded by the government of India.

- *Seed-stock*: Defective products at the OEM's manufacturing plants are purchased by the remanufacturers.
- *Exchange offers*: Used products are often collected in exchange offers where old ones are replaced by new products or new models with lower sales price. This is a popular practice by big retail houses, particularly during stock clearance at the end of a season.
- *Warranty returns*: Used products may be returned during the warranty period due to technical failures and replaced by new ones. This is another popular practice by manufacturers of machines, parts, white goods, and so on.

There exists a set of approaches for the acquisition of returns, the most popular of which are take-back, buyback, off-lease or off-rent, auction, seed-back, exchange offers, and warranty returns.

Return forecasting is an important activity both for RTS and RTO strategies of remanufacturing. Production planning in this business is affected by the quantity and condition of the returns and their timing of acquisition. It is more critical for RTO, as the decision on accepting an order is entirely influenced by the availability of sufficient supply of cores. The supply is also determined by the identification of the right type of sources and mode of cores acquisition, as listed earlier. If it is a case of CLSC and the products are with customers under a lease contract, the OEM knows in advance the timing and quantity of cores expected to be acquired, although their conditions and quality are still not exactly known. So returns forecasting in this situation is not a big issue. Moreover, if the OEM is implementing *extended producer responsibility* policy, it is supposed to keep track on the product after it was sold. It is the responsibility (both physical and financial) of the OEM to plan any meaningful post-end-of-life treatment of the product. The manufacturer may take the full responsibility itself or may outsource it to the *producer responsibility organization (PRO)*. The OEM may include the associated costs in price of the product itself. In Europe (particularly in Germany), *Green Dot* is one of the trademarks representing certified PROs and these PROs take care of the supply of returns to remanufacturers (OEM in this case). So the management of returns is relatively easier in such a situation.

Moreover, it is to be noted that even if the extended producer responsibility policy is not applicable, some OEMs may outsource or create a long-term contract with external recycling partner for core acquisition exercise. For example, Samsung has its take-back and recycling program known as *STAR*. The customers who are interested in getting rid of the used mobiles may arrange the pick up by informing Samsung by telephone or email (ewasterecycling@samsung.com). Samsung's recycling partner collects the returns free of cost from the customer's residence. Alternatively, customers may drop it at specific e-waste bins installed by Samsung across the country near the service centers.

Incidentally, the traditional time series analysis may not give the correct result in forecasting of returns. Thus, alternatively, they are to opt for *cause-effect modeling*, like multi-variate regression analysis. *The primary influencing variable for forecasting future returns is the sales of new products.* There is a time lag for conversion of the new products to returns or cores, which usually occurs at the end of its useful life at customer's end. We may estimate the probability of availability of returns at a particular point of time. *This may be modeled as a transfer function involving the sales and the time lag.* The primary attention of the forecaster is formulation of this transfer function, which may follow the *Box and Jenkins method of time series analysis or Bayesian estimation.* Some other factors (both quantitative and qualitative) may also be considered in modeling of returns forecasting for better result. These are frequency of failures, work environment at the customer's end, end of PLC in the market because of the entry of new models or better substitute for the existing product, stringent laws for end-of-life of products, technology development, customer's emotional attachment to the product, and of course, the price of remanufactured product in comparison to the new product. Price prevailing in secondhand or unstructured market may also affect the forecasting process.

> *Unlike the traditional demand forecasting in production planning, returns forecasting may be estimated applying cause-effect relationship or multivariate statistical model using sales of new products as the primary influencing variable accompanied by several other market-focused and technology-focused variables as the independent variables.*

Remanufacturers need to take the most appropriate decision while buying back the used product from customers for right timing and offering the right price to be negotiated with the customers. The following paragraph shows how a cost-benefit analysis may be carried out in case of deciding on the buyback time of used products.

4.3.1.1 Cost-Benefit Analysis on Buyback Decision

The management of RL starts from the decision on core acquisition, which is the raw material for any product recovery operation. *Experts opine that the scarcity of cores of good quality at an acceptable price at right time seems to be the crucial limiting factor restricting the popularity of product recovery business worldwide.* Inherent uncertainty in recovery rate from the cores makes the acquisition planning quite complex. Figure 4.8 displays the managerial activities or decisions under the management of RL in general and how the uncertainty factors influence them at every stage.

In this context, we will now discuss how analytically we may address the acquisition planning problem considering the decision-making on estimation of appropriate buyback time of the used products.

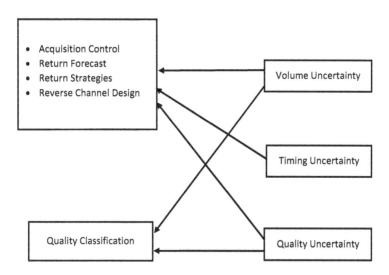

FIGURE 4.8   Activities in core acquisition and reverse logistics management.

Let us see the inherent conflict existing in this decision problem. *If the buyback occurs earlier, naturally the user or customer of the product will accept higher price offered by the remanufacturer, whereas if the buyback occurs at the later stage of the product's life, the user may be happy even with a lower buyback price. However, the older product will require more replacement by new parts and repairs, and thus, the cost of remanufacturing will increase.* How do we solve this trade-off analytically? With this research question, a research project was initiated and the author of this book took part in this two-member project as the project guide. The brief outcome of the research study is outlined here, although the detailed project report is available in the research report Mondal and Mukherjee (2006b).

The whole project was based on the core acquisition management considering a case of a multinational photocopier company, which offers a unique service policy to its customers. The company has its factory and organizational establishment in India, and it has already created a popular brand image in the photocopier sector. It is virtually enjoying a monopolistic market because of non-existence of any significant competitor in the market.

Customers of its product get free servicing (other than the cost of consumables) to all new products sold till its end-of-life (approximately ten years). The company opted for remanufacturing through buybacks of used products so as to avoid the cost of servicing, which exponentially increases over time. The following assumptions are taken into consideration in this model development.

- The model takes care of the interest of OEM, which has initiated remanufacturing of its own products.
- The cost on inventory carrying of the used products is ignored.
- The servicing cost and buyback price are deterministic parameters following a mathematical function.

- None of the cost or revenue parameters are influenced by inflation or discount rates.
- The remanufacturing cost and revenue collected from recyclers by selling the unrecovered parts are stochastic parameters because of influence of the random recovery from the returns.

Let us assume that after the period of t from the date of sale of the new product, the used product is bought back by the OEM. The scale of t is considered within 0 to 1 as a continuous variable. Here $t = 0$ means the sale of new product and $t = 1$ means its end of life, after which nothing can be recovered. In this analysis we consider the following parameters.

1. *Recovery rate*: This is a probabilistic parameter based on reliability theory and product deterioration. Recovery at period t may be expressed as a linear function $\propto (t) = 1 - \beta t$. Here β is a random number showing the unknown rate of deterioration of product quality.

2. *Remanufacturing cost*: This cost decreases with increase in recovery or increases with deterioration (non-recovery) over time. Let this cost be expressed as $R(\propto) = K_f + K_p \beta t$, where K_f is fixed cost, representing the resources for creating and maintaining the remanufacturing facility. $K_p \beta t$ or $K_p (1 - \propto)$ is the variable portion of the remanufacturing cost, depending on the condition or non-recovery (deterioration) of parts from used products. The higher the recovery, the lower the remanufacturing cost.

3. *Resale value as revenue from recyclers*: Revenue generated by selling the unusable or unrecovered parts/components to the recyclers may be expressed as $V(\propto) = K_d (1 - \propto) = K_d \beta t$. It is proportional to the unrecovered or non-usable parts of the returns or cores.

4. *Service cost*: Service cost at t may be simply expressed as $S(t) = K_s t$, where K_s is the rate of the use of resources for servicing a unit. It increases with time or life of using the product. So savings of service cost by buyback of the used product at t will be $S_s(t) = \int_t^1 S(t)dt = \int_t^1 K_s t\, dt = \frac{K_s}{2}(1 - t^2)$.

5. *Buyback price*: Buyback price may be expressed as $B(t) = K_1 - K_2\, t$, where K_1 is the price of new product and K_2 is the rate of loss of value as perceived by the customer because of using this value for his or her own activities. The older the product bought back by remanufacturer, the lower the price acceptable to the customer.

So the net return (i.e. $N(t)$) gained by the OEM (also the remanufacturer) at the time t after the buyback is shown as follows.

$$N(t) = V(\propto) + S_s(t) - B(t) - R(\propto) = K_d \beta t + \frac{K_s}{2}(1 - t^2) - K_1 + K_2 t - K_f + K_p \beta t$$

$$= K_2 t - \frac{K_s}{2}t^2 - (K_p - K_d)\beta t + \left(\frac{K_s}{2} - K_1 - K_f\right)$$

Now in terms of expected value of $N(t)$,

$$E(N(t)) = K_2 t - \frac{K_s}{2} t^2 - E(\beta)(K_p - K_d)\beta t + \left(\frac{K_s}{2} - K_1 - K_f \right)$$

We may also note that β is the only random parameter in the right-hand side expression. Let us now maximize this expected net return. In this pursuit, let us consider classical optimization method. We differentiated the $E(N(t))$ function and got its second derivative clearly negative. So $E(N(t))$ is a concave function in t, and there exists an optimal t^*, which maximizes the $E(N(t))$. By equating the first derivative to zero, we get the optimal buyback time.

$$t^* = (K_2 - E(\beta)(K_p - K_d))/K_s$$

Here this t^* value gives some new insights on role of servicing of OEM in this buyback exercise. As K_2 represents the value used up by the customer while using it and $E(\beta)(K_p - K_d)$ is the expected deterioration, so the numerator of t^* shows the remaining value of the product till t^* period at the user's end. This is the outcome of servicing by the OEM. Moreover, K_s is the rate of investment for servicing. Thus, other than the optimal buyback time, it also reflects the *service effectiveness (SE)* of the OEM. This SE is further divided into *recovery index (RI)* and *service index (SI)*.

$$\text{RI} = \left(K_2 - E(\beta)\left(K_p - K_d \right) \right)/K_2 \text{ and } \text{SI} = K_2/K_s. \text{ Thus, SE} = \text{RI} \times \text{SI}.$$

Here we may perhaps explain the phenomenon by considering that better user's management may improve RI, and the OEM may improve the service efficiency by better SI. SE shows the overall effectiveness of service. We may logically conclude the following:

- If either or both RI and SI increase, the SE increases and OEM may delay the buyback process cost-effectively.
- Higher investment on servicing (i.e., K_s) without significant creation of user's value means poorer SI.
- Low non-recovery or more recovery (high numerator of RI) is not within one's control, but if it happens, the buyback may be delayed.

These conclusions have been numerically shown and proved with an example in our published report (Mondal and Mukherjee, 2006b). In fact, **three important business concepts and performance measures, SE, RI, and SI, have been developed by the research project of Mondal and Mukherjee (2006b).** Another important fact related to these findings is that OEM can reduce the buyback price by improving its service performance by delaying the buyback.

> *Buyback time for acquiring cores may be decided by optimizing the two conflicting parameters—buyback price and remanufacturing cost. If buyback is done at a later stage of life of the product, the buyback price will be lower, but remanufacturing cost will be higher. It is just reverse, if the cores are bought back early.*

4.3.2 INVENTORY MANAGEMENT IN REVERSE LOGISTICS

Reverse logistics (RL) encompasses the bi-directional flows of materials. However, like any logistics process, in RL we also consider the fact that sometimes the materials are on the move (i.e., distribution in both the directions) and some other times they are at rest (i.e., keeping inventory). Here we are going to discuss on inventory management in RL. The following are the unique characteristics of inventory management in this situation.

- Returns flow is uncertain.
- Returns quality and recoverability are uncertain.
- Inventory of returns, serviceable inventory, and inventory of newly procured parts are to be managed.
- Supply and demand are to be matched, both of which are not exactly known beforehand. Of course, in RTO, order is placed before the remanufacturing activities and so the demand is known.
- Remanufacturing lead time is not known due to unknown disassembly efficiency and product recovery rate.
- It is difficult to estimate the unit shortage cost of returns, as the economic impact of lost sales or lost profit cannot be estimated without the knowledge of the recovery from returns.

In remanufacturing business, the following four types of inventory are actually to be maintained by the concerned organization.

- *Inventory of returns* before disassembly
- *Inventory of recovered parts* or components before reassembly
- *Inventory of new parts* procured for replacements
- *Inventory of remanufactured products, often termed as serviceable inventory*

Inventory of returns and serviceable inventory are the two types of relevant inventory that demand special attention of the organization in inventory management in reverse logistics. Serviceable inventory actually includes recovered parts after disassembly along with remanufactured ones, if the OEM is simultaneously carrying out manufacturing and remanufacturing of its products, as recovered parts may be used as input materials for manufacturing. However, recovered parts after proper inspection are also sold to the market. New parts procured from outside, either as mandatory parts for remanufacturing business or as replacement of non-reusable items (e.g., plastics, adhesives, fasteners), are also treated as inventory. These may also be considered as serviceable inventory.

Figures 4.9 and 4.10 show simple frameworks of returns inventory (R) and serviceable inventory (S) in a closed-loop supply chain. Figure 4.9 represents the business model showing a situation, where the manufacturing and remanufacturing processes are catering to the needs of two separate markets. It may be noted that the markets may not be completely independent. In case of shortage in Market II, the

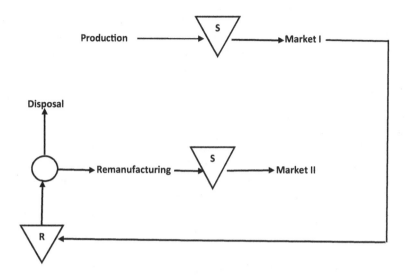

FIGURE 4.9 New and remanufactured products sold in separate markets.

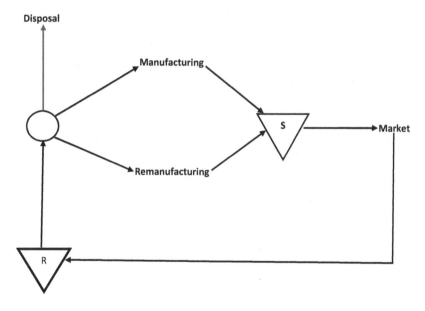

FIGURE 4.10 New and remanufactured products sold in same market.

customers may resort to purchasing the new products from Market I. Of course, the price difference has a role in this decision. Figure 4.10 is the framework for a case, where the same market is used for selling new and remanufactured products. Customers may be interested in remanufactured products because of lower price with maintenance of same quality like the new ones. Here the customer is to consider the acceptable trade-off between the price saving and sacrificing the new product or even a new model. In other words, customers are to decide how much perceived quality deterioration (old vs. new product) is acceptable to gain the advantage of one unit of price saving. In both the diagrams, the serviceable inventory is meant for a new or remanufactured product ready for marketing. When OEM remanufactures its own products, it resolves two conflicts. First, it is to decide on optimal share of manufactured and remanufactured products to meet the demand of market by techno-economic cost-benefit analysis. Accordingly, once the OEM receives the cores, it is to decide whether to go ahead with the remanufacturing process or to keep it as returns inventory. Secondly, the OEM is to decide on maintaining a particular type of stock, whether it is better to keep the returns or cores as it is in returns stock or to keep them as serviceable inventory after remanufacturing. Here the conflict exists between the difference of inventory-holding costs (holding cost as serviceable inventory being higher than that of returns inventory) and gain in responsiveness (due to remanufacturing lead time).

4.3.2.1 Quantitative Analysis in Inventory Planning

In case of independent remanufacturing, the inventory planning is relatively easier. However, as the supply of returns is partially uncontrollable, it cannot be treated as decision variable like "how much to procure" for raw materials in the manufacturing business. However, in case of CLSC, the OEM may control the acquisition process by fluctuating buyback prices. Nevertheless, optimal lot size modeling may also be applicable in the remanufacturing process. EOQ-type models may be applied in CLSC with its traditional set of assumptions (Minner and Linder, 2004).

Let us consider that OEM is engaged in remanufacturing its own products and there is a constant and deterministic rate of demand D. The used products or returns are also received at a constant rate R, which is also known beforehand. If we consider the fact that all products are not returned for remanufacturing because of their disposals and damaged condition, the condition $R \leq D$ is quite valid. In fact, R/D represents the reusability of a product. The stock or inventory of serviceable parts or whole products is replenished by remanufactured, manufactured, or newly procured items. Because of extensive inspection and quality control, the remanufactured items are treated as like new items. Further, there are inventory-holding costs associated with keeping inventory in the two stocks. Let h_r and h_s be the unit-holding costs per annum for returns and serviceable items, respectively. Logically the condition $h_s \geq h_r$ is also valid here. On the other hand, like any process-focused activity, the manufacturing and remanufacturing operations are also associated with some fixed or setup costs. Let the fixed or setup costs be F_r and F_p for remanufacturing and manufacturing operations, respectively. Setup costs are independent of quantity of production, and so the bigger the lot size, the lower the unit cost of production. But this will give rise to more holding costs of serviceable inventory. Traditionally, optimal manufacturing lot size is computed based on this conflicting issue. Here we are interested in computing lot size of both manufacturing and remanufacturing operations of OEM.

Let us assume that all returns are remanufactured and replenishments of both the inventory are made instantaneously—that is, no lead time is associated with them, which is a known assumption in EOQ-type models. The manufacturing batch serves to satisfy the net demand—that is, the gap between demand and the returns. This is also termed as *netting approach*, popular among European researchers on remanufacturing. Two cases have been proposed in this EOQ modeling. In first case, it is proposed that each manufacturing lot Q_p is followed by K number of remanufacturing setups, each having same lot size (Q_r). In this situation the following optimal lot sizes may be computed.

$$Optimal\, Q_p^* = \sqrt{\frac{2(D-R)F_p}{h_s(1-R/D)+h_r\,R/D}}$$

$$Optimal\, Q_r^* = \sqrt{\frac{2DF_r}{h_s+h_r}}$$

$$Optimal\, K^* = (\frac{R/D}{1-R/D})(\frac{Q_p^*}{Q_r^*})$$

Readers can easily understand the adjustments on the traditional EOQ model. The manufacturing lot size takes care of the net demand (i.e., $D - R$) and the weighted holding cost in the denominator. The remanufacturing lot size, on the other hand, is meant for meeting the customer demand, and it affects the inventory-holding costs both as inventory of returns and that of serviceable. Teunter (2001) proposes just an opposite case. Here a single remanufacturing batch Q_r is followed by P number of same Q_p lots of manufacturing. The resultant models are the following:

$$Optimal\, Q_p^* = \sqrt{\frac{2DF_p}{h_s}}$$

$$Optimal\, Q_r^* = \sqrt{\frac{2RF_r}{h_s R/D+h_r}}$$

$$Optimal\, P^* = \left(\frac{1-R/D}{R/D}\right)\left(\frac{Q_r^*}{Q_p^*}\right)$$

It has been proved that the first case is applicable for a high R rate—that is, if R/D tends to 1, whereas the second case is more appropriate for a low R rate—that is, if R/D tends to 0.

Let us now relax some of the assumptions applicable for EOQ-type model formulation. The primary relaxation is consideration of stochasticity in demand and return rate estimation. In case of traditional inventory control of manufacturing operations, there are two types of review schemes or policies for procurement decision, when the exact demand rate is unknown. These are the continuous review system/Q system/ (s,Q) policy and the periodic review system/P system/(R,S) policy. In case of CLSC, the OEM faces difficulty in deciding on quantity of manufacturing and remanufacturing, if both the return and demand rates are not exactly known, and also there exists some lead times of two production operations. Here we will refer to the simple model mentioned in the work of Mukherjee (2002). Let us consider the existence of the following new set of assumptions relevant to the model.

- Single item or type of product is considered.
- No setup or fixed cost is considered for manufacturing or remanufacturing; only linearly variable costs are used in model development.
- Demand and return rates are estimated as stochastic parameters.
- Once the returns are received, it is decided whether to dispose of or to go ahead with remanufacturing keeping in view cost-effectiveness.
- The demand of the customers are satisfied from the serviceable stock made up of newly produced products and remanufactured products.
- There are lead times for manufacturing and remanufacturing, but the values are known beforehand.
- The OEM has implemented push or RTS and MTS strategy.

Decision Variables

p_t = Quantity to be manufactured in period t
r_t = Quantity to be remanufactured in period t
d_t = Quantity to be disposed in period t

Parameters

D_t = Stochastic demand of serviceable products in period t
R_t = Stochastic return of used products in period t
t_p = Production lead time
t_r = Remanufacturing lead time
c_p = Unit production cost; $c_p \geq 0$
c_r = Unit remanufacturing cost; $c_r \geq 0$
c_d = Unit disposal cost; $c_d \geq 0$
h_r = Inventory carrying cost of returned products per unit per period
h_s = Inventory carrying cost of serviceable products per unit per period
v = Shortage cost of serviceable products per unit per period (mainly lost sales)
$S_{r,t}$ = Stock on hand of returned products at the end of period t
$S_{s,t}$ = Stock on hand of serviceable products at the end of period t

Model

Cost of operations and relevant activities: $C(p_t, r_t, d_t) = c_p.p_t + c_r.r_t + c_d.d_t$

Inventory-holding costs of returned products: $I_r(S_{r,t}) = h_r.S_{r,t}$

Inventory-holding costs or shortage costs for serviceable products:

$$I_s\left(S_{s,t}\right) = h_s.S_{s,t} \text{ for } S_{s,t} \geq 0$$

$$= -v.S_{s,t} \text{ for } S_{s,t} \leq 0$$

To minimize the total expected costs over the whole planning horizon T periods:

$$F = Expected\ on\ D, R\left[\sum_1^T\left\{C\left(p_t, r_t, d_t\right) + I_r\left(S_{r,t}\right) + I_s\left(S_{s,t}\right)\right\}\right]$$

Subject to satisfaction of the following system relations:

$$S_{r,t} = S_{r,(t-1)} + R_t - r_t - d_t$$

$$S_{s,t} = S_{r,(t-1)} + p_{t-t_p} + r_{t-t_r} - D_t$$

where $p_t \geq 0, r_t \geq 0$ and $d_t \geq 0$

The relationships among the cost parameters also may help in taking some decisions based on simple logic.

1. $c_p < v$, as v represents loss of unit sales, and also $c_r < v + c_d$, which justifies remanufacturing in place of disposal of returns. Both are the justifications for the two production operations.
2. If net disposal is cheaper than the inventory-holding cost of serviceable products—that is, if $h_s > c_d - c_r$ or $c_d < h_s + c_r$ —then it is better to dispose the returns instead of remanufacturing. This will reduce the flow to inventory of serviceable products.
3. If $c_r - c_d < c_p$, we may justify remanufacturing of returns.

4.3.2.2 Simple Policies for Inventory Management and the Application of MRP

The model described in the previous section (4.3.2.1) is a dynamic stochastic model. Even in case of exclusive manufacturing management, getting an optimal solution in this situation is really difficult. In fact, this leads to the creation of policies like Q-system and P-system. In CLSC, the complexity is little more, as we are to take care of both returns inventory and serviceable inventory, and also it is more so because of nonzero lead times. However, if we assume $t_p = t_r$, we may propose a simple set of inventory control policies with practicable implications. Here we assume that we may dispose returns, if they are not cost-effective.

Situation I: No Stock-Keeping of Returned Products

- It is a two-parameter (S_l, S_u) policy. These are the lower and upper limits of inventory position.
- If the inventory position of serviceable inventory is less than S_l, remanufacture the returns once received without disposal and also produce up to S_l.
- If the inventory position is more than S_l and less than S_u, then also remanufacture any returns received without disposal, but do not produce.
- If the inventory position of serviceable inventory exceeds S_u, do not remanufacture any returns, rather than dispose them down to S_u or sell them to recyclers and do not produce.

Situation II: With Stock-Keeping of Returns

- It is a three-parameter (S_l, S_m, S_u) policy. S_m is the third limit of inventory position.
- If inventory position of serviceable is less than S_m, remanufacture the available returns up to S_m.
- If the inventory position of total returns and serviceable is less than S_l, produce up to the S_l level.
- If total inventory position exceeds S_u, we dispose returns down to the level of S_u.

During the later part of 20th century and early 21st century, lot many research reports were published by European researchers in the area of stochastic modeling of returns and serviceable inventory planning. Researchers developed and tested two control policies similar to continuous review system or (s, Q) policy. These are push-based and pull-based policies.

Push-Based Policy

Whenever, Q_r, quantity of remanufacturable products, are available, the batch is pushed through remanufacturing facility. The manufacturing production of Q_p batch size is initiated, wherever the serviceable inventory position touches s_p or lower than that level. This stands for the reorder point, and thus, the policy may be known as (s_p, Q_p, Q_r) policy.

Pull-Based Policy

Remanufacturing activity only starts if the serviceable inventory position drops to the s_r level (reorder point for remanufacturing), and the remanufacturing batch should increase the serviceable inventory position to the S_r level (remanufacture up to level). On the other hand, manufacturing production is initiated when the inventory position touches down to the s_p level or less and quantity of manufacturing should be Q_p (manufacturing lot size). This policy may be named as (s_p, Q_p, s_r, S_r) policy. It may further be noted here that s_p is generally lower than s_r, which means remanufacturing (more cost-effective than manufacturing) is prioritized. In this situation, the safety stock of $(s_r - s_p)$ takes care of demand during remanufacturing lead time.

MRP-based systems have also been attempted to manage inventory during remanufacturing. However, this modified MRP is not based on parts or components procured or manufactured per required bill of materials (BOM). Actually, the modified MRP makes use of "reverse" BOM per the processing time in disassembly and the quantity of recovered or released components or parts. *This MRP poses difficulty because of three primary reasons—uncertainty in disassembly process (time and quantity of components or parts), recoverability, and the timing, quantity, and quality of return flows.*

> *Four types of inventories are managed in reverse logistics—inventory of returns or cores, inventory of recovered parts or components, inventory of new parts for replacement, and inventory of remanufactured products. Traditional approaches of inventory planning from manufacturing process like EOQ, push-based or pull-based policies, MRP, and so on are equally applicable in RL or CLSC management, but with some modifications.*

4.3.3 REVERSE-DISTRIBUTION-BASED NETWORK ANALYSIS IN REVERSE LOGISTICS

Reverse distribution is the set of activities for transporting the used products from their sources to the facilities where the remanufacturing takes place. It may originate at the disposer's or user's place, retail stores, collection centers, and so on. It is an important component of RL other than acquisition and inventory management. If we consider the reverse distribution as an open-loop or independent process applicable primarily for independent or third-party remanufacturers or third-party logistics service providers, it is relatively simpler and somewhat similar to forward distribution networking from source to destination, with some differences because of the uncertainty in acquiring returns, their packaging, and in some cases, the design of vehicles. The closed-loop supply chain (CLSC) process involves a combination of forward and reverse distribution in one business model, if the OEM is carrying out product recovery operations on its own products along with the existing manufacturing activities.

The management of bi-directional distribution process includes taking the following decisions:

- Location selection of facilities in the network
- Capacity planning of facilities
- Identification of appropriate linkages between sources of acquisition and facilities and among facilities at different levels of the distribution network
- Assessment of depth and width of distribution networks
- Decision on centralization or decentralization of facilities, like warehouses and testing centers.
- Integration of forward and reverse distribution involving route-level decision-making, vehicle scheduling, and so on.
- Selection of modes (mostly decisions related to multi-modal network) for distribution.

If forward distribution is considered to have few-to-many, or divergent, structure, reverse distribution may be treated as a representation of many-to-few, or convergent, network structure.

The management of this convergent network structure requires consideration of different set of issues. Two most important ones are as follows:

1. The issue is whether the few remanufacturing facilities have enough space and resources to receive and remanufacture the huge returns from various

sources. If it is insufficient, the remanufacturer is to become selective in deciding on the quantity and sources of returns. *Should the criteria be the cost of inbound logistics, buyback price, or estimated uncertainty in the timing, quality, or quantity of cores acquisition?* Or is it something else?

2. As the sources of core collection differ in terms of work environment and mode of handling at the user's end, the timing, quantity, and quality of the cores will vary drastically. *So there will be variation in recovery rates and rejection rate or disposability.* This will affect the planning of reverse transportation and mode of transport. Moreover, *the parts-matching problem* also may arise because of the variation in types of product models. Uncertainty on inputs create difficulty in planning for *remanufacturing lot size*. Variation in recovery rate may also create *variation in prospective customers of the items recovered*. For example, the outputs from remanufacturing plant may be a complete remanufactured product, a set of remanufactured components or some recovered parts, depending on the actual recoveries occurred, depending on the condition of the returns. So it may be sold to customers of the product, manufacturers, suppliers of parts, and so on. *How do we plan for reverse distribution network incorporating the uncertainties and resultant effect?*

Some other issues as mentioned in the following list may also be considered while managing distribution network in CLSC involving a bi-directional flow of items.

- Nodes of the forward distribution channel normally includes warehouses, distributing centers, and wholesalers, whereas those of reverse distribution are testing or inspection centers, and disassembly centers. So normally the channels are separate.
- The forward distribution network may be further divided into two directions with different sets of channels, if the markets for manufactured and remanufactured products are separated.
- If the manufacturers are reusing the containers or recovering them after some repair or extra fitting, the CLSC may use the same fleet of transport for bi-directional distribution and routes may be same. This is quite a common practice for the distribution of soft drink bottles. The reverse distribution of empty bottles may be followed by the forward distribution of filled bottles using the same fleet of trucks.
- In CLSC management, the sharing of manufacturing and remanufacturing lot sizes is often integrated with closed-loop flow management while developing a comprehensive decision-making model.

Fleischmann et al. (1997) identified three key factors that are responsible for efficient management of reverse distribution (or CLSC):

1. *Actors in reverse distribution (or CLSC) management*:
    It may be the traditional members of forward channel (e.g., OEM, retailers, distributors, logistics providers) and some members in reverse distribution (cores collectors, in-charge of testing centers, etc.). Independent parties

exclusively looking after materials recovery or NGOs may also be included as actors.

2. *Exclusive facilities and their locations*:

   A reverse distribution network is designed to decide on appropriate locations for specific functions meant for collecting, testing, sorting, repair, and so on. Early testing and disposals may reduce the transport costs (non-value-additive) of non-recoverable products. Further, if the facilities like testing centers contain expensive machines and other resources, they should be very few in number with high rate of utilization. They are also to be centrally located. Of course, the outbound transportation costs on these cases are also to be taken into account in these decision-making situations. Further, it is to be noted that the outputs from a separation or distributing center are to be subsequently distributed to various destinations. These may be spare parts suppliers, distributors of parts, manufacturing or remanufacturing plants, company-specific retail stores, service centers, or repair centers.

3. *Relations between forward and reverse distribution channels*:

   CLSC is usually not meant for recycling, as the output from recycling operations are often sent to other industries, as by recycling the output will be a completely different product or set of various items, which are often used in different industries. Integration of forward and reverse distribution is applicable for remanufacturing and reusing used products. Although transportation modes for both the channels may be the same, material-handling, loading, and unloading mechanisms will be different for new and used products.

Matching of movements, flows, and transport modes between forward and reverse logistics are crucial, while planning for distribution networking of a CLSC. Uncertainty on several parameters are the sole cause of complexity in RL and CLSC management.

4.3.3.1 Analytical Modeling for CLSC Management

Various types of analytical models had been applied for formulation of CLSC process and its management. *Most of the deterministic models are mixed-integer linear programming (MILP)*, although some are mixed-integer non-linear programming (MINLP) models as well. Problems under uncertainty are actually more realistic than the deterministic ones. The uncertain parameters may be the demand, returns, delivery time, returns acquisition time, quality of returns, carbon emission rate, and recoverability rate. These are often modeled applying a fuzzy approach, robust optimization, or stochastic programming.

Let us formulate the basic closed-loop network design considering the decisions on facility location and bi-directional flow of items, as proposed by Fleischmann et al. (2004). Here the OEM intends to identify the potential locations for plants and warehouses for forward distribution and those for test centers in reverse flow of used products. As it is a case of CLSC, the same manufacturing plants do have the remanufacturing facilities. So as to simplify the modeling framework, we are here ignoring the lead time of various processes and its resultant impact on inventory cost issues. The model attempts to assess the impact of optimal forward flow of manufacturing products and reverse flow of used products. Here we consider all impacts on annual basis. The model details are described in the following.

Decision Variables

Y_i^p = Binary variables indicating opening the plant at location I, where $i \in I$, representing the set of alternative plant locations.

If $Y_i^p = 1$, the plant is located at location i; $Y_i^p = 0$, otherwise.

Y_j^w = Binary variables indicating opening the warehouse at location j, where $j \in J$, representing the set of alternative warehouse locations.

If $Y_j^w = 1$, the warehouse is located at location j; $Y_j^w = 0$, otherwise.

Y_k^t = Binary variables indicating opening the test or inspection center at location k, where

$k \in K$, representing the set of alternative test center locations.

If $Y_k^t = 1$, the test center is located at location k; $Y_k^t = 0$, otherwise.

X_{ij}^p = Quantity of products to be distributed from plant i to warehouse j (in product units).

X_{jl}^w = Quantity of products to be distributed from warehouse j to customer l (in product units). Here $l \in L$, representing the set of all customers.

X_{lk}^r = Quantity of returns to be distributed from customer l to the test center k (in used product units).

X_{ki}^t = Quantity of recoverable returns to be distributed from test center k to the plant i (in used product units).

S_l = Amount of unsatisfied demand or shortage created at the end of customer l.

Cost Parameters

F_i^p = Annualized fixed cost, if a plant is opened at location i.

F_j^w = Annualized fixed cost, if a warehouse is opened at location j.

F_k^t = Annualized fixed cost, if a test center is opened at location k.

C_{ij}^p = Sum of unit production cost at plant i and its transportation cost to warehouse j.

C_{jl}^w = Sum of unit product-handling and storage cost at warehouse j and its transportation cost to customer l.

C_{lk}^r = Sum of unit transportation cost of returns from customer l to test center at k along with cost of testing, inspection, and disposal, if any, at test center k.

C_{ki}^t = Sum of unit transportation cost of recoverable returns from test center at k to plant at i along with cost of reprocessing or remanufacturing at plant i.

C_l^s = Unit shortage cost at customer l end.

Other Parameters

D_l = Annual demand from customer l.
R_l = Annual possible returns collected from customer l.
α = Average percentage recoverable products from returns like recovery rate.
P_i = Annual production capacity at plant i.
W_j = Storage and handling capacity of warehouse at j.
T_k = Annual testing and inspection capacity at test center k.

Objective Function

Let us minimize total annual costs involving location and bi-directional distribution decisions.

$$\text{Minimize } z = \sum_{i \in I} F_i^p Y_i^p + \sum_{j \in J} F_j^w Y_j^w + \sum_{k \in K} F_k^t Y_k^t + \sum_{i \in I} \sum_{j \in J} C_{ij}^p X_{ij}^p + \sum_{l \in L}$$

$$\left(C_l^s S_l + \sum_{j \in J} C_{jl}^w X_{jl}^w \right) + \sum_{l \in L} \sum_{k \in K} C_{lk}^r X_{lk}^r + \sum_{k \in K} \sum_{i \in I} C_{ki}^t X_{ki}^t$$

Constraints

1. Consideration of shortage at the customer's end:
$$\sum_{j \in J} X_{jl}^w + S_l = D_l \text{ for each } l \in L$$

2. Balance constraints at various nodes of transshipment:
The equation at each warehouse as follows.
$$\sum_{i \in I} X_{ij}^p = \sum_{l \in L} X_{jl}^w \text{ for each } j \in J$$

The total plant production includes both the remanufactured and new products.
$$\sum_{k \in K} X_{ki}^t \le \sum_{j \in J} X_{ij}^p \text{ for each } i \in I$$

Disposals are carried out at test centers to reject the returns that are not worthy of recovery.
$$\sum_{i \in I} X_{ki}^t \le \alpha \sum_{l \in L} X_{lk}^r \text{ for each } k \in K$$

3. Capacity constraints at each facility and linking facility location decision with flow variables:
$$\sum_{j \in J} X_{ij}^p \le P_i Y_i^p \text{ for each plant } i \in I$$

$$\sum_{i \in I} X_{ij}^p = \sum_{j \in J} X_{jl}^w \le W_j Y_j^w \text{ for each warehouse } j \in J$$

$$\sum_{l \in L} X_{lk}^r \le T_k Y_k^t \text{ for each test center } k \in K$$

4. Y_i^p, Y_j^w and $Y_k^t \in \{1,0\}$ and binary
$$X_{ij}^p, X_{jl}^w, X_{lk}^r, X_{ki}^t, X_{ki}^t \text{ and } S_l \ge 0 \text{ and continuous}$$

This is a simple MILP model, and the problem may be solved applying any available software package. Once we get the optimal solution, some additional information is also directly obtained as shown here:

Annual appropriate estimate of the quantity of manufacturing at the plant: $i = \sum_{j \in J} X_{ij}^p - \sum_{k \in K} X_{ki}^t$

Estimate of the total annual disposal at the test center: $k = \sum_{l \in L} X_{lk}^r - \sum_{i \in I} X_{ki}^t$.

Key Learning

- The management of reverse flow of returns from location of its source to the location for product recovery operations is known as reverse logistics (RL) management, although the e-commerce sector also uses this term for return of defective and undesirable items from customers.

- The management of both forward flow of marketable (serviceable) products and reverse flow of recoverable products is popular as closed-loop supply chain (CLSC) management. Here, the OEM is taking care of remanufacturing operations of its own products along with its manufacturing activities following the cradle-to-cradle business model.

- REVLOG is a globally known research project on RL by a European research group comprising six European universities. It was sponsored by the European Union and conducted from 1997 to 2002. The outcome of the research was a generation of extensive knowledge and concepts on RL and development of relevant quantitative techniques for solving various related decision problems.

- The management of RL is a combination of managing three essential components: returns acquisition, inventory, and reverse distribution.

- Returns may be acquired by various schemes—take-back, buyback, off-lease or off-rent, auction, seed-stock, exchange offers, and warranty returns.

- For buyback acquisition, the timing for buying back may be optimized. If the used product is bought back later, the buyback price reduces, but the remanufacturing cost increases. Optimization of buyback time gives rise to development of a performance index named the **service effectiveness** of the OEM.

- There exist four types of inventory items in RL. These are inventory of returns before disassembly, recovered parts before reassembly, new parts to be fitted during reassembly, and remanufactured products for selling.

- In a deterministic situation, optimal manufacturing and remanufacturing lot sizes may be determined, applying modified EOQ models. Inventory review policy-based systems, applicable for uncertain situations, may also be developed depending on whether pull-based or push-based policy is being followed in remanufacturing management.

- Uniqueness of reverse distribution is because of its many-to-few characteristics, unlike the few-to-many model of forward distribution.

- The MILP model may be applied in optimization of facility location decision along with determination of quantity of bi-directional distribution in CLSC management.

> *Prior to the discussion session, it is expected that student groups will be formed. Now each of these questions may be discussed among the group members. The objective of the discussion session is to encourage students to think threadbare and explore all related issues, not arriving at the answer or solution to the problem,*

Discussion Questions

1. How do you characterize the product recovery options—refurbishing, remanufacturing, cannibalization, and recycling—in terms of their impacts on the original product and managing its remnant value after its use? How do you measure their suitability? Discuss on the industrial, technical, behavioral, or economic factors that may dictate this decision-making process?

2. In order to combine the benefit of both cost and responsiveness, a remanufacturer may opt for RTS-cum-RTO strategy like the postponement strategy in manufacturing management. How do you plan for location of facilities of different activities in remanufacturing operation and for designing the layout for proper implementation of this strategy?

3. How does industrial ecology contribute to achieving sustainability? What is its relevance to circular economy?

4. Explain the three types of recycling with practical and industry-related examples. Which of the three seems to be most effective in terms of product recovery and sustainability?

5. Compare DFE, DFR, and DFD from the product recovery viewpoint. Differentiate with reference to the areas of focus.

6. What challenges are encountered by Indian factories in managing product recovery operations? How do we overcome them?

7. Why are "matching demand with supply" and "parts-matching problem" so crucial in PRM? Discuss with industrial examples, sharing your experience in industries, if possible.

8. Do you agree with the outcome of the study in Section 4.2.1.? If not, please share your arguments and discuss.

9. The primary hurdles in achieving the efficiency and effectiveness in remanufacturing management is because of the existence of uncertainties on various issues—cores acquisition (quantity, quality, and time), recovery from returns, disassembling time, and forecasting of demand for remanufactured products. Share your expertise in developing the strategies for managing these uncertainties.

10. What is the primary difference between RL and CLSC? What conditions are to be fulfilled in implementing CLSC by the OEM?

REFERENCES

Amezquita, T., Hammond, R., Salazar, M., & Bras, B. (1995). Characterizing the remanufacturability of engineering systems. Proceedings 1995 ASME Advances in Design Automobile Conference, Boston, MA. DE-Vol 82, 271–278.

Benyus, J. M. (1997). *Biomimicry: Innovation Inspired by Nature*. New York: William Morrow.

Deng, Q-W., Lian, H-L., Xu, B-W., & Liu, X-H. (2017). The resource benefits evaluation model on remanufacturing processes of end-of-life construction machinery under the uncertainty in recycling price. *Sustainability*, *9*, 1–21, 256; doi:10.3390/su 9020256.

Ferrer, G. (1996). Product recovery management: Industry practices and research issues. Working Paper 96/55/TM. INSEAD, France.

Fleischmann, M., Bloemhof-Ruwaard, J. M., Beullens, P., & Dekker, R. (2004). Reverse logistics network design. In: Dekker, R., Fleischmann, M., Inderfurth, K., & van Wassenhove, L. N. (Eds.). *Reverse Logistics: Quantitative Models for Closed-Loop Supply Chains*. Berlin: Springer, 64–94.

Fleischmann, M., Bloemhof-Ruwaard, J. M., Dekker, R., Laan, E. van der, van Nunen, J. A. E. E., & van Wassenhove, L. N. (1997). Quantitative models for reverse logistics: A review. *European Journal of Operational Research*, *103*, 1–17.

Frosch, R. A., & Gallopoulos, N. E. (1989). Strategy for manufacturing. *Scientific American*, *261*(3), 144–153.

Guide, V. D. R. Jr. (2000). Production planning and control for remanufacturing: Industry practice and research needs. *Journal of Operations Management*, *18*, 467–483.

Guide, V. D. R. Jr., Jayaraman, V., & Linton, J. D. (2003). Building contingency planning for closed loop supply chains with product recovery. *Journal of Operations Management*, *25*, 259–279.

Guide, V. D. R. Jr., Srivastava, R., & Spencer, M. S. (1997). An evaluation of capacity planning technique in a remanufacturing environment. *International Journal of Production Research*, *35*(1), 67–82.

Guide, V. D. R. Jr., & van Wassenhove, L. M. (2001). Managing product returns for remanufacturing. *Production and Operations Management*, *10*(2), 142–155.

Hawken, P., Lovins, A., &. Lovins, H. L. (1999). *Natural Capitalism: Creating the Next Industrial Revolution*. Boston, MA: Little, Brown & Company.

Inderfurth, K., & Langella, I. M. (2008). Planning disassembly for remanufacture-to-order systems. In: Gupta, S. M., & Lambert, A. J. D. (Eds.). *Environment Conscious Manufacturing*. Boca Raton, FL: Taylor and Francis Group, 387–412.

Inderfurth, K., & Mukherjee, K. (2008). Decision support for spare parts acquisition in post product life cycle. *Central European Journal of Operations Research*, *16*, 17–42.

Jiang, J. (2006). Remanufacturing and engineering sustainable development, March 2016. https://www.researchgate.net/publication/296618781. Accessed on 14 August 2020.

Lund, R. (1998). Remanufacturing: An American resource. Proceedings of the fifth International Congress Environmentally Conscious Design and Manufacturing. June 16 and 17, 1998. Rochester Institute of Technology, Rochester, NY.

Lyle, J. T. (1996). *Regenerative Design for Sustainable Development*. New York: Wiley.

McDough, D., & Braungart, M. (2002). *Cradle to Cradle: Remaking the Way We Make Things*. New York: North Point Press.

Minner, S., & Linder, G. (2004). Lot sizing decisions in product recovery management. In: Dekker, R., Fleischmann, M., Inderfurth, K., & van Wassenhove, L. (Eds.). *Reverse Logistics: Quantitative Models for Closed Loop Supply Chain*. Berlin: Springer, 157–179.

Mondal, S. (2006). Economic analysis on product recovery/remanufacturing in Indian market—an exploratory study. Unpublished PhD dissertation. Department of Management Studies. IIT (ISM) Dhanbad. India.

Mondal, S., & Mukherjee, K. (2006a). An empirical investigation on the feasibility of remanufacturing activities in Indian economy. *International Journal of Business Environment, 1*(1), 70–88.

Mondal, S., & Mukherjee, K. (2006b). Buy back policy decision in managing reverse logistics. *International Journal of Logistics Systems and Management, 2*(3), 255–264.

Mondal, S., & Mukherjee, K. (2012). Simulation of tyre retreading process—an Indian case study. *International Journal of Logistics Systems and Management, 13*(4), 526–539.

Mukherjee, K. (2002). Reverse logistics for product recovery—issues and models. *Vision,* Special issue on Supply Chain Management, 141–149.

Mukherjee, K., & Mondal, S. (2009). Analysis of issues relating to remanufacturing technology—a case of an Indian company. *Technology Analysis and Strategic Management, 21*(5), 639–652.

Muris, L. J., & ho Filho, M. G. (2015). Production planning and control for remanufacturing: Exploring characteristics and difficulties with case studies. *Production Planning and Control.* doi:10.1080/09537287.2015: 1091954.

Pauli, G. A. (2010). *The Blue Economy: 10 Years, 100 Innovations 100 Million Jobs.* Taos, NM: Paradigm.

REVLOG. (1998). The European working group on reverse logistics. www.fbk.eur.nl/OZ/REVLOG. Accessed on August 2020.

Rodale, R. (1983). Breaking new ground: The search for a sustainable agriculture. *Futurist, 17*(1), 15–20.

Rogers, D. S., & Tibben—Lambke, R. S. (1999). *Going Backwards: Reverse Logistics, Trends and Practices.* Pittsburgh, PA: Reverse Logistics Executive Council.

Sakai, S., et al. (2014). An international comparative study of end—of—life vehicle (ELV) recycling systems. *Journal of Mater Cycles Management, 16*, 1–20.

Seitz, M. A., & Peattie, K. (2004). Meeting closed–loop challenge: The case of remanufacturing. *California Management Review, 46*(2), 74–89.

Simic, V. (2013). End–of–life vehicle recycling—a review of the state-of-the-art. *Technical Gazatte, 20*(2), 371–380.

Stahel, W. (1994). The utilization focused service economy: Resource efficiency and product-life extension. In: Richards, D. J., & Allenby, B. R. (Eds.). *The Greening of Industrial Ecosystems.* Washington, DC: National Academic Press, 178–190.

Teunter, R. H. (2001). Economic ordering quantities for recoverable item inventory systems. *Naval Research Logistics, 48*(6), 484–495.

Thierry, M., Solomon, M., Vannunen, J., & van Wassenhove, L. (1995). Strategic issues in product recovery management. *California Management Review, 37*(2), 114–135.

Vongbunyong, S., & Chen, W. H. (2015). *Disassembly Automation, Sustainable Production, Life Cycle Engineering and Management.* Cham: Springer International Publishing.

Part IV

Measures and Assessment
of Sustainability

5 Performance Assessment of Sustainable Operations Management

5.1 SUSTAINABILITY PERFORMANCE MANAGEMENT

Performance management is an essential domain of managerial activities and decision-making in management of any business unit. Any business setup aspires to manage its resources, process, and system in order to maintain and enrich its value-additive activities both efficiently and effectively. Here by efficiency, we mean "getting more from less resources" or better utilization. Whereas effectiveness in this context represents higher satisfaction, and survival and growth of the business unit in future. In fact, in case of supply chains these efficiency and effectiveness issues are dealt with the two distinct areas of supply chain performance: cost and responsiveness to customers. This gives rise to two distinctly different supply chain strategies for gaining competitive advantage—efficient strategy and responsive strategy. Of course, hybrid strategies like postponement strategy or delayed product differentiation are also adopted by supply chains for taking advantage of these two extremely opposite strategic orientations.

Activities Meant for Performance Management

Performance measurement is an important component of performance management, as "you cannot manage what you cannot measure." Activities under this domain of management are carried out at two levels—long term and short term.

1. **Long-term or relatively static:**
   These activities are for development of performance measurement system or scorecards. This measurement system is often used in strategy or plan formulation. The following are the required steps.

   a. Identification of indicators or criteria for measuring the performance. These should be closely related to the goal(s) of the organization or the supply chain. They should satisfy the expectations of all relevant stakeholders. The performance indicators at the process level or departmental level should be generated from the goals at corporate level and should have proper vertical and horizontal alignments. KPIs are identified at all levels.
   b. Units and scales of measuring the criteria are to be determined so that appropriate scorecards are developed. Kaplan's balanced scorecard is an example of such a scorecard.

DOI: 10.1201/9780429195600-9

    c. Correlations among these criteria are to be assessed. Any negative correlation between the criteria reflects conflicting characteristics, whereas the criteria with positive correlations may be clubbed together, if possible.

2. **Short-term or dynamic activities**:

This set of activities represents the actual use of the performance measurement systems. Practically it is the monitoring and control phase. The following activities are normally carried out in this phase.

    a. Actual performance is measured after a fixed period (annually, monthly, or even daily) depending on monitoring and control plan. This requires communication of actual outcomes from relevant operational levels for their subsequent comparison with the desired outcomes.

    b. The current performance is compared with the target or expected performance. This is more like a benchmarking approach for performance improvement.

    c. The outcome of the comparison is used for taking corrective actions or modification of target values.

Benefits of a Performance Management System

Any performance management system is expected to contribute to effective management of an organization or supply chain in various ways.

1. It supports the integration of performance indices with strategic goals, which naturally helps in goal achievement through required activities.
2. It helps in planning and decision-making at all levels for the realization of local targets and objectives.
3. It creates a basis for communication among all processes and activities. Integration becomes effective through execution of performance management system.
4. Strategic goals are linked with operational goals. All stakeholders are expected to be satisfied because of linking of their goals with all the processes.
5. Effective monitoring of process outcomes is made possible.
6. The performance management system (PMS) is activated on the basis of a representative framework meant for dynamic control of all the processes.
7. The implementation of the benchmarking approach is made possible by establishing the PMS.

5.1.1 INFLUENCING FACTORS OF SUSTAINABILITY PERFORMANCE MANAGEMENT (SPM)

By using the phrase **"sustainability performance,"** we mean here the performance management with special focus on sustainability or incorporating sustainability as one of the key representatives of organizational performance. Sustainability performance of any industrial organization or a supply chain may be perceived and measured with diverse perspectives and viewpoints. It may be the inclusion of the *triple* *bottom line* concept in capturing various dimensions or may also be creating a platform of measurement by integrating the performance indicators representing various levels of the organizational structure or entities of supply chain. Here by levels of management we mean the responsibility centers, which are supposed to achieve some levels of performance. These may include an individual, a facility, a machine shop, a plant, an organization, a nation, or even an international body comprising diverse interest groups. There are other perspectives of managing performance of sustainability, which attempt to model the business processes and their performance using input and output parameters or causes and effects of issues pertaining to sustainability. The perspectives differ in terms of value system, areas of focus, specific purposes, scopes for performance measurement (target fixing, benchmarking, accounting, reporting, etc.), and relevant parameters.

Performance measurement of sustainability, once incorporated in PMS, becomes a manifestation of comprehensive form of commonly used PMS in organizations. As discussed in detail in Chapter 1, sustainability involves new set of stakeholders other than the business-related ones by inclusion of physical environment (ecosystem) and society at large. So sustainability performance management (SPM) encompasses both business (economic, technical, and operational) and non-business (environmental and societal) indicators meeting the requirements of the enlarged set of stakeholders.

Here, the performance issue of sustainability may be first understood as an integration of various factors or criteria of measurement capturing all relevant aspects and dimensions because in any performance management system the primary focus is always on understanding the mutual support and conflict among all relevant measurement criteria. However, like any other management approaches, success of any proposed SPM is assessed by the degree of its acceptance by the business community and stakeholders or by its popularity worldwide.

> *Sustainability performance management (SPM) enables the organization measure the performance considering both business (economic, technical, and operational) and non-business (environmental and societal) indicators.*

In this pursuit, it is always meaningful to initiate the discussion on SPM by identifying the critical issues or activities that surround the management of industrial operations keeping sustainability as the strategic goal. Subsequently, better insights may be obtained by exploring their mutual influence and relationships. Literature review and interactions with business executives give rise to creating a long list of various activities and issues which influence management of green operations initiatives. But a cumbersome list of 25-to-30-odd such issues is really not manageable at all for further study and analysis. So a list of manageable eight critical factors are proposed here which is an outcome of a research project conducted by the author of this book along with his project team (Choudhury et al., 2017). The input data for this study is the responses from executives of Indian SME sector. The interdependence of these factors is the base for discussion on sustainability-inclusive performance management. The following eight critical factors or issues influence each other in various degrees for managing sustainable operations or even green supply chain.

1. Green design: This factor includes the design issues for products, packages, and the production process, which reduce the environmental impact.
2. Supplier's environmental collaboration: It covers all possible collaborations and involvement of suppliers relating to sustainability. It may cover partnership, information sharing, joint decision-making, and collaborative R&D with suppliers.
3. Customer's environmental collaboration: It includes activities and issues pertaining to collaboration and interactions of the supply chain with the customers on implementation and maintenance of sustainability. It may cover mutual understanding, joint decision-making (particularly in B2B), and information sharing. It would be more meaningful if this firm is a supplier to a manufacturing unit (i.e., customer in this supply chain).
4. Government regulations and support: This critical factor includes all taxes and penalties, environmental regulations, subsidies, and assistance, which reflect the role of regulating authority in managing green supply chain. It may be motivator or demotivator as the case may be.
5. Performance measurement practice: It includes external reporting (e.g., GRI), environmental monitoring, and green auditing. Various assessment mechanisms on environmental performance are covered under this factor.
6. Top management commitment: The managerial activities like formulation of goal and environmental strategy, fund allocation, and rewards and incentives are parts of this factor.
7. Organizational resources and capabilities: This includes financial resources, expert team, advance technology, and human resources of the firm.
8. Reverse logistics: This factor addresses all issues relating to acquisition of used products, remanufacturing and recycling, and forward and reverse distribution.

Among these eight important factors, governmental regulations as coercive force and governmental support as an enabling force surely trigger the process of green supply chain management practices, and they play the roles of powerful external influencers. This influence gets strengthened (or weakened) by better (or worse) management and use of resources of the organization along with high commitment of corporate management. Greenness of supply chain is achieved once other supply chain members are collaborated and involved in this pursuit. Complete green supply chain demands green practices in managing activities of all supply chain members along with green product design and reverse logistics. The success of this sustainability orientation is measured by performance measurement system established in the organization.

The research study is further extended to critically explore the role of these factors as driving or influencing tools in green supply chain practices. The following conclusions may be drawn.

1. Governmental regulations and support, top management commitment, and organizational resources and capabilities seem to have maximum power of

influencing other factors. In other words, other five factors are directly or indirectly driven by these three factors.

2. The factors like performance measurement practice, green design, and reverse logistics are influenced or affected by other factors. We may even conclude that the outcome from activation of other factors or activities are reflected on these three factors.

These two conclusions of the study are expected to play very crucial role in management of green supply chains for SME sector of India. The study shows that the factors may be grouped in four levels based on their direction of influence among themselves. The factors under the same level are influencing each other in both ways. The factors in lower-level drive or influence the factors in higher levels, but reverse is not true.

So the Figure 5.1 clearly justifies the fact that for sustainability in the Indian SME sector the most critical external driving forces are governmental acts, rules, incentives, penalties, and so on, and internal drivers are commitment of corporate management and organizational culture and capabilities. All the other factors and related activities are pushing the organization to sustainability, which is ultimately reflected by the result of performance measurement.

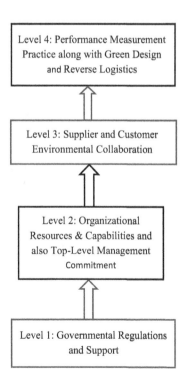

FIGURE 5.1 The multilevel relationship model among critical factors in managing green supply chain practices in the SME sector.

This was developed applying the interpretive structural model (ISM), which depicts the influences among the factors. This study empirically and scientifically demonstrates that the performance measurement system is the most important factor in green operations management, as all other factors influence it collectively.

5.1.2 Two Viewpoints on SPM

Let us now address the sustainability issues in performance management in two ways. First, it will be the identification of the performance metrics or indices on sustainability along with other indicators of business performance, and second, we will try to extract insights from the practices commonly carried out by business units and accepted by regulating authorities. The former initiates building of the performance management system on the basis of the appropriate performance indicators. On the other hand, the later primarily focuses on the business practices for performance management of sustainability prevailing in various sectors of economy, which are already accepted by the society and business community. These two viewpoints combine both theoretical and practical modes and thus provide meaningful insights on various issues related to SPM comprehensively.

5.1.2.1 Performance Indices or Indicators

Researchers and practitioners proposed various viewpoints in order to make a comprehensive list of performance measures incorporating all factors of business, including sustainability. The viewpoints do differ depending on the stakeholders considered in the framework, factors included in measuring the performance, and sustainability-related activities within the organization and outside, which influence and are influenced by the outcomes of business processes.

Organizations often define key performance indicators (KPIs) in order to prepare final scorecard of their performance, which are further used for fixation of targets meant for benchmarking and planning and subsequently for control of all organizational processes and functions. Identification of KPIs for any organization demands critical study on needs of key stakeholders, inputs from strategic goals, SWOT outcome and futuristic vision of the corporate management. In present context, these KPIs may be grouped under two sets of indicators.

The first set is of the traditional KPIs, which focus on operations leading to reduction of various cost items, increase of sales revenue, and enhancing the capability of coping up with the dynamic environment and growth potentiality of the organization. Thus, it includes

- operational efficiency (productivity, cost reduction, shortening of lead times, waste reduction, etc.),
- effectiveness (customer satisfaction, delivery time, product quality improvement, after-sales services, etc.),
- quality enhancement of business processes,
- agility, and
- innovation and creativity.

The second set of KPIs may be generated on the basis of the capability and contribution of the organizations measured in terms of the factors other than economic and operational ones. These factors are the impacts on outside environment other than those which directly result in financial gain or advantage. In other words, this calls for exclusive consideration of sustainability-related factors affecting physical environment and society. These KPIs include but are not limited to

- carbon footprint generation (including Scope 1, 2, and 3 of emissions during the product life cycle),
- consumption of energy (energy efficiency, replacement of thermal energy by other non-thermal sources, etc.),
- supply chain miles or food miles (transportation or logistics-based emissions, waste generation, and environmental degradation),
- excess consumption of materials (conservation of scarce material resources) and waste reduction and reuse (recycling or remanufacturing to convert the product or components as good as the new ones), including other external environmental impacts like biodiversity, flora and fauna, deforestation, and so on,
- social impacts (gender equity, wage rationalization, responsibility towards society, environmental, social, and governance (ESG) issues, etc.),
- safety and security factors (for in-house employees, customers, other affected individuals, particularly in the mining industry, thermal power plants, chemical plants, nuclear power plants, transportation, etc.), and
- health-related factors (generation of hazardous or toxic materials and gases in the business process.

Sustainability-related factors that give rise to the creation of performance framework for a business organization primarily include carbon emissions, energy consumption, excess material use, safety and security factors, social impact, health-related issues, biodiversity affecting factors, impact on fauna and flora of surroundings, and deforestation.

Experts from both practitioners and research domains address the sustainability performance measurement from various viewpoints as mentioned in the following.

5.1.2.1.1 Performance Measurement Indices on the Basis of the 3Ps Concept

On the basis of the 3Ps of sustainable development, we may consider the following performance indices:

1. *Profitability*—It may be measured by indices of financial efficiency (operating profit, net cash flow, return on assets/investment, resource utilization, etc.) and that of customer satisfaction (delivery speed, product availability, product reliability, etc.). Sometimes these two sets may be conflicting in

nature—for example, reduction of delivery time versus reduction of costs or resource (inventory) utilization versus product availability.

2. *Social or human welfare*—It may be measured in terms of benefits offered to employees (equity in wage structure, job security, gender equity, leave policy, workers' participation, etc.), safety (safety index, accident cost, compensation, accident prevention, etc.), and contribution to society (employment potentiality, corporate social responsibility compliance, etc.).

3. *Environment-friendliness*—It may be measured by unwanted waste generation, various emissions (carbon dioxide or GHG emissions, unwanted waste liquid generation, noise or vibration generation, industrial or urban solid waste generation as pollutants, etc.) and resource consumption (utilization of limited resources, generation of renewable resources, generation of substitutes for nonrenewable resources, reduction of energy and materials consumption, implementation of lean production system, etc.).

This three-dimensional measurement of sustainability performance seems to be the most popular performance measurement metrics considered in corporations or even in ministerial level of national governance. Obviously, this is essentially a reflection of TBL. Progressively, other welfare dimensions may also be suitably added to it, like industrial or occupational safety factors (fatality indices, accident proneness, post-accident compensation, provisions for life insurance, etc.) and risk of health hazards (industry-specific or job-specific). Traditionally, organizations mostly prefer the economic (or financial) bottom line for assessing the business performance with a profit-fetching goal. The next preference may be the environmental bottom line, and then the third bottom line of social performance is considered as the last performance measure, mostly reflected by social responsibility and social responsiveness.

Judicious integration of these three bottom lines is still a complex task, attracting continuously the attention of practitioners and researchers. The complexity is primarily twofold. The first one is the problem of quantifying all measurement indices, and the second one is the integration issue. The quantification issue is most prominent in case of measuring the social dimension of sustainability. For example, if the attempt is made to establish triple-bottom-line performance measurement at the national governmental level, we may find difficulty in measuring the level or degree of community involvement as social responsiveness. The social bottom line factor also includes the indicator depicting how the community is supporting the culture of indigenous groups or poverty-stricken areas. Most of the large industrial setups, like the hydel power projects or mining projects may result in rehabilitation and resettlement of local people. This means there is enough chance of cultural conflicts, resistance to change and difficulty in mixing with the new social community. It is really problematic to quantify these factors of social sustainability. Here the difficulty still remains in scaling of measurement ascertaining the degree or intensity of involvement. In these situations, of course, we may look for some surrogate or proxy indicators, like the monetary contribution to charitable organizations for the benefit of the community or of the people in social programs. The second reason of complexity is the integration of these factors, each of which is oriented toward a particular direction, and most often they are conflicting in nature. This integration may be achieved also in two ways. The first one is overall optimization-based framework. If possible, we may try to optimize the performance, which is the toughest form of integration. This actually means judicious and acceptable compromise among the factors under the three dimensions for maximum achievement of corporate goal. This may also incorporate the compulsions and restrictions at the national policy level in the integrated model of optimization. Moreover, the relative importance of the factors and/or trade-offs among them are also considered in this optimization process for meeting the overall goal of the organization, which is really difficult to estimate and quantify objectively. The second option for this integration is relatively easier, and it is prioritization-based integration. It may be explained as a multistage process of optimization (pre-emptive priority levels), considering the environmental protection as the top priority and profitability as the least priority in case of sustainability performance at a national level. Once the performance measure of higher priority is maximally considered for comparison or benchmarking, the dimension of lower priority is given due consideration. But the primary concern is how to create the priority levels among the criteria, which is simultaneously acceptable by corporation, society, and the concerned governmental agency. In case of corporations, the tendency would be consideration of financial benefits as the criterion of top priority, which will make the sustainability-focused performance analysis simply futile, as the environmental protection and social responsibility will be of very low importance in depicting the overall performance.

5.1.2.1.2 Performance Indicators Representing Managerial Efforts in Maintaining Sustainability

This classification is primarily focusing on how much the managerial activities of the organization are likely to capture the sustainability in the plan and control functions and what may be its ultimate impact. Various guidelines have already been proposed in this pursuit, and industrial organizations are already practicing them. SPM, when used for the whole supply chain, is often termed as sustainable supply chain performance management (SSCPM) or green supply chain performance management (GSCPM).

We may here identify three classes of indicators of organizational performance representing managerial efforts primarily on environmental management by the organization.

1. *Managerial performance indicators*: These represent the managerial efforts in protecting and improving the surrounding environment. These are the drivers for protecting the environment, which are reflected in the formulation of organizational goals, policies, strategies, and plans. It may include the costs or budgeted amount targeted in dealing with environmental issues, goals, or targets fixed for environmental planning by the organization and time spent in dealing with these activities.

2. *Operational performance indicators*: These indicators are measured by the intensity of activities or operations relating to environmental management, like the use or consumption of materials per unit product, frequency of preventive maintenance (for improved efficiency and machine availability),

and average fuel consumption of vehicle per kilometer. These may be understood as the execution of corporate policies, plans, and budgetary controls (as shown in previous set of indicators) at the operational level. Moreover, these simultaneously represent both the economic efficiency and efficiency in dealing with environmental issues while conducting organizational operations.

3. *Environmental (direct) indicators*: These indicators are the direct environmental impacts or effects on local environment. These may be the generation of air, water, or solid pollutants; frequency of smog due to emissions from thermal power plants; chemical or petrochemical plants; and acid rains.

It is quite understandable that these three indicators are linked by the cause-effect or enablers-results relation model. Better managerial performance causes improved operational performance, which is reflected on environmental impacts.

5.1.2.1.3 Performance in Terms of Responding to Various Drivers

In this context, the most crucial question is why a business being a profit-making organization will be interested in achieving sustainability goal instead of meeting financial targets. Is it the pressure from society or compulsory governmental directive or demand from market or customers? Does the pressure represent compulsory compliance, or is the compliance voluntary?

Latif et al. (2020) studied various pressures in manufacturing sector of Pakistan, which actually act as enablers in adopting effective environment management techniques like environment management accounting (EMA). Sustainability performance may be measured by how positively the organizations are reacting to these enablers. The study on these enablers is based on the application of both institutional theory and stakeholder theory. The theoretical framework of the institutional theory assumes that organizations are embedded in a web of values, norms, rules, and beliefs that guide their behaviors and practices, whereas the stakeholder theory proposes engagement of stakeholders to strengthen the sustainability-focused performance and ultimately to gain the competitive advantage in the market. In case of any change in values or norms at the stakeholder's level (e.g., government, customers, suppliers), the companies are likely to suffer, if they ignore these changes and miss the opportunity of reacting and adopting them in their activities in right time.

These drivers or enablers for sustainability are classified under three groups.

Normative drivers: Various pressures may be exerted by internal and external stakeholders. These pressures are not meant for compulsory compliance but trigger the voluntary activities in order to meet these pressures. For example, green supply chain adoption of sustainability by a manufacturer automatically puts pressure on other members of the chain to perform in a sustainability-friendly manner. Sometimes this performance may also be measured by the intensity of response to media perception on the sustainability-related activities of the organization. Corporate social responsiveness (as practiced by several Indian companies nowadays) or corporate green image may be two such examples of corporate policies and strategic goals. Performance under this category enables companies to improve their green image, impression, and perception of external stakeholders.

Coercive drivers: These drivers are powered by governmental agencies or regulating authorities. These are mostly mandatary and compulsory. These include acts, laws, and directives by the government related to pollution, solid waste management, resource or water conservation, recycling and reuse, and disposal. This group of drivers also covers directives of MoEF (now renamed MoEFCC) and restrictions imposed by state and central pollution boards. Coercive pressure may reinforce or modify the essential framework of the environmental protection and legislative mandates of organizations. Moreover, coercive pressure may also originate from global customers, foreign investors, and transnational institutions, particularly in case of globalized business environment. We may refer to the fact that European customers exerted coercive pressure to globalized Indian companies for adoption of TQM and getting ISO 9000 certification. These are the most powerful drivers. Organizations are to plan for developing their sustainability-focused strategies and operational activities as the mandatory reaction to this pressure. The sustainability performance indicators may be established on the basis of these factors and the relevant prioritization.

Mimetic drivers: Organizations are indirectly pressurized by sustainability-oriented action of any successful organization and tend to mimic them. This case of imitation often occurs through comparison with the competitors or as a result of any benchmarking process. The origin of this pressure lies in the competitive behavior of the organizations and in building survival strategies in an open and globalized market. This pressure often plays a very crucial and effective role in improving sustainability management performance of organizations. Let us suppose that an organization in a developed country has implemented an environment-friendly technology. It is quite possible that organizations of same sector in developing countries will soon be pressurized to opt for similar technology. So ultimately, this leads to improved environmental management not only in the organization of the developed country, but most probably also in the same industrial sector of other developing countries. Technology transfer and adaptation may also be applicable in this pursuit. Mimetic drivers do not pressurize organizations toward carrying out any mandatory activity. However, any strong mimetic driver may also be converted to coercive one, once governmental agencies get influenced by it. Actually, this leads to creation of an act, rule, or directive for its implementation in national level. Installation of a particular type of pollution arrester in the factory, construction of water treatment plant for reuse of wastewater, and remanufacturing of the firm's own products are some of the examples of mimetic drivers.

Organizations are forced to perform environment-friendly, when they face coercive pressure from powerful government and partners, and normative pressure from industry associations, trade associations, media and other social actors. Otherwise, they may be punished (by payment of penalties, carbon taxes, compensations) as decided by the judicial authorities. Non-law-related punishments may be loss of brand value or tainting the green corporate image. They may also be isolated by some stakeholders and members of the business environment.

> *The indicators in this performance measurement framework are also expected to reflect managerial, operational, and environmental performance, and at the same time, they should respond to coercive, normative, and mimetic drivers.*

They may even lose the opportunity of acquiring better external resources and market share. Proper response to mimetic pressure helps in developing suitable strategy in an industry or work environment while competing with the peer group members, particularly on the basis of green image factor.

5.1.2.2 Performance Measurements Based on Business Practices

Earlier the business community used to believe in a wrong notion that any focus on issues outside the business is not beneficial to the profit-making organization. So any activity relating to contribution to the environment and society used to be treated as non-business performance, philanthropy, or benevolent action. But now the industry practitioners could understand that efforts on better performance for making an industrial organization an environment-friendly one automatically improves business performance of the organization in following two ways.

First, the organization tends to improve its financial performance by availing the new opportunity of sustainability-oriented programs as this plays the role of an additional competitive advantage to have an edge over the competitors in the market. This is also supposed to add values to almost all business activities. However, some gains can only be achieved in long run. These are the higher market share, improvement in growth rate and enhancement of corporate brand value.

Second, a lot of improvement in operational performance will be experienced by the organization, like quality improvement, waste reduction, energy and resource conservation, and lean production system in the manufacturing process, because of sustainability-oriented practices. Thus, focus on sustainability leads to overall improvement of an organization's performance, both in terms of effectiveness and efficiency or growth and cost.

5.1.2.2.1 Green Supply Chain Management (GSCM)
Practices and Performance Indicators

Green supply chain is the integrated virtual organization (managing the flow in the network of all separately owned business units in the supply chain), which intends to characterize the management processes of the whole supply chain by green orientation. The management practices of GSCM encompass operations from green purchasing to reverse logistics, involving all entities like suppliers, manufacturers, distributors, retailers, all service providers, and even customers.

Let us suppose that the sustainability-focused performance of the supply chain is measured simply by three popular dimensions—environmental, economic, and operational. The green performance may be measured as cost reduction, quality improvement, waste reduction, emission control, speedy delivery, and new eco-friendly product development. Each of these performance indices demands coordination of all members of the supply chain. On the other hand, these are also the elements required for gaining competitive advantage in the market. Thus, green performance and competitive advantage may be achieved by carrying out some GSCM practices, which, in general, cover some internal environmental management practices, investment recovery (reuse and disposal of used products in eco-friendly way), and cooperation and involvement of critical supply chain members (relationship management with suppliers, distributors, retailers, etc.).

Readers will have significant insights on the prevailing green practices in various economic or industrial sectors from the following brief reports collected from published literature.

5.1.2.2.1.1 US Industrial Sector By the later part of previous century, when sustainability was not significantly popular among corporations, Walton et al. (1998) conducted a study among business units of US and identified the following five GSCM practices in industrial sector.

1. Design of products with environment-friendly materials
2. Environment-friendly process design for production
3. Improvement of operational processes at the supplier's end
4. Evaluation of processes and activities of suppliers
5. Internal logistics process

Further, Hervani et al. (2005) later on identified four other GSCM practices, which primarily focus on making the supply chain process green:

1. Green purchasing
2. Green manufacturing and material management
3. Green distribution marketing
4. Reverse logistics

So from these two lists, we find the following classes of practices, which we may term as GSCM practices.

1. Eco-product-design
2. Environment-friendly processes at the supplier's end
3. Green operational processes, like purchasing, manufacturing, inventory, and all logistics operations (inbound, outbound, and in-factory logistics)
4. Reverse logistics for product recovery

5.1.2.2.1.2 Indian Automobile Sector In this context, the readers may also have some glimpse of GSCM practices in the Indian automobile industry based on a small survey on five automakers and five auto-component manufacturers (Jain and Sharma, 2012). The respondents gave their opinions on the existing management practices and activities at operational activities, which are meant for implementing and maintaining green supply chain in automotive sector of India. We may conclude that 22 GSCM practices may be identified for the Indian

TABLE 5.1

GSCM Practices in the Indian Automobile Sector

Groups of GSCM Practices	Management/Operational Practices under Each Group
1. Internal practices	• Controlling toxic emissions at the plant level* • Monitoring resource consumption and pollution* • Organizing employee training program on sustainability/greenness • Creating opportunity for cross-functional cooperation
2. Green activities in all functional areas	• Application and use of information technology* • Green transport • Green purchasing • Green warehousing • Green manufacturing*
3. Eco-design of products	• Eco-labelling of products for making customers aware of environmental consciousness of manufacturer in designing the product* • Eco-designing considering all phases of life cycle • Customers' involvement in designing of environment-friendly products
4. Life cycle assessment and product recovery	• Monetary value recovery from used products* • Reverse logistics • Management of product life cycle
5. GSCM performance assessment and resource allocation	• GSCM performance standards, evaluation of green performance, and green effectiveness • Resource allocation*
6. GSCM budgeting and implementation of EMS	• Green as a driver in the SCM agenda • Regular performance of environmental audit • Priority of GSCM during budget preparation • Environment R&D* • EMS ISO 14001 certification

(*) Most popular GSCM practice

automobile sector, which may be classified under six major groups, as shown in Table 5.1. The six groups of practices are internal environmental practices, green activities in all departments or functions, green design, life cycle assessment and product recovery practices, performance assessment, and inclusion of GSCM in annual budget (proactively fund allocation for achievement of greenness). It is further noticed that out of 22 GSCM practices, eight are most popular ones among the factories under this sector, which are spread over all the groups or types of GSCM practices. The popular practices are earmarked by asterisk in Table 5.1.

5.1.2.2.1.3 Brazilian Automobile Sector Almost similar GSCM practices are prevalent in the automaking sector of Brazil. As mentioned in published report of Drohomeretski et al. (2014), the Brazilian research team studied three automakers and collected data during one year (2012 to 2013). These car manufacturers from southern Brazil have been studied thoroughly. Each of the automakers has recruited either a manager or a supervisor, who will be solely responsible for implementing sustainable practices matching with the environmental policy of the company. Each of the automakers maintains its goal of restricting the consumption of water, resources, and energy, and the sustainability is very closely aligned with the production strategy of each of the manufacturers. The management practices of green supply chains of these automakers are somewhat similar to that of Indian ones, like the following:

- *Internal processes*—primarily implementation of an environmental management system along with ISO 14001 certification
- *Supplier management and green purchasing*—training, supplier development programs, insistence of ISO 14001 certification, and sustainability-focused conditions during supplier selection
- *Green packaging*—use of long-lasting materials like plastisol, use of materials with low environmental impact (both during production and use by customers), and use of pallets made of recycled materials
- *Internal, external, and reverse logistics*—use of material-handling equipment with low consumption of natural resources (e.g., replacing forklifts fueled by liquified petroleum gas with those run by electricity), optimization of transport time within factory and external travel time, reduction of emission and consumption of resources, effective system of return of defective items, and rework, repair, and disposal in reverse logistics
- *Clean production*—well-defined system of monitoring and restricting the use of natural resources in product design and production technology, water recycling through water treatment stations and reuse of treated effluents for various activities in the plant and other non-production areas, and technology with an effective waste disposal mechanism
- *Eco-design in new product development*—reduction of the number of components in the new product design, consideration of the product disposal system after the end of its life, use of recycled materials, and consideration of the low consumption of materials during its use at the customer's end

Additionally, the study outcome also reveals the fact that with varied levels of maturity the automakers are following lean manufacturing system in their production process. It can also be justified that JIT, TQM, lean, and agile manufacturing are the potential antecedents to the successful GSCM practices.

5.1.2.2.1.4 Chinese and Japanese Industrial Sector Q. Zhu and his co-researchers carried out various research projects on green supply chain management (GSCM) practices and their impacts on performance of organizations in China and Japan. Zhu et al. (2008) first conducted a survey on 200 Chinese organizations for this purpose. As an extension to this survey, Zhu et al. (2010) once again carried out another similar study involving Japanese manufacturers. They further analyzed the GSCM practices and resultant performance based on the survey outcomes from both the studies—that is, the business activities of both

China and Japan in this context. Zhu and his co-researchers contacted 12 large Japanese firms for this study, out of which nine firms ultimately got involved in this research project. This sample includes four chemical manufacturers, one petrochemical firm, two producers of electrical and mechanical appliances, one electrical manufacturer, and one producer of food items. The main objective of this study was to understand the effort of Japanese firms in implementing GSCM practices and impacts of the practices on business performance. Subsequently, the study was extended to compare the scenario with those of Chinese firms (report of 2008). The study was initiated with the data collection on GSCM practices and on the corresponding performance indicators of the organizations from both the countries. This empirical study has been conducted using sets of structured questions both for the sustainability practices and performance. Two sets of questions were considered for data collection—that is, 21 questions on GSCM practices and 17 questions for data on performance. Each of these questions is an alternative GSCM practice or a performance indicator. The business executives, who are the respondents of this survey-based study, were supposed to respond to each of the structured questions in a 1–5 scale. In the case of questions about GSCM practices, the scale represents the five values ranging from "not practicing" (1) to "actively practicing" (5) by the organization. On the other hand, the performance data may be collected from the responses on a scale showing the range from "not at all significant" (1) to "significant" (5).

Direct involvement of industrial practitioners or executives in data collection adds enough of practicality to this study, and thus, it is expected that the study will be quite meaningful to the managers practicing GSCM or planning to practice the greenness in operations. Let us now discuss on the various green practices and performance indicators that have been considered in this study.

There exist five groups of GSCM practices. These groups and the respective practices are as follows:

Internal Environmental Management Practices (Seven Types of Practices)

1. Commitment for GSCM at the corporate level
2. Support of mid-level managers for a green supply chain
3. Cross-functional cooperation across the organization for an environment-focused management
4. Total quality and environmental management
5. Strict compliance with environmental directives and guidelines with inhouse auditing programs
6. ISO 14001 certification.
7. Implementation of the environment management system.

Green Purchasing Practices (Five Types of Practices)

1. Providing environment-friendly designs to suppliers for items to be procured
2. Supports to suppliers for managing environmental issues
3. Audit of suppliers' internal environment management programs

4. Mandatary ISO 14001 certification
5. Persuasion of suppliers to pressurize second-tier suppliers to implement environment-friendly practices

Customer Interactions on Environmental Issues (**Three Types of Practices**)

1. Interactions and involvement on eco-design of products
2. Interactions and involvement on cleaner production
3. Interactions and involvement on green packaging

Eco-Design (**Three Types of Practices**)

1. Product design with minimum use of materials and energy
2. Product design with the reuse, recycle, and recovery of materials, parts, or components using design for disassembly or remanufacturing
3. Product design with avoidance or elimination of hazardous items

Investment Recovery (**Three Types of Practices**)

1. Sales of excess materials or inventory (slow moving) by making the process lean
2. Sales of scrap and excess used materials
3. Sales of unused capital equipment lying idle for long

Similarly, Zhu and his co-researchers have also identified 17 performance metrics of sustainable operations or green supply chain. The following list of indicators collectively represents the overall performance of the organizations practicing GSCM. This list is again grouped under three sets of performance indices.

Environmental Performance (**Six Types of Measures**)

1. Reduction of air pollution
2. Reduction of wastewater
3. Reduction of solid waste
4. Reduction of use of toxic, hazardous, or harmful materials
5. Reduction or even elimination of environmental accidents
6. Improvement of overall environmental condition of facilities or manufacturing plant

Financial or Economic Performance (**Five Types of Measures**)

1. Reduction of costs of input materials
2. Reduction of waste generation and waste treatments
3. Reduction of costs of waste disposal
4. Reduction of energy consumption
5. Reduction of fines, taxes, and penalties due to pollutant generation and accidents caused by poor handling of environment-related issues

204 Sustainable Operations Management

Operational Performance (Six Types of Measures)

1. Reduction of scrap generation (due to lean processes)
2. Improvement in inventory management
3. Improved capacity utilization (efficiency in operations)
4. Improvement in product quality
5. More scope of adding new product lines (better research and development)
6. Improvement of in-time deliveries of goods (due to elimination of non-value-additive activities)

Logically, we may conclude that better operational performance leads to higher value of the environmental and financial performance metrics. The outcome of this study provides meaningful insights on GSCM practices and performance of manufacturing organizations. The researchers also carried out a comparative analysis of GSCM efforts and the performance between Japanese and Chinese organizations. The survey report of Chinese firms was published earlier (Zhu et al., 2008).

Large corporations in Japan have actively implemented internal environmental management practices, although their external environmental management practices are primarily limited to green purchasing and investment recovery along with eco-design. It may be noted that internal management practices of the environment in Japan are more intensive than in China. In terms of other GSCM practices, both countries are active at more or less similar levels. It may be opined that most of the global manufacturing companies are keen in implementing internal environmental management practices for GSCM implementation. Incidentally GSCM as a business model has been implemented in Japan for a long time, and so the expertise of Japanese firms in these practices is much greater compared to other countries, particularly in Asia. India is yet to come up to that stage of implementation of GSCM practices. Nevertheless, close interactions with customers and activating their cooperation are yet to be practiced by Japanese firms. Chinese corporations also implement internal environmental management practices. However, survey result shows that Chinese firms take help of customer cooperation in clean production technology and green packaging. This enables the firms meet the requirements, particularly, in case of exporting the products to the Western world. This is also applicable in situations, when the Chinese firms play the role of suppliers to the external manufacturers.

These GSCM-based activities have resulted in significant performance improvement. Overall impact on Japanese firms is clearly more prominent than Chinese ones. Performance of Japanese firms shows higher than significant status (i.e., value 4 in a 1–5 scale) in almost all measures of performance of an organization. It may be due to the maturity of Japanese corporations in working with some guidelines, directives, or policies relating to sustainability. Among the three types of performance measures, Japanese firms perform remarkably well in environmental and financial measures. Their excellence in terms of reduction of scrap generation, improvement in capacity utilization, and product quality is also quite prominent. However, operational performance of Japanese corporations is less significant compared to other two performance dimensions. Environmental performance of Chinese firms do not display significant improvement after implementing GSCM practices.

Because of excellent internal environmental management practices, Japanese firms could achieve significant environmental performance, particularly in measures like reduction of air pollution. Further, in this pursuit, Japanese firms could even improve financial performance by reducing input material costs, decreasing fines or taxes because of accidents involving environmental issues, and improving the quality of products through GSCM practices. Incidentally, Chinese firms could not achieve these improvements significantly. In fact, Chinese firms are yet to gain financially. However, experts opine that better financial performance is more like a long-term outcome of sustained use of GSCM practices, which is now experienced by Japanese corporations. On the other hand, GSCM practices have not led to significant operational performance among corporations of both countries. However, because of Japan's rich legacy in lean production and TQM, their performance on products' quality promotion, capacity utilization, and reduction of scrap rate is somewhat significant.

The general conclusion drawn from the study outcome is that better environmental performance may be achieved by effective implementation of internal environmental management practices, green purchasing, eco-design, and investment recovery (including the policies on reuse, recycle, and recovery). However, better operational performance demands close interaction with external stakeholders, like suppliers and customers. on sustainability-related activities. Positive outcomes on both environmental and operational performance surely improves financial performance of the organization.

5.1.2.2.1.5   Green Practice Implementation in a Multi-Tier Framework   The practical implications of 3Ps-based sustainability dimensions in managing performance should be understood by industrial organizations with futuristic perspective. Corporations are to incorporate the Sustainable Development Goals (SDGs) per the 2030 Agenda for the benefit of both the planet and society in future. So all the organizations need to formulate the strategic goals accordingly and the organizational performance in terms of sustainability is to be mapped systematically down to operational target level. A three-tier approach may be adopted considering the levels, like strategic goals, key performance indicators, and quantitative assessment, similar to what was developed by Hristov and Chirico (2019). This demonstrates top-down management hierarchy for implementing strategic goals to actions. This hierarchical set of parameters will be applicable for all three sustainability dimensions—environmental, social, and economic. The quantitative evaluation, being the last level, shows the scale of assessing the achievement of each strategic goal through the corresponding performance indicator(s). Once it is known, management can fix the annual target of achieving a performance criterion for each function and department. This framework depicted in Figure 5.2 is expected to translate the SDG achieving strategies into achievable targets at operational level.

Corporations have other options of managing performance of GSCM, which is much more holistic and which takes care of various aspects of business activities in a comprehensive manner (Kazancoglu et al., 2018). The tree-like structure of performance criteria gives logical representation of performance assessment system. Practically it looks like the hierarchical structure similar to the analytical hierarchy

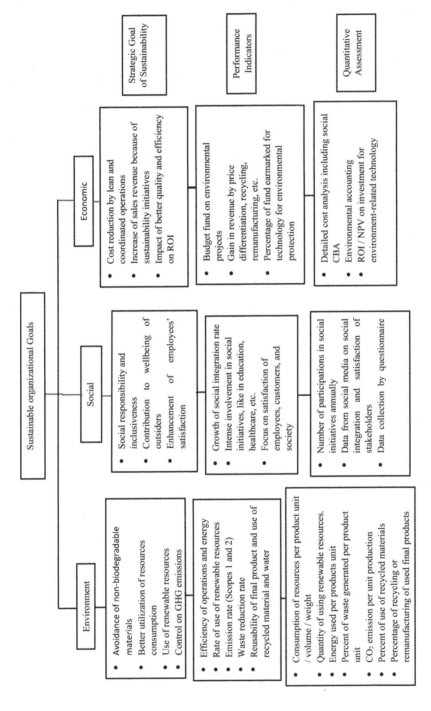

FIGURE 5.2   A tree-tier approach for meeting strategic goals on three dimensions of sustainability.

TABLE 5.2

Holistic Performance Assessment System

Main Criteria (or Indicators)	Sub-Criteria	Measures
Environmental performance	• Decreasing emissions • Decreasing energy coal consumption • Decreasing business waste • Decreasing environmental costs • Increasing environmental revenues	• GHG emissions • Energy utilization ratio • Solid waste generation • Disposal costs • Revenue from green products
Economic/financial performance	• Cost oriented • Revenue oriented	• Manufacturing and transportation costs • Average profit from green products
Operational performance	• Increase in quality • Increasing efficiency • Improving green manufacturing • Improving green packaging • Improving green/eco-design	• Customer rejection rate • Capacity utilization • Waste reduction and use of pollution monitoring equipment • Use of recyclable materials in packaging • Quantity of recycled or recyclable materials in new design
Logistics performance	• Improving green logistics • Improving revenue logistics • Improving green purchasing	• Eco-driving for lower fuel consumption • Reusing, remanufacturing, and/or recycling via reverse logistics • Purchase of environment-friendly materials
Organizational performance	• Improving green image • Incorporating environmental management • Green information system	• Reduction of environmental accidents • Commitment from managers and employees • Exchange of environmental information with suppliers and partners
Marketing performance	• Increasing customer satisfaction • Improving cooperation/ collaboration with customers • Marketing measures	• Out-of-stock situation for green products • Sharing sustainability-related goals with customers • Number of new customers for green products

process (AHP) model. This framework for performance assessment is hierarchically divided into three levels—main criteria (or performance indicators), sub-criteria (sub-indicators under each indicator), and measures (along with scale of measurement). The uniqueness of this proposed approach as shown in Table 5.2 is that all indicators or criteria are actually quantifiable. Further, this has a focus on organizational operations in managing the manufacturing and related processes with least importance to social dimension. Six main criteria (or indicators) are the main representation of the performance of any organization. The main indicators are environmental performance, financial performance, organizational overall performance, and operations, logistics, and marketing performance. Actually, each sub-criterion gives rise to several optional measures and management may select any of them based on its suitability in a given situation. Here, in order to simplify the presentation, we are including only one measure for each sub-criterion.

> *The common set of GSCM practices, which are primarily practiced world-wide, are internal environmental management practices, eco-design of products and processes, internal and external green logistics, green procurement, reverse logistics, ISO 14000 certification, reduction of input materials and energy consumption, and limit of waste generation.*

5.1.2.2.1.6 General Conclusions from Studies on GSCM Practices and Performance Measures The following is some of the general comments and learnings on GSCM practices and performance measurements due to sustainability in operations.

1. Five types of management actions are required for implementing GSCM.

   - Developing internal management practices within the organization for managing environment in the form of strategies, planning, and control
   - Activating suppliers' management for green or environment-friendly sourcing
   - Managing customer involvement in product design, cleaner production, and packaging to make the activities greener
   - Considering eco-design principles in product design
   - Making the process more cost-effective by lean production process, reselling scrap materials, and reusing items (applying principles of product recovery management like recycling, remanufacturing, refurbishing, etc.)

   Further, there should be cooperation and involvement of external stakeholders in all processes of supply chain. For example, new product design should have activities involving eco-design, green procurement, and support of suppliers and customers in planning and control of sourcing, production, and transportation processes.

2. It is advisable that manufacturing firms should first initiate internal environmental practices and subsequently motivate suppliers to incorporate greenness in both production and supplies of their inputs. Thus, green sourcing will be activated in the manufacturing plant. Then suggestions from customers may also be assessed for eco-design and packaging of the products. Customers' involvement is more important in case of globalized manufacturing organizations and in B2B business model. Moreover, GSCM will only be completed once the organization introduces the policies of reselling scrap or extra items and reusing, recycling, or remanufacturing products or components. This is product recovery process. This also enables generation of additional financial revenue by the organization.

3. Primarily there are three types of performance metrics of the organization or the supply chain as a whole, which are expected to reflect the impacts of GSCM practices.

   • Environmental performance is to be measured by reduction of various types of pollutants, better material conservation (particularly for scarce materials), and reduction of using hazardous material. These are the direct outcomes of GSCM practices.

   • Operational performance shows the capability of the organization in the efficient production of better-quality products for selling at lower prices. This performance simultaneously reflects how sustainable the operational processes (production, transportation, etc.) are.

   • Financial performance may be measured by cost savings due to leaner production process and lower material consumption. These savings are due to implementation of sustainability-oriented GSCM practices. It may also be the gain in increased sales revenue from green-conscious market. Financial outcome may often be considered as the net result of other two performance dimensions.

   • There may be the fourth type of performance, which is social performance. It shows the social responsibility and responsiveness of the organization. These include organization's policies relating to expanding societal relationships and contribution. Often this performance causes better image building, improvement in financial outcome and it even drives for better environmental performances

4. It is often found that the larger organizations with higher value of corporate brand image are more active in implementing GSCM practices. The reasons perhaps are their prominent existence in industry, importance of maintaining corporate image, and huge risk of loss of brand value. Moreover, the organizations are supposed to have relatively better infrastructure and reserve funds to initiate new management practices. In other words, it is because of larger implications of normative enablers on this organization and also the fact that the effects of coercive drivers like governmental mandatory directives are more visibly and prominently traced, these organizations are keen to implement management practices involving sustainability.

5. There should be focus on scrap and waste management along with the reuse or recycling of these items. Handling used products or components seems to represent an extended set of economic activities in addition to managing the business of primary manufacturing activities. The main drivers for these activities are cost or financial benefits, reduction of pollution by reducing load of manufacturing new products, delaying disposals, value recovery, corporate image, and social responsiveness or responsibility.

6. The most significant benefits in terms of organizational performance are primarily reflected on environmental and financial performance measures. The immediate impacts of GSCM can be measured directly on the reduction of pollutant emissions, better efficiency of facilities, energy use, and decreased use of materials in designing products. In fact, better financial

performance satisfies the internal stakeholders (shareholders), and environmental performance upgrades the condition of external stakeholders (eco-system and society).

7. Implementation of internal environmental management practices essentially requires support of a higher management level, as they mostly represent corporate policies and strategies, which are subsequently followed by corresponding plans and schedules of activities applicable for middle and lower management levels. Although we are discussing five types of different efforts for implementing GSCM separately, most of them are interdependent. Internal environmental management practices control most of the remaining four types of GSCM efforts. Green purchasing and eco-design are closely related. The outcome of customers' interactions surely contributes to the eco-designing of the products. On the other hand, performance measures are also similarly connected among themselves. For example, eco-design with lean design and implementation of lean processes leads to better utilization of process capacity, or better operational performance, which ultimately results in cost savings or, in other words, financial gains. If organizations perform well in environmental management (i.e., by reducing pollution and avoiding the use of hazardous items and waste generation), it surely saves the fines or taxes relating to pollution issues and the input consumption in manufacturing process, and thus, there is financial improvement.

8. Financial performance is the most powerful driver for implementing sustainability in operational processes or supply chain management. Most of the manufacturers can be motivated, if all types of resulting performance measures are translated into financial benefits. Of course, the firms are compelled to implement environment-friendly practices, if government enacts any mandatory act, law, or rule on sustainability. It may further be noted that very often financial betterment may not be achieved in the short term as a result of GSCM practices. Normally, financial gains may be obtained only after long time span in future and this may be prominent, if the customers are green-conscious. The implementation of the lean manufacturing system is quite popular among manufacturers, particularly in assembly lines like in the automobile industry. But this is primarily to achieve the goals of cost reduction or efficiency improvement, not for intensifying sustainability.

9. Experts opine that if the manufacturers attempt to obtain ISO 140001 certification of an environmental management system (EMS) and maintain it, they are also expected to perform well in cost reduction, reduction of wastes, and shortening of lead times. This EMS certification can never be achieved without implementation of GSCM practices, and it also demands environmental management practices among their suppliers.

Key Learning

- Survival and growth of any business unit depend on aligning the performance with the strategic goal(s) and subsequently monitoring and controlling the activities per the performance management system of the organization.

- Performance management and related activities may be considered in two levels—long-term and short-term.
- The sustainability performance management (SPM) is a little more complex than traditional performance management system because of its inclusion of impacts on environment and society, the essential components of sustainability.
- Eight critical factors have been identified from an empirical study taking inputs from the Indian business sector, which affect the management of sustainable operations. Further, the study reveals that the performance measurement practice is one of the three factors representing the ultimate impact of other critical factors.
- Sustainability-based performance may be explained in two different ways— first, by identifying the appropriate performance indicators and, second, by studying the green business practices prevailing in various industrial sectors.
- Using the basic three dimensions under 3Ps or TBL seems to be the most popular performance indicators practiced by business units, which are practicing sustainability. However, integration of these dimensions is still an unsolved problem.
- Performance indicators may also be classified as managerial performance, operational performance, and their impact on environment. Other indicators may also be identified as the performance against three types of drivers: normative drivers, coercive drivers, and mimetic drivers.
- On the other hand, industrial executives in real life follow various green practices. The role of GSCM practices and their performance have been elaborately explained taking sectoral examples in the US, Indian, and Brazilian automobile sectors and the Chinese and Japanese industrial sectors.
- Because of multidimensional nature of various parameters involved in the performance management of green operations, it is also possible to capture the relevant parameters in a hierarchical framework. Two such frameworks are shown here. The first one maps the SDG goals from Agenda 2030 (in other words, TBL) to operations target level through corresponding KPIs and measurements. The second framework models the performance assessment by a three-level structure originating from six main criteria of assessing the sustainability performance.

5.2 POPULAR PRACTICES FOR ENVIRONMENTAL PERFORMANCE MANAGEMENT

As sustainability is deepening its footprint in industrial management across the globe, several tools and practices are gaining popularity among them for measuring the performance of sustainable operations. Industrial executives of this new world order are supposed to be aware of these practices. However, these are somewhat different in terms of their use. Some are *reporting system in a predefined performance framework*, whereas others are *single index-based approach, scorecard-based approach*, and *process-based* or *system-based approaches*. The following is the list of some such popular practices.

* Global Reporting Initiative (GRI)
* ISO 14000 Environmental Management Standards
* Dow Jones Sustainability Indices
* Air quality index
* Environmental impact assessment (EIA)
* EFQM model
* Composite index on sustainable development
* Finance-focused performance index

5.2.1 GLOBAL REPORTING INITIATIVE (GRI)

The GRI may not be primarily treated as a system of performance management of sustainability. The primary application of GRI is global reporting. This is done using world's most widely used standards for sustainability reporting, known as GRI Standards. The GRI is an independent international standards organization formed with the support of the United Nations Environmental Program (UNEP). It enables any organization (large or small, private or public, or from a sector of any type) to understand and report the impact of its business activities on the environment, people, and economy of the organization. The standards are prepared and reported in a universally accepted common language and in a comparable and credible manner, adding transparency to the declaration. Naturally, these reports become quite relevant and meaningful to all possible stakeholders, like investors, governmental agencies, stock markets, policy-makers, and the society at large. It may also be treated as a channel of communication with the outside world on sustainability issues. GRI reporting means that the organization is taking the full responsibility of the authenticity of report and the impact of the business activities of the organization. The GRI's headquarter is in Amsterdam, the Netherlands. The mission of GRI is to enable organizations to become transparent and responsible for all the activities having impacts on sustainability dimensions—people, planet, and profit. The standards do cover various related topics ranging from anti-corruption to water management, biodiversity to occupational health and safety, taxes to emissions. Incidentally, GRI reporting is not mandatory to date. The Global Sustainability Standards Board (GSSB) has the sole responsibility for setting the world's first globally accepted standards for sustainability reporting (i.e., GRI Standards). GSSB members are representatives of various domains of expertise relevant to sustainability and the standards incorporate perspectives of all related stakeholders.

The following benefits may be derived from proper and systematic GRI reporting.

* Better sustainability performance
* Better risk management
* Improved involvement and relations with stakeholders

- Improved impression and credibility in market as a committed and responsible corporate entity
- Better sustainability goals and more effective goal achievement strategies
- Improved performance management for sustainability

GRI Standards were first developed in 1997, which came to light in full version in 2000. After several updates, the latest version of the GRI Standards of 2016 is being used now. These standards were made by a non-profit organization named as CERES (Coalitions for Environmentally Responsible Economies). The GRI Standards allow third parties to assess the activities and impacts of business sustainability. The GRI Standards are produced in accordance with the guidelines included in ISO 14000, ISO 26000, and OHSAS 18001 for health and safety.

GRI 100 series are considered to be applicable universally for all organizations for preparing the sustainability reports.

GRI Standards 101 is the first step in preparation for all standards, considered as Foundation 2016 reporting.

GRI Standards 102 includes an overview of the organization and its activities. It is thus the standards of General Disclosure 2016 report. The organization is to disclose the name, ownership, market being served, scale of operations, and information on its supply chain.

GRI Standards 103 is meant for reporting on how the organization is managing the generation of economic, environmental and social impacts as a result of its activities and operations. It is known as Management Approach 2016 reporting. The organization is to explain and declare the management approaches on each area of management (material topic), including policies, responsibility, resources, and goals in detail along with the evaluation of each approach and its impacts.

After reporting based on universal GRI 100 series, each organization is to disclose the activities and impacts on each material topic through GRI 200 (economic), GRI 300 (environmental), and GRI 400 (social) Standards.

We must note that GRI Standards represent a general framework for depiction and declaration of business activities and impacts on sustainability in a universally acceptable and standardized manner. It is also a form of reporting to others and accepting the responsibility of impacts (good or bad).

In the recent past, the topics of tax, human health and safety, and water management had been included in GRI standards as extensions on the three dimensions of the 3Ps. These had been launched in 2018 and 2019.

5.2.1.1   GRI 207: Tax 2019

This enables firms declare the tax payments along with the management approaches relevant to that. This standard supports maintenance of transparency on tax contribution of the firm to the national economy and society at large.

5.2.1.2   GRI 403: Occupational Health and Safety (OHS) 2018

This additional standard takes care of reporting on OHS-related events and OHS management approaches, like policies for implementing preventive measures and promoting health among workers. Its role has become very significant during last three to four years because of the COVID-19 pandemic.

5.2.1.3   GRI 303: Water and Effluent 2018

This standard represents the policy, strategy, and planning for water management along with its consumption, discharge, and reuse in addition to the impacts of this water treatment on the local communities.

Readers can visit the website www.globalreporting.org to get some new information on revised Universal Standards. Actually, since its launching in 2016, the GRI reporting system experienced many modifications from time to time. Here, let us discuss on some revisions of the GRI Standards on its structure and content. This is primarily to incorporate the forward-looking policy of GRI, which reflects the emerging trends in managing sustainability globally. Interested readers may also read "a short introduction to the GRI Standards," available in the same website of global reporting. The details of revised Universal Standards and guidance for GRI reporting are now freely available for downloading. It is expected to be applicable for reporting from 1 January 2023. This, in fact, justifies the forward-looking policy of GRI.

Revised GRI Standards is a modular system of interconnected standards. The GRI Standards comprise three series of standards: the GRI Universal Standards, the GRI Sector Standards, the GRI Topic Standards. Each standard actually contains the format of **disclosures**. Disclosures do include **requirements** (must report or disclose) and **recommendations** (relevant information but not necessarily to be reported). Figure 5.3 exhibits an exemplified framework of this revised GRI Standards. It consists of the following levels of standards.

5.2.1.4   GRI Universal Standards

These standards are applicable for all organizations and the following three standards are included under this series.

- *GRI 1, or Foundation 2016*: As the foundation, this standard (like the GRI Standard 101 of 2016) outlines the purpose and explains the concepts and use of GRI Standards. Disclosure requirements are elaborated under this standard.
- *GRI 2, or General Disclosures 2016*: It includes the general disclosures on structure, activities, and policies of the organization, giving fairly good idea on its profile and expectations on reporting.
- *GRI 3, or Material Topics 2016*: This explains the steps to be followed to specifically report on activities relating to all relevant material topics and their impacts. The impact on various materials topics may be understood once the Sector Standards are considered under this broad category.

5.2.1.5   GRI Sector Standards

This class of standards is expected to identify the specific disclosure requirements unique for an industrial sector. Incidentally, the GRI Sector Standards are not yet

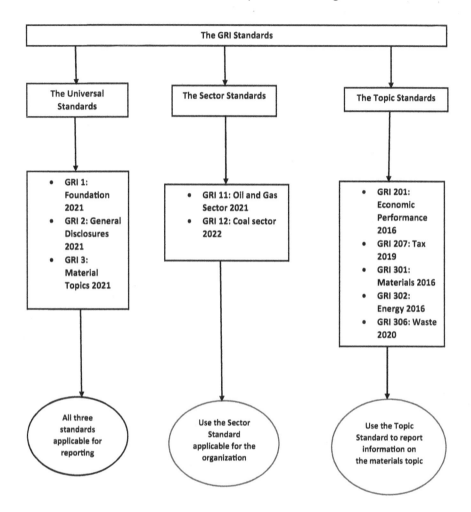

FIGURE 5.3   A sample framework of Revised GRI Standards.

fully developed and available for reporting by organizations. Because of its proactive policy, GRI is planning to develop 40 sector standards that prioritize the impact on environment and people. The prioritized sectors are oil and gas, agriculture, coal, mining, aquaculture, and fishing. Each sector standard shows the sector's characteristics and its activities in terms of their impacts. The disclosures under this sector should match with the requirements under each relevant material topic.

5.2.1.6   GRI Topic Standards

Under this set of standards, we find the GRI guidance in reporting the relevant information in detail for material topics of interest. Various topics have been added over the time to enrich GRI reporting with detailed guidelines. The material topics under this set of standards include occupational health and safety, tax, energy, and economic performance. Topics may have varied levels of importance for different sectors.

GRI Standard reports may be published in electronic or paper-based format and may be accessible to others. So it may be included in webpages, annual reports, or business journals for accessibility to various stakeholders. It is further expected that there should be a **GRI content index** in the report for better information, traceability, credibility, and transparency.

However, it cannot logically and effectively integrate all dimensions for the development of a policy or strategy acceptable to all stakeholders, nor can it provide any approach for improvement of the activities toward some desired direction.

It may further be noted that GRI is a mode of declaration, disclosure, and reporting through globally acceptable formats. The businesses are thus taking ownership and responsibility of the consequences of the business processes operated and managed by them.

> *The Global Reporting Initiative (GRI) is a system of sustainability reporting accessible to the globe, based on the standards set by the Global Sustainability Standards Board (GSSB). The GRI was established with the support of United Nations Environmental Program (UNEP) in 1997.*

5.2.2   ISO 14000 Environmental Management System (EMS)

Out of so many other sustainability issues, like people issues, depletion of non-renewable natural resources, and water conservation, many organizations today are mostly concerned about three main environmental issues: pollution or waste generation, climate change or global warming, and ozone depletion.

An environmental management system (EMS) may be defined and understood in different ways primarily depending on whose perspective is being considered during implementation of the EMS. If it is from an organizational perspective, then this may be treated as an important component of the overall management system. EMS thus gets influenced by the organizational structure, processes, resources, practices, planning, and procedure while developing, implementing, reviewing, monitoring, and maintaining the environmental policy and corporate policies of the organization. The implementation and effective management of an EMS is expected to reduce the negative impact of organizational activities on the environment. The driving forces for integrating EMS with the management system of an organization may be coercive or mandatory because of existing regulatory pressure from the government or from external stakeholders. It may also be discretionary, which often ultimately results in economic benefit or increase in brand value.

ISO 14000 and its series offer various globally recognized ways to support organizations implement EMS. The standards under ISO 14000 may be broadly classified into three classes: organizational evaluation (e.g., ISO 14001, ISO 14004), environmental auditing (e.g., ISO 14010, ISO 14011, ISO 14012, ISO 14030, ISO 14031), and product evaluation (e.g., ISO 14040, ISO 14020, ISO 14060).

ISO technical committee ISO/TC 207 is primarily responsible for developing ISO 14000 family of standards, which is meant for environmental management. To date, by and large, 21 international standards have been developed under this family, and it is still evolving. The committee was established in 1993 to work on sustainable development standards and related documents. It has close relations with more than 30 international organizations for consultation during developing and updating the standards. Moreover, during this process, the ISO/TC 207 consulted closely with ISO/TC 176—that is, the technical committee responsible for developing the ISO 9000 family of quality management standards.

Under the ISO 14000 series, ISO 14001 is the most recognized and popular framework for the **environmental management system**. It helps in guiding the organization for better overall environmental management. It has been implemented by more than half of the ISO members. ISO 14004 provides **additional guidelines and explanation** in EMS implementation. An auditing standard such as ISO 19011 is equally applicable in EMS for **environmental auditing**. ISO 14031 helps in **evaluating environmental performance** of organizations. The standards provide supports in selecting suitable performance indicators so that actual performance may be assessed based on these criteria. ISO 14020 addresses various means for **environmental labelling and declarations**. ISO 14040 provides relevant guidelines on the **concepts, principles, and the steps for application of life cycle assessment (LCA)**. ISO 14063 is meant for **environmental communication** with guidelines for making relevant links to external stakeholders. ISO Guide 64 guides the designers and manufacturers on **environmental aspects relating to product standards**.

As the standards are evolving, new standards get published in every year. The website www.iso.org provides new updates. On the basis of ISO 14000 family 2009, a new standard like ISO 14045 is going to provide guidelines and principles on **eco-efficiency assessment**. In fact, standards of this type do focus on specific assessment tools. The ISO 14051 standard includes principles and framework of **material flow cost accounting (MFCA)**. MFCA is a management tool, which may be used to achieve better resource and energy utilization by reducing the cost associated with their consumption. MFCA critically analyzes the manufacturing and logistics processes and measures the flow and inventory of both materials and energy based on physical units. Subsequently it is converted to the essential component of production cost. Actually, this contributes to **environmental management accounting (EMA)**. ISO 14064, ISO 14067, and ISO 14069 together provide valuable guidance in quantification and calculation of carbon emissions, GHG release, and their reporting. ISO 14064 parts 1, 2, and 3 are international **greenhouse gas accounting and verification standards**. ISO 14065 standards are meant for the **accreditation** of organizational bodies that may undertake **greenhouse gas validation and verification**. ISO 14067 and ISO 14064 help with **carbon footprint assessment, study, and reporting, including GHG accounting** for products, services, and supply chain.

ISO 14005 provides guidelines for the **phased implementation of EMS** to facilitate the SME sector. ISO 14006 will provide guidelines on **eco-design**. ISO 14033 guides in **communicating environmental information** effectively. ISO 14066 will specify the **competency requirements for GHG validators and verifiers**.

It is once again reiterated that because of its nature of continuous development, interested readers may consult the ISO webpage to get the up-to-date program of standards developed by ISO/TC 207. The ISO 14000 series is developed for execution of EMS-related standards following the principles of plan-do-check-act (PDCA) cycle, which is the core principle of quality management system of ISO 9000.

Although ISO 14000 certification reflects the green consciousness of an organization and offers guidelines for EMS, it cannot directly confirm improvement of environmental performance of the organization, neither can it be treated as a benchmarking approach for assessing the gap of performance. An organization is to develop its own performance management system for achieving sustainability using the set of criteria matching with its strategic goals and existing systems and processes. However, the guidelines of ISO 14000 standards immensely help in establishing the EMS and relevant strategies, plans, and control measures.

> *The family of international standards under ISO 14000 covers standards on organizational evaluation, environmental auditing, and product evaluation relating to environmental management. It offers guidelines and supports to organizations to implement an environmental management system (EMS) and organizations accept its role in improving environmental performance.*

5.2.3 Dow Jones Sustainability Indices (DJSI)

The Dow Jones Industrial Average (DJIA) is a widely followed index used for assessing blue-chip stocks of US. DJIA is price-weighted index that tracks some large publicly owned companies trading in New York Stock Exchange and NASDAQ. Similarly, Dow Jones Sustainability Indices (DJSI) are float-adjusted market capitalization weighted indices that measure the performance of companies selected with ESG (environment, social, and governance) criteria using best-in-class approach. As these are the companies listed in the stock exchange, the economic criterion is automatically taken care of. The DJSI are a set of indices (family of indices) meant to be used by prospective investors, who recognize and value sustainable business practices. The DJSI allow the creation of portfolios of companies that fulfill certain sustainability criteria better than majority of their peers within a given industry. This family of DJSI was first launched in 1999 as the first universal benchmark of sustainability. DJSI are meant for sustainability-conscious investors, representing the top 10% of the largest 2,500 stocks of S&P Global Broad Market Index on the basis of sustainability criteria. For example, the top three holdings of the DJSI in September 2019 were Microsoft, Johnson & Johnson, and Visa.

The DJSI family comprises global, regional, and national benchmarks like DJSI World, DJSI North America, DJSI Europe, DJSI Asia Pacific, and DJSI Korea. Interested readers may refer to the DJSI website, www.spglobal.com/esg/performances/indices/djsi-index-family. The DJSI World index consists of top companies on sustainability practices worldwide, classified under various industry groups like automobiles and components, banks, capital goods, consumer durables and apparel, consumer services, energy, and transportation. Based on documents on DJSI methodology (www.spglobal.com/spdji/en/documents/methodologies/methodology-dj-sustainability-indices.pdf) of January 2021, the current geographical classification of DJSI may be expressed in two levels. In the first level, it is divided into DJSI World (with the scope of total world), DJSI Regions (some specific segments of the globe), and DJSI Countries (some specific countries). DJSI World is further classified in the second level into Dow Jones Sustainability World, World Enlarged, and Emerging Markets. DJSI Regions include DJSI Asia/Pacific, Europe, and North America, whereas DJSI Countries include Dow Jones Sustainability Australia, Korea, Korea Capped 25%, and Chile. India is allocated by benchmarks in all the three indices of DJSI World (i.e., in ESG).

The S&P Global Corporate Sustainability Assessment (CSA) for any company is meant for getting insights on its sustainability performance compared to peer companies. The DJSI index of a company is S&P Global Environment, Social, and Governance (ESG) score calculated by SAM's annual CSA. Usually, the CSA process is initiated in April each year, and new scores are released in September. The result of the annual CSA process is the creation of an "assessed universe" for each member of the DJSI family. Indians would be happy to know that Dow Jones Sustainability Indices 2020 rank Hindalco Industries Limited as the world's most sustainable aluminum company. Among 61industries, Hindalco is the sole Indian company recognized as an "Industry Leader." Hindalco, the metals flagship of Aditya Birla Group, got the total score of 75 points in ESG based on DSSI CSA against the industry average of 51. Its excellence has been reflected in the environment, social, and governance (ESG) dimensions. The ESG performance criteria include customer relationship management, human capital development, climate strategies, biodiversity, environment and social reporting, and water-related issues. In 2019, its score in DJSI Emerging Markets was only 60 points, with its third rank globally under aluminum industry category. Thus, there is considerable improvement during one year in sustainability-related activities.

> *Dow Jones Sustainability Indices (DJSI) are float-adjusted market capitalization weighted indices that measure the performance of companies with ESG (environment, social, and governance) criteria using best-in-class approach.*

5.2.4 AIR QUALITY INDEX (AQI)

Among several indices available as the performance indicators of sustainability, the Environmental Performance Index (EPI) seems to be one of the globally recognized indices to gauge the performance of environmental sustainability being maintained by a country (epi.yale.edu). The first EPI score is computed and then the rank of the country is determined out of 180 countries. The EPI score is quantitative assessment under two main aspects—health (health hazards due to poor environmental condition) and ecosystem vitality. The EPI research team carries out the study to offer a scorecard to each country in terms of environmental performance, which further provides the practical guidance to achieve the sustainability targets in future. The EPI is computed primarily on 11 issue categories, which is further spread to 32 performance indicators. Air quality is one of the categories (under the health aspect) comprising three performance criteria—household solid fuel, $PM_{2.5}$ exposure, and ozone exposure.

Per EPI 2020, India ranks 168 (out of 180), with EPI score 27.6 (100 is the best). In terms of air quality, India's ranking is worse (179) and its EPI score for this category is only 13.4. The EPI research team works both in Yale and Columbia, and interested readers may read the relevant literature (Wendling et al., 2020) for detailed information on EPI. However, we are not going to elaborate more on EPI, but the discussion here will be focused on AQI, which is a more popular sustainability index, particularly in India. However, we should not forget that air quality is an important component of environmental sustainability, and here lies the relevance of initiating discussion with EPI.

The air quality index (AQI) is an index meant for measuring air quality only, not any other environmental or sustainability parameters. It reflects the harmful content of the polluted atmosphere with the primary objective of warning people about the danger and helping government or local administration in taking corrective measures. It may be treated as an effective tool of communication to people about the current air quality through monitoring and updating air quality of a geographical area or region. AQI was first formulated by the US Environmental Protection Agency (EPA) in the 1970s to check whether it is safe to breathe the air or not. Soon it became popular in various countries and cities around the world. The national air quality index in India was launched on 17 September 2014 in New Delhi under the Swachh Bharat Abhiyan. It aggregates complex air quality data of various pollutants and translates it to a single number, known as AQI. It includes the following eight pollutants.

1. Particulate matter (size less than 10 μm) or PM_{10}
2. Particulate matter (size less than 2.5 μm) or $PM_{2.5}$
3. Nitrogen dioxide (NO_2)
4. Ozone (O_3)
5. Carbon monoxide (CO)
6. Sulfur dioxide (SO_2)
7. Ammonia (NH_3)
8. Lead (Pb)

The website of the Central Pollution Control Board (www.cpcb.nic.in/National-Air-Quality-Index) includes all relevant information on AQI. The AQI in India is calculated based on the average concentration of a particular pollutant measured over a standard time interval (24 hours for most of the pollutants, eight hours for carbon monoxide and ozone). The air quality data are collected and the AQI is calculated by CPCB in India.

Actually, the effect on human health depends not only on the concentration of harmful pollutants in air but also on the duration of exposure. The overall AQI for a full day may be worse than each individual hourly air quality index. Average

TABLE 5.3

AQI Ranges and Categorization Based on Pollutant Measurements or Their Concentration Ranges

AQI Category	AQI Score Range	PM$_{10}$	PM$_{2.5}$	NO$_2$	O$_3$	CO	SO$_2$	NH$_3$	Pb
Good	0–50	0–50	0–30	0–40	0–50	0–1.0	0–40	0–200	0–0.5
Satisfactory	51–100	51–100	31–60	41–80	51–100	1.1–2.0	41–80	201–400	0.6–1.0
Moderately polluted	101–200	101–250	61–90	8oo1–180	101–168	2.1–10	81–380	401–800	1.1–2.0
Poor	201–300	251–350	91–120	181–280	169–208	11–17	381–800	801–1200	2–1–3.0
Very poor	301–400	350–430	121–250	281–400	209–748	18–34	801–1600	1201–1800	3.1–3.5
Severe	401–500	430+	250+	400+	748+	34+	1600+	1800+	3.5+

Note:
1. CO is measured in mg/m^3 and other pollutants in μg/m^3
2. 2-hour average values for PM$_{10}$, PM$_{2.5}$, NO$_2$, SO$_2$, NH$_3$, and Pb and 8-hour average values for CO and O$_3$.

measurement benchmarks for one-hour, eight-hour, and 24-hour intervals provide more accurate air quality data, which in turn allows people, the local administration, and the government to take better, corrective health and safety decisions. The national AQI of India ultimately computes a single value as overall air quality and matches it with a range of values in six AQI categories to declare the air quality of the region. These six categories are good, satisfactory, moderately polluted, poor, very poor, and severe, in order of best to worst air quality. Table 5.3 displays the categories of AQI showing the range of the scores for each category along with those for their components meant for individual pollutants.

The AQI computation is made with the input data of ambient concentration values of air pollutants and their likely health impacts (termed as health breakpoints). At the first stage, air quality sub-index is computed for each pollutant. This AQI sub-index is a linear function of various concentrations. For example, the PM$_{2.5}$ sub-index may be computed with 51 at the concentration of 31 µg/m^3, 75 at the concentration of 45 µg/m^3, and 100 at the concentration of 60 µg/m^3. The linear combination of various concentrations is the category of a sub-index. The category of the worst sub-index will show the AQI category.

So long as the AQI is below 50 (i.e., good), it may be concluded that during exposure to outside environment, people are breathing fresh or clean air. The region is quite safe, and people may spend more time outside and stay active. The following risks are expected for other categories of AQI.

Satisfactory (51–100): Minor breathing discomfort is expected among people, particularly the sensitive ones.
Moderately polluted (101–200): Breathing discomfort is expected among people with lung conditions (e.g., asthma) and heart diseases.
Poor (201–300): Breathing discomfort is expected among most people with prolonged exposure.

Very poor (301–400): Respiratory illness is possible for most people, demanding immediate consultation with physicians.

Severe (401–500): The air is harmful even to healthy people, and there are serious impacts to people with existing diseases, which demands immediate
consultation with physicians and hospitalization, if required.

In this context, it is to be mentioned that the toxic level of air pollution in and around
Delhi has been creating quite a menace during last three to four years, with AQI
records hovering around poor, very poor, or severe. The following are some of the
reasons, which cause generation of this dangerous air quality in Delhi and NCR.

1. Farmers burn rice stubbles in Punjab, Haryana, and Uttar Pradesh, the
   immediate neighboring states of Delhi. The wind carries all the pollutants
   (mostly particulate matters—i.e., $PM_{2.5}$ and/or PM_{10} in this case), which
   mix intensely with the air of the Indian capital.
2. Vehicular emission is another primary cause of air pollution and smog. The
   CPCB also declared that traffic emission is the main contributor to air pollution of this region. This emission primarily causes generation of NO_2 and CO.
3. During the winter season or pre-winter period, because of stagnant wind, the
   dust particles and pollutants in the air do not move in the air, creating smog.
4. Large-scale construction in Delhi and NCR is another culprit for major dust
   (PM) and air pollution. This is also because of increasing population in
   Delhi, Noida, Gurgaon, and so on.
5. Industrial pollution (e.g., brickkilns, coal-based power plants) and waste
   generation due to industrialization also create garbage dump. These
   immensely contributes to air pollution. These significantly emit SO_2 to air.

As an urban entity, New Delhi seems to create highest ambient particulate matter
among other cities of India. When annual average $PM_{2.5}$ across India was around
58 µg/m³, Delhi's average was recorded near 98 µg/m³. Interestingly, China, the most
populated country, with huge growth potential in industrialization and urbanization,
has improved its air quality a lot in recent years. Its air quality is much better now.

Incidentally, the nationwide lockdown significantly helped in improving air quality in India. The first countrywide lockdown was announced on 20 March 2020 for
21 days due to COVID-19. Immediately there were drastic reduction or elimination
of industrial activities, civil constructions, and movements on the road. This has
its impact on air quality. The $PM_{2.5}$ concentration in Kolkata and Delhi reduced by
34.52% and 27.57%, respectively. Similarly, in Mumbai, Chennai, and Hyderabad,
the $PM_{2.5}$ concentration reduced by 19.28%, 5.4%, and 3.99%, respectively (Singh
and Chauhan, 2020).

This is quite visible from real time AQI data (from AQI_Bulletin_20220505.pdf
of webpage cpcb.nic.in accessed on 6 May 2022) of some Indian cities as snapshots.

Good—Shillong, Srinagar, Agartala, Guwahati
Satisfactory—Varanasi, Patna, Kolkata, Kozhikode, Hyderabad, Chennai,
Bangaluru

Moderate—Udaipur, Visakhapatnam, Ujjain, Mumbai, Lucknow, Nagpur, Ghaziabad, Delhi, Kanpur

Poor—Jodhpur

These data show that the air quality of most of the towns or cities are either moderate or satisfactory, which means that the situation is not very dangerous now, and there may be only few cases of minor breathing discomfort. This is clearly an impact of restrictions on business activities and movements and closing of restaurants, malls, and other establishments due to the pandemic. Thanks to COVID-19, only a couple of cases are showing poor air quality, including the city like Jodhpur, where there is expected to be breathing issues in case of prolonged exposure to the environment. The good sign of the restrictive industrial and logistics activities is quite visible with many examples of cities with good air quality. But incidentally, most of them are from North East region of India, which, in general, provides Indians a relatively cleaner environment.

It may be noted that Delhi experienced relatively cleaner air (satisfactory) during the COVID-19 lockdown period (particularly during first three phases, March–May 2020) as industrial activities almost came to a halt and traffic movements declined drastically. However, the air quality of Delhi again started deteriorating as lockdown was withdrawn, sometime in October 2020. Ironically, because of the existence of this viral infection in varied intensity, people are breathing clean air and are required to wear a mask.

Now I would like to draw the attention of readers to this glaring real-life example of the impact of the conflicting two dimensions of sustainability (TBL), economy and environment.

The air quality index (AQI) is an index meant for measuring air quality, one of the several environmental parameters. It reflects the harmful content of the polluted atmosphere, with the primary objective of warning people about the danger and helping the government or local administration in taking corrective measures. The national air quality index of India was first launched in 2014. Its scores are classified under six categories.

How much sacrifice is the country going to tolerate on economic development in order to improve upon the environment, or vice versa? What is the acceptable trade-off? Is it possible to quantify it optimally? It may be treated as an interesting food for thought.

5.2.5 ENVIRONMENTAL IMPACT ASSESSMENT (EIA), ENVIRONMENTAL MANAGEMENT PLAN (EMP), AND STRATEGIC ENVIRONMENTAL ASSESSMENT (SEA)

An environmental impact assessment (EIA) is broadly conceptualized by United Nations Environment Programme (UNEP) as a process of evaluating the likely

environmental impacts of a proposed project or development plan ready to be implemented, taking into account interrelated socio-economic, cultural, and human health
impacts, both beneficial and adverse.

Its primary aim is to predict the environmental impacts at an early stage in project planning and design. Its subsequent intention is to explore the possible ways and
means to mitigate the adverse impacts and modify or restructure the project accordingly. So EIA enables the project management incorporate possible changes for better
consequences on the surroundings and society, and most probably for avoiding subsequent costs because of environmental disasters. It is essentially not a measurement
of index like AQI, but rather it is the outcome report of a process of assessing the
possible impacts of a proposed project or a plan of development. These impacts cover
physical, economic, socio-economic, human health, and cultural aspects of the surroundings. The impacts may be beneficial (positive) or adverse (negative or harmful)
in nature. In this context, it is to be mentioned that the environment management plan
(EMP) is to be prepared well in time to plan for the mitigation of harmful impacts
before the occurrence of the consequences with optimum utilization of resources. The
EIA needs to be integrated with all the stages of a project, from exploration and planning, through constructions, operations, decommissioning, and beyond site closure.

Till the 1970s there was no systematic or formal process for assessing environmental impacts for any project. Once the National Environment Policy Act (NEPA)
1969 was implemented in the US, the EIA was introduced by the NEPA in 1970 as
a mandatory practice in the US. During the 1970s, guidelines for the mode of public
participations and standard methodologies (tools for impact analysis; e.g., checklist,
table, network) were developed and introduced. Australia, New Zealand, and Canada
started following the footsteps of the NEPA by the mid-1970s. Other industrialized
and also developing countries like France, the Netherlands, the Philippines, Germany, and Colombia started implementing the EIA by the early 1980s. In Europe,
the EIA Directives (85/337/EEC) is in force since 1985 for private and public projects. In 1989 World Bank made EIA a compulsory condition for getting funds.

Globally, various multilateral environmental agreements (MEAs) contributed to
improvements in the EIA's legal, policy, and institutional arrangement. For instance,
in Espoo Convention on Environmental Impact Assessment in a Trans-Boundary
Context (1991), the general obligations on the EIA for vulnerable projects drew
attention of all parties of the globe. Principle 17 of the Rio Declaration (in 1992)
calls for use of the EIA as a national level tool for assessing adverse environmental
impacts of projects. Agenda 21 of the Rio Declaration clearly emphasized the use of
appropriate methodologies for impact assessment of any forthcoming developmental
activity on environment and society.

In India the very necessity of assessing the environmental impacts was first felt
in 1976–1977, when the Planning Commission of India requested the Department
of Science and Technology to examine the environmental impact of the newly proposed river valley projects. This is the first administrative initiative of studying the
environment-related elements in any big project of India. Slowly it gained popularity.
Subsequently, the Public Investment Board insisted on study of environmental issues
for getting its approval. Till 1993, the environmental study used to be an administrative effort, not any mandatary activity with legislative support. In 27 January 1994,
the Ministry of Environment and Forest (MoEF) included the EIA in the process

of Environmental Clearance (EC). The EIA notification became mandatory for the modernization or expansion of any existing or running project or for setting up any new project listed in Schedule 1 of the notification. Although since 1994 more than 12 amendments have been made in the EIA notification of 1994, the EIA is always statutorily supported by the Environment Protection Act 1986. The following are some important milestones on legislative changes of the EIA notification.

1. EIA notification was made in 1994.
2. First amendment on EIA notification was made in 2006.

   - Two categories of projects are declared for decentralization of EC activities for ease of process or quicker clearance.
     Category A: national-level appraisal by the Impact Assessment Agency (IAA) and the Expert Appraisal Committee (EAC)
     Category B: state-level appraisal by the State-Level Environment Impact Assessment Authority (SEIAA) and the State-Level Expert Appraisal Committee (SEAC)
   - Introduction of four stages of the EIA cycle: screening, scoping, public hearing, and appraisal. Category A projects directly move to mandatory environmental clearance by passing screening process. Category B is further classified during screening into B1 (mandatory EIA) and B2 (EIA not required).
   - EC, for some pollution-prone projects, like mining, thermal power plants, river valleys, infrastructure, chemical plants, paper factories, foundries, and electroplating operations, is mandatory.

3. EIA 2006 is further expected to be amended and draft EIA notification 2020 has been prepared and still under scrutiny. The following are some of the proposals included in the draft.

   - Notice period of public hearings was reduced from 30 days to 20 days.
   - Many projects are exempted (B2) from mandatory EC and public scrutiny, like oil, gas, and shale exploration; hydel projects up to 25 MW; and small cement plants.
   - After EC, each project is to adhere to the rules laid down in the EIA and is to submit the compliance report annually.
   - It excludes reporting of violation and non-compliance to the public for public scrutiny.
   - It also proposes post-facto clearance, which allows projects to apply for clearance after it has sufficiently progressed in operations.

Some of the draft proposals of EIA 2020 raise questions on the scope of public consultation and upper hand of bureaucracy.

Although the EIA procedure may vary from country to country, the following are some of the stages normally included in initiating and conducting the EIA process.

1. *Screening*: The EIA is project-specific and is essentially a process-oriented endeavor. This initial step is the selection of the project or developmental work to be assessed fully or partially, keeping in view its criticality in

the organization. Some CBA analysis is necessary, as the EIA needs some funds and time for carrying it out. Sometimes the EIA is required as the situation demands or due to some public pressure. Any stakeholder of the organization may also ask for the EIA for some reason. Of course, if it is required for EC, this step is redundant.

2. *Scoping*: This includes focus areas for assessing the environmental impacts that are relevant to the EIA because of legislative compulsion, international convention, expert opinion or advice, or public or societal pressure. This also shows the terms of reference for carrying out this assessment along with the boundary of the area to be covered and the time limit of the study.

3. *Assessment and evaluation of impacts*: In this stage of the EIA, the likely environmental and social impacts (both beneficial and adverse or harmful) of the proposed project are identified and predicted. Subsequently, the significance of these impacts is evaluated at the local, national, and global levels.

4. *Mitigation of adverse impacts*: Under this stage of the EIA, actions are recommended to reduce severity, mitigate, or if possible, eliminate the potentially adverse consequences (both environmental and social) of the project or the developmental activities under study.

5. *Reporting*: The outcome of the EIA or environmental impact statement (EIS) is communicated in an appropriate format to the persons or offices concerned or to EC.

6. *Review*: During this stage, the EIA report is examined, keeping in view the purpose and scope (from scoping) of the assessment. Here there may be participation from outside agencies or people, if necessary.

7. *Decision-making*: This includes the final decision on approval of the project—that is, whether the project is approved, rejected, or to be modified, bringing in some changes in resources, facilities, or processes.

8. *Post-commissioning*: Once the project is approved and commissioned, there should be required checks and balances through monitoring. It is to be ensured that the impacts should not exceed the permitted limits or violate the legal standard norms. Mitigations should be assessed and relevant control measures should be implemented during the whole project life cycle.

In this context, it is advised that the EIA report should be submitted during the feasibility study stage of project management or project life cycle. This will enable any recommendation on modification of the project plan to be incorporated at the early stage of the project life cycle. Otherwise, it will increase environmental costs or will delay the project implementation. In fact, the EIA process should be integrated with project management cycle for better effectiveness and overall result.

The International Institute for Sustainable Development (IISD), located at Winnipeg, Canada, provides a lot of learning materials on EIA. Its contribution on impact assessment methods is quite noteworthy (iisd.org/learning/eia/eia-7-steps/step-3-impact-assessment-and-mitigation, retrieved on 12 May 2022). The following assessment methods either singly or in combination may be applied in the EIA process.

5.2.5.1   Expert Judgment

This method intends to carry out the assessment on the basis of domain (preferably multidisciplinary) expertise on environmental parameters (e.g., water, soil, air, biodiversity, and communities) and also on technicalities of the project. The method may be applicable, when sufficient data are not available for quantitative analysis, and of course, predictive analysis may not be feasible in that case. Expert judgment may also be used as a complementary method during final interpretation of the impact(s).

5.2.5.2   Quantitative models

Here quantitative models are developed for impact predictions formulating the cause-effect relations. The development of air dispersion models for predicting emissions and pollution concentration at some locations because of installation of a coal-fired power plant may be a typical example of this method. Another example may be ecological models to predict the existence of aquatic lives resulting from discharge of toxic materials to river. Here by aquatic lives, I mean microorganisms, plants and animals living, growing, and often found in water bodies. Social impacts may also be predicted in the same manner, but it is relatively more complex.

5.2.5.3   Cumulative Impact Assessment

Most often, environmental and social impacts from multiple sources lead to creation of a cumulative effect having more serious implications to surroundings and people. Examples of situations covered by this type of assessment include increase in pollution concentrations in a water body having outlets from more than one factory, reduction of waterflow in a watershed due to multiple withdrawals, and depletion of a forest because of several industrial activities in an area. As these situations often involve multiple stakeholders, intervention of governmental agencies may be required to conduct a comprehensive study on cumulative environmental impact and its serious implication.

5.2.5.4   Matrices and Interaction Diagrams

These are the common tools to represent the overall impact by matrices or tables. Rows may be considered as the stages of a project life, like development, operations, and closure, whereas the columns are all the elements of environment and societal dimensions that are likely to be affected by the project implementation. The content or the elements of the matrix are the quantities of various impacts in a particular stage, such as emission amount, water extraction, forest depletion, and displacement of families. The impact may also be measured qualitatively, like high, moderate, or low. In this context, the Leopold matrix became popular when it was first used in 1971 (normally 100×88 matrix size). Table 5.4 shows the use of matrix model for EIA of projects in the coal mining industry, where site preparation, mine construction, coal/overburden production, and so on are the relevant processes which are likely to generate environmental and social impacts during the coal mining project. The columns of the matrix represent the processes, and the rows represent the indicators of both environmental and social impacts. The column "Weight" includes the relative importance of sustainability indicators in mining industry. The house of sustainability (HOS), developed by the author (Mukherjee, 2011), also shows the similar

TABLE 5.4

An Example of EIA Format for an Open-Cast Coal Mine Project

Sustainability Criteria	Weight	Site Preparation	Mine Construction	Coal/OB Extraction	Coal/OB Transport	Coal-Handling Plant	Workshop
Land Use (Soil Erosion)							
Ecology and Forest							
Air Quality							
Water Quality							
Noise Pollution							
Ground Vibration							
Social Issues (Resettlement and Rehabilitation)							
Cultural Issues							
Fauna and Flora							

assessment method with additional scope of capturing the inter-process interactions. HOS has been elaborately discussed in Chapter 3. Interested readers may refer to that in order to look for the impact assessment model.

5.2.5.5 Rapid Impact Assessment Matrix (RIAM)

RIAM is a systematic approach using the qualitative set of data for analysis of impact of projects on four aspects—physico-chemical, biological, human, and economic. On the basis of a baseline, RIAM identifies the change (positive and negative) brought in by the mitigation process and evaluates various options contributing substantially to the monitoring stage of EIA. The comparison of values on the four aspects and subsequent actions at the later stages of EIA are the value additions of the RIAM method.

5.2.5.6 Battelle Environmental Evaluation System

It is a quantitative model based on the weighted average method. Environmental impacts are first classified under four categories—ecology, pollution, aesthetics, and human interest. Now each category is expected to be composed of some parameters, each of which is measured by an indicator. In this model we intend to assess the environmental issues on those four categories in terms of marginal deterioration because of implementation of the project. Conceptually it is somewhat similar to the two earlier methods (matrices and RIAM). Indicators are represented by a scale ranging from 0 (poor quality) to 1 (excellent quality). Moreover, the importance of parameters (indicators) are quantified by a total of 1,000 points. The mathematical model may be formulated as follows:

$$ENI = \sum_{i=1}^{n} W_i x_i^1 - \sum_{i=1}^{n} W_i x_i^2$$

where

ENI: Environmental net impact after the project is implemented

W_i: Relative importance or weight of parameter or indicator i

n: Total number of indicators

X_i^1: Value of environmental parameter In term of indicator i after the implementation of the project

X_i^2: Value of the environmental parameter in term of indicator i before or without the implementation of the project

EIA reports are quite significant for projects from most polluting industries. Various components of construction, commissioning, and operations, which are likely to affect environment and society, are first identified. The format of the EIA outcome in Table 5.4 represents a case of highly polluting open-cast mine project.

As mentioned earlier, EIA norms have been well-framed with legislative support in developed countries. There exists active involvement of competent authority, governmental agencies, and affected people since early stage of EIA implementation. However, there is a lack of formal legislative support in many developing countries. In India, although it has legislative support, but there is still limited participation of people and NGOs.

Similar to the EIA, there is another term known as the environmental management plan (EMP), which is more like a plan or strategy for managing the environmental parameters and for mitigation of the adverse impacts identified in the EIA. The difference between EIA and EMP is highlighted in Table 5.5.

As EIA is focusing primarily on assessing the adverse environmental impact because of the project operations (preferably before starting the project), EMP strives for the effective management of the facilities and resources for the mitigation of adverse environmental impacts. Any organization may carry out an EIA as a mandatory activity for project implementation or due to pressure from society or other stakeholders or even as a strategic move for image building. An EMP may be designed at the beginning of the project life, by integrating it into the project management activities. An EIA should be included in the EMP process for making environmental management plan effective.

Along with the EIA and the EMP, there exists another globally accepted concept in environmental management, known as the strategic environmental assessment (SEA). It is a form of strategic assessment of environmental impact with a broader scope of study. SEA refers to systematic analysis of environmental impacts in case of implementing various development policies, plans, programs, and other strategic actions. This leads to the extension of aims, principles, and scope of EIA toward upstream, involving a strategic decision-making process beyond the project level. So the unique strength of SEA is the opportunity of exploring other options

TABLE 5.5
Characteristics of EMP and EIA

Environmental Management Plan (EMP)	Environmental Impact Assessment (EIA)
• This represents development of a plan in an organization for effectively managing environmental parameters.	• This is a structured document showing possible environmental impacts for a proposed project.
• It is meant for administrative and management activities	• This is a formal document for reporting not only to the management of the project but also to competent authority at governmental level.
• It makes use of detailed information on adoption and implementation of management strategy over a time frame, allotting possible resources for the best possible outcome.	• It is carried out at the early stage of project planning for getting approval through EC of the proposed project.
• It takes into consideration all possible resources and best utilization of the resources with proper fund management.	• It is a mandatory requirement.
• It may be developed at any stage of the project—during operations or even at the time of decommissioning.	• Focus is mitigating adverse consequences by identifying and analyzing the environmental impacts.
• It is more like a strategic decision-making process.	
• It may use the data from EIA reports.	
• It may be included in a strategic plan of an organization or may be conducted for meeting some specific requirements with contextual importance.	
• It may also be developed as the activity subsequent to EIA reporting so as to plan for mitigating the adverse impacts.	
• Focus is managing resources to meet the goal of improving environmental parameters of the surroundings.	

involving generation and evaluation of various alternatives to be considered at higher levels of decision-making. Although it is difficult to identify a globally established definition of SEA, the one proposed by Sadler and Verheem (1996) seems to be referred to quite often in various literature. It defines an SEA as a systematic process for evaluating the environmental consequences of a proposed policy, plan, or program initiative in order to ensure that they are fully included and appropriately addressed at earliest stage of decision-making on par with economic and social consideration.

If we consider 4Ps as the management decision-making stages in an organization—policies, programs, plans, and projects—an EIA is primarily meant for assessing the environmental impacts once the project is selected as the feasible and viable option for implementation. On the other hand, an SEA is more on assessment and analysis of the futuristic 3Ps and is more like a proactive approach. It may be noted that EC is the last phase before starting any project execution. Table 5.6 exhibits how an SEA is characteristically different from an EIA. The key features of an SEA may thus be summarized as the following:

TABLE 5.6
Comparison between SEA and EIA

Characteristics of SEA	Characteristics of EIA
• Occurs at the early stage of the decision-making cycle	• Occurs at the end of the decision-making cycle
• Proactive approach to a development proposal of the organization; may be at the strategic level	• Reactive approach to project development; ready for implementation
• Analyzes implications on environmental issues of sustainable development for futuristic proposals or plans	• Only assesses the environmental impacts
	• No scope of consideration of other options
• Evaluates all possible optional proposals from an environmental perspective	• Has limited view only on environmental impacts of the project proposal already selected
• As it is futuristic, a warning signal emits in case of disastrous environmental impacts expected in future from the strategic proposals under consideration.	• Focuses only on mitigation and reduction of environmental impacts
	• Limited scope with detailed description on impacts
• Focuses on achieving objectives and meeting targets with a broader perspective of maintaining the green image of the organization	• Well-defined process with limited dimensions and parameters related to the project management and environmental issues
• Broader scope and lower detailing for creating an overall framework	• It has the standard agenda of identifying the environmental deterioration due to the specific project.
• Complex process involving various aspects, passing through a multistage process with the objective of PPP formulation	
• Three-dimensional sustainability agenda considered in its ultimate goal	

1. SEA is a formalized, systematic, and comprehensive process, not a tool or technique.
2. SEA is likely to improve policy-making at the strategic level.
3. SEA contributes to overall sustainable development, not simply identifying the environmental impacts because of organizational activities.
4. SEA is objective-oriented, and the process starts at the beginning of the decision-making cycle.
5. Like other strategic approaches, SEA integrates policy- and strategy-making with operational- and project-level activities.

Incidentally, SEA had limited development and implementation till 1990. However, after 1990 a number of developed countries like Canada and Denmark adopted SEA for policy, plans, and programs separately from EIA legislation and procedure. Historically Europe is always more active in adopting sustainability-related approaches and techniques. Some European countries introduced SEA through some reforms or adaptation of EIA legislation. Of course, SEA adoption is growing and has now become relatively popular among big corporations. SEA may be applicable to an entire sector at the macro level (e.g., implementation of a national policy, energy policy, scheme for intense industrialization) or to a geographical area (e.g., regional development plan, rural electrification) or to an organization (e.g., strategic expansion plan, automation policy). SEA is proactive, and it may also incorporate EIA as a tool for assessing impacts of implementation of a scheme, plan, or program. SEA is a continuous process, and it assesses cumulative effects and identifies the implications of sustainable development. Perspective of SEA is wider, and it may lead to improvement or maintenance of environmental quality of the surroundings. This may be a geographical area, region, or organization itself.

> *An environmental impact assessment (EIA) has been conceptualized by the United Nations Environment Program (UNEP) as a process of evaluating the likely environmental impacts of a proposed project or development plan ready to be implemented, taking into account the socio-economic, cultural, and human health impacts, both beneficial and adverse. In 1994 the Ministry of Environment and Forest (MoEF) included the EIA in the process of environmental clearance (EC).*

> *Environmental Management Plan (EMP) is a plan or strategy for managing the environmental parameters and for mitigation of adverse impacts identified in EIA.*

> *Strategic Environmental Assessment (SEA) may be defined as a systematic process for evaluating and managing the environmental consequences of a proposed policy, plan or program initiative in order to ensure that they are fully included and appropriately addressed at earliest stage of decision-making on par with economic and social consideration.*

5.2.6 European Foundation for Quality Management (EFQM) Model

The European Foundation for Quality Management (EFQM) is the brainchild of 14 companies of Europe (British Telecom, Bosch, Bull, Electrolux, Fiat, KLM, Nestle, Philips, Renault, etc.) originated in 1988 and founded in 1989. Its mission was to push European organizations to excellence and to maintain that sustainable business excellence in future. The EFQM excellence model is a non-profit organization comprising members from industries, academic institutions, research institutions, and so on, the number of which is growing annually touching around 1,000 now from various countries of the globe (see www.efqm.org). Any organization interested in performance improvement is to create a system or framework for getting required data for EFQM.

It is a practical tool that indicates the position of an organization in the path of excellence. It also helps determine the shortcomings and appropriate actions necessary to achieve the target level of excellence. This excellence model makes use of a circular system (named RADAR [result, approach, deployment, assessment, and refinement]) of measuring and monitoring performance in all key areas of the organization using the model. There exist some popular models that are specifically meant for achieving continuous improvement in any organization. While Deming's PDCA cycle has been extensively used in the development and deployment of quality policies, DMAIC added the required rigor to Six Sigma projects. Similarly, RADAR (EFQM excellence model) is primarily used for assessment of organizational performance.

The performance is assessed by a score and it is used according to the RADAR method. The components of RADAR may be simply explained as follows:

- Determine and fix the expected **results** or targets to be achieved in near future.
- Formulate a set of interrelated, feasible, and effective **approaches,** which are expected to deliver the mentioned results.
- Now **deploy** or implement the approaches using available resources and facilities.
- Lastly, **assess** and **refine** the deployed approaches through proper monitoring and analysis of the actual results achieved.

The cycle of RADAR continues (see Figure 5.4), and as mentioned earlier, the EFQM system is implemented and operationalized following the continuous cycle of RADAR similar to Deming cycle (plan-do-check-act).

The basic presumptions of the EFQM model are quite simple and these cover the following considerations.

- Any performance management system is never static in nature, and it is built on the objective of moving or pushing it forward continuously to maintain continuous improvement.
- The model framework of EFQM is equally divided into two main components—causes or **enablers** or **inputs** and effects or **results.** Each of the components is having equal weight or importance in performance scoring.
- All elements of these two components should be influenced and affected by learning, creativity, and innovation. This essentially reflects dynamicity of the process and betterment of organizational performance.

The prominent focus of this model is achieving customer satisfaction, employee satisfaction, and noteworthy contribution to the society. The EFQM excellence model is based on the principle (Uygur and Sumerli, 2013) that excellent results are essentially reflected on organizational performance, satisfaction of both customers and employees, and society. This emphasizes the fact that evidence of excellence is never restricted to financial results, but it has a wider connotation, such as loyalty of customers, sense of belongingness and motivation of employees, and impression or even

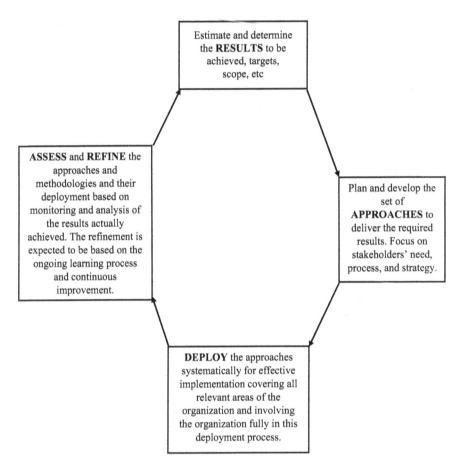

FIGURE 5.4   RADAR matrix cycle.

loyalty of the society at large. The organization can deliver excellent results with the help of its policy and strategy, employees, resources, and processes. The original EFQM model was reviewed several times like in 1997, 1999, 2003, 2010, and even thereafter. The EFQM model has been renamed during review of 1999 as the EFQM excellence model. RADAR logic is the heart of EFQM for maintaining its dynamic continuity. The results of RADAR is similar to the results of EFQM and then other components of RADAR are representing the elements of inputs or enablers of EFQM.

The model framework has been developed on the basis of the following principles (Olaru, 2011).

- Achieving balanced result
- Taking responsibility for a sustainable future
- Adding value to customers
- Nurturing learning, creativity, and innovation

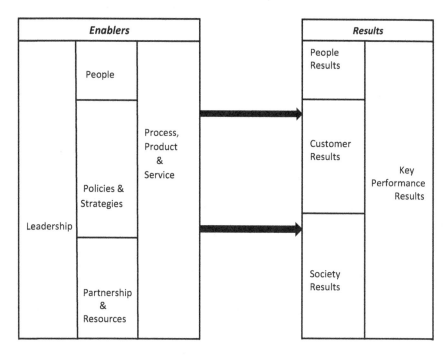

FIGURE 5.5   The EFQM excellence model.

- Leading with vision, inspiration, and integrity
- Managing by process
- Succeeding through people

The model considers collection of two sets of elements as inputs and outputs (results) of business performance. So there are nine elements in total. Five input elements lead to delivery of four result elements. All these factors are guided by learning, creativity, and innovation as shown in Figure 5.5.

The EFQM excellence model judiciously aligns the need of the stakeholders with the business performance. This model shows a focus of taking economic and social responsibility (environmental impacts may be considered indirectly in process, product, people, and results) by the concerned business. The 2010 version of the EFQM model introduced assignment of following weights on the factors or elements of EFQM, which are used for assessing the performance of any business unit.

Enablers:

- Leadership—10%
- People—10%
- Strategy and policy—10%
- Partnership and resources—10%
- Process, products, and services—10%

Results:

- People results—10%
- Customer results—15%
- Society results—10%
- Key performance results—15%

Here we are considering same weights of enablers and results (50% each). Performance of an organization may also be assessed out of 1,000 scores, taking 500 each for enablers and results.

In the process of updating, EFQM includes all contemporary issues in every year. EFQM 2020 model emphatically addresses sustainability by incorporating United Nations Sustainable Development Goals (SDGs) and European ethics values. EFQM 2020 considers three dimensions in model framework development: **direction** (*why*), **execution** (*how*), and **results** (*what*). The EFQM 2020 has the criteria with stronger adaptation of guidance points compared to the previous version (EFQM 2013), which was quite popular among organizations. These are purpose, vision and strategy, and organizational culture and leadership. The study also shows that there exists a relation between business excellence and sustainability (Jankalova and Jankal, 2020). The fundamental idea of the EFQM 2013 excellence model may be stated as the integration of sustainability concept with the organizational core strategy, value chain, process design, and resource allocation process. The current version (EFQM 2020) more strongly emphasizes this fundamental idea of capturing sustainability in model framework. People (stakeholders and leaders) and process components of enablers and results are primarily contributing to sustainability in organizational performance. SDGs are also incorporated in some criteria used for assessment in EFQM. Some other criteria of EFQM 2020 reflect relevant concepts, like sustainable levels of performance, sustainable future, and sustainable value. Although explicit description of the three sustainability dimensions (environment, social, and economic) may not be available in EFQM model framework, most of the elements of the model framework (particularly in 2020 version) include references of sustainability. Social dimension is relatively more prominent in this model. Society results and people reflect social responsibility, whereas key results are more oriented toward achieving economic sustainability. The EFQM model is operationalized by filling up questions on all the elements of the model. The score is further utilized for maintaining continuity using RADAR approach, as mentioned earlier. So this calls for comparison with best-of-the-class performance like benchmarking method for improvement. Goals and indicators are also aligned with organizational requirements and legal demands. Implication of EFQM modeling lies on continuous improvement (PDCA cycle), result orientation, and involvement of stakeholders, processes, and people.

The European Foundation for Quality Management (EFQM) excellence model is essentially a model for improvement of business performance satisfying all stakeholders. It has five enablers and four result elements. EFQM 2013 and EFQM 2020 addressed sustainability as an essential component in business performance and included it in model framework.

5.2.7   COMPOSITE INDEX ON SUSTAINABLE DEVELOPMENT

Although there exist several popular methods for assessing the performance of sustainable business activities, it is still somewhat difficult to identify a single composite and integrated index on sustainable-development-oriented performance of any business unit. An attempt has been made by Krajnc and Glavic (2005) to capture all dimensions of sustainable development mathematically so as to develop an integrated index of performance. The following steps are proposed for assessing the performance of a business unit after a particular time period (e.g., annually), which integrates various relevant factors.

1. A hierarchical structure is first developed following the analytic hierarchy process (AHP). The performance index for contribution to sustainable development is represented by the *triple bottom line*, which considers three dimensions—economic, environmental, and social. In the next level of hierarchy, each dimension is further represented by relevant indicators. The input data for this analysis are the values of the indicators at the time period.
2. The indicators are divided into positive indicators (increasing value gives rise to positive impact on sustainability) and negative indicators (increasing value gives rise to negative impact on sustainability). The values are further normalized for making them comparable.
3. For the aggregation of these values, the pairwise comparison method of AHP is applied. Then per hierarchy the aggregation takes place first by creation of sub-indices at the dimension level and then at the composite performance level. This leads to the computation of relative importance or weights of indicators and also sub-indices.
4. First, the indicators are aggregated with weights of indicators and their values within each dimension, and we get sub-indices, and subsequently, the ultimate sustainable development-oriented performance of an organization is computed at a time period (year) by combining the values of sub-indices and the weights. All the weights are the outcome of applying paired comparison approach in AHP.

The essence of this approach lies on the following principles of computation.

1. Aggregation of various indicators of measuring sustainability on performance primarily depends on the relative importance (weights) of the indicators in terms of their contribution to the overall performance and the performance achieved by the organization at that time period in terms of indicator values. Here the aggregation is made by weighted sum of the values.
2. Weights of the indicators are computed applying AHP in two-level analysis.
3. The values of indicators are separated as positive and negative indicators. Considering the symbols j $(j = 1, 2, and 3)$ for each of the three sustainability dimensions and i $(i = 1 to n within a dimension)$, the values of indicators are normalized by the following formulae.

$$NX_{ijt}^{+} = \frac{X_{ijt}^{+} - miniX_{jt}^{+}}{maxiX_{jt}^{+} - miniX_{jt}^{+}}$$

$$NX_{ijt}^{-} = 1 - \frac{X_{ijt}^{-} - miniX_{jt}^{-}}{maxiX_{jt}^{-} - miniX_{jt}^{-}}$$

Where,

X_{ijt}^{+} & X_{ijt}^{-} = Values of ith positive and negative indicators within j^{th} dimension at t^{th} period

NX_{ijt}^{+} = Normalized value of ith positive indicator within j^{th} dimension at t^{th} period

NX_{ijt}^{-} = Normalized value of ith negative indicator within j^{th} dimension at t^{th} period

$miniX_{jt}^{+}$ & $maxiX_{jt}^{+}$ = Minimum and maximum values of ith positive indicator over last several periods under study.

$miniX_{jt}^{-}$ & $maxiX_{jt}^{-}$ = Minimum and maximum values of ith negative indicator over last several periods under study.

4. Let us now compute the composite index of sustainable development at period t ($X_{sd,t}$), as an integration of weighted values of sustainability sub-indices ($X_{sj,t}$). It is a step-by-step procedure following AHP. Values of sustainable sub-indices, $X_{sj,t}$ for j^{th} dimension, may be computed as:

$$X_{sj,t} = \sum_{ij} W_{ijt} * NX_{ijt}^{+} + \sum_{ij} W_{ijt} * NX_{ijt}^{-}$$

$$\sum_{ij} W_{ijt} = 1; and\ W_{ijt} \geq 0\ for\ each\ t$$

where

W_{ijt} = Weight of the corresponding i^{th} normalized indicator within j^{th} dimension at period t with the assumption that the level of importance of ith indicator (within the perspective of jth dimension) may change over time

Thus, ultimately the integrated composite index of sustainable development at period t ($X_{sd,t}$) may be assessed as follows:

$$X_{sd,t} = \sum_{j} W_{jt} * X_{sj,t}$$

Here, W_{jt} = Weight of the jth dimension as the relative importance in terms of sustainable development of the organization at period t with the assumption that the level of importance of jth indicator may also change over time.

This approach helps detailed analysis of the situation (combining sensitivity analysis along with the presented exercise). It enables inter-company comparison of performance on sustainability and use of benchmarking concept for improving this performance. Interested readers may go through the study (Krajnc and Glavic, 2005) to have the required insights on practical application of this approach and detailed analysis thereafter. Two large-sized global oil companies have been studied—BP (petroleum and petrochemical groups, primarily as downstream business activities) and Royal Dutch/Shell Group (exploration and production of crude hydrocarbon items, primarily as upstream business activities). The data during 2000–2003 have been used for this analysis considering four economic, six environmental, and four social sustainable development indicators. The case study focused report also includes investigation on the trend of composite index values and its practical implications.

Here the aggregation process of all relevant indicators of sustainability is quite effective because of the consideration of their positive and negative impacts and their normalization before computing the composite index. Nevertheless, the model fails to capture the interdependence and correlation among the indicators. This interdependence plays very crucial role in selecting the right strategy for achieving sustainability. A typical example of such interdependence of indicators is the negative correlation between the cost of manufacturing and the carbon emission from manufacturing activities.

> *The composite index on sustainable development may be developed by integrating various sustainability indicators with their degrees of importance.*

5.2.8 FINANCE-/COST-FOCUSED PERFORMANCE INDEX

Any business activity is actually enabled by expectation of financial outcome. Thus, finance-focused performance assessment, perhaps, is quite logical. Every business unit is supposed to meet the investors' financial expectations for its survival and growth. So converting all measurement indices to financial elements may be a practical way of devising a single index for performance assessment. But how do we do that in case of sustainable-development-focused business performance? Translating environmental and social contribution to financial units is a real challenge to any business organization. For both the dimensions, it is not possible to quantify all the related indicators and to translate them in financial terms. Penalty or carbon tax savings or savings of water inputs by its recycling are directly quantifiable measurement units. Same is true for extra cost of investing for equipment meant for arresting pollutions or cost of resettlement and rehabilitation of families displaced for production activities. But some other critical issues cannot be quantified so easily, like increase of sales exclusively due to new green image of the corporation, cost savings due to better negotiation power with supply chain partners, and improvement of productivity because of lowering the risk of environment-related occupational health hazards. Effectiveness of inter-organizational comparison is another questionable issue, as the scale and model of conversion may not match universally.

Incidentally, this approach is relatively less popular among practitioners compared to other approaches mentioned earlier, like GRI, AQI, and ISO 14000. However, those who are really interested in expressing sustainability-based performance in financial terms may make use of the following three techniques for performance measurement:

1. Use of the DuPont pyramid
2. Use of the economic value-added approach
3. Use of direct and indirect cost (shadow price/opportunity cost)

5.2.8.1 DuPont Pyramid Approach

If the environmental and social contributions of the business unit are represented in terms of costs and revenues, ROI or ROA may be computed applying traditional DuPont pyramid. The model developed by DuPont Corporation emphasizes on step-by-step computation for detailed assessment of a company profitability. Its advantage is the opportunity of separately focusing on the relevant parameters like cost, sales, asset utilization, and equity-debt combination in asset development and conversion for the improvement of profitability (ROI, ROA, or ROE). The DuPont pyramid represents the combination of three components of financial and operational performance in measuring return on equity (ROE) as shown here:

ROE = (net income/sales) × (sales/average total assets) × (average total assets/average shareholders' equity)

Most of the management experts and practitioners suggest the existence of a possible link between profitability and sustainability, although any clear and foolproof evidence on the causal relations is yet to be established. At several occasions it has been found that corporations with sustainability enabled management can have better financial results. There is a strong belief in the business circle that pollution-reduction efforts or energy-saving technologies, along with socially responsible strategies, enhance brand image and have a measurable effect on financial performance. As the DuPont model clearly segregates the ultimate profitability into net profit margin, total asset turnover, and financial leverage, the possible impact of sustainability-focused performance particularly on first two components may be translated into financial performance as contribution to profitability. We may refer to the examples like reduction of material inputs in product design, reduction of energy consumption, and increase of asset utilization or elimination of movement of empty vehicle in logistics planning, which simultaneously contribute to environmental sustainability and cost reduction.

The model developed by the DuPont Corporation emphasizes on step-by-step computation for detailed assessment of a company profitability separately focusing on the relevant parameters, like cost, sales, asset utilization, and equity-debt combination in asset development and conversion for the improvement of profitability (ROI, ROA, or ROE), which essentially represent financial and operational performance of the organization.

5.2.8.2 Economic Value-Added Approach

Economic value-added (EVA) approach is conceptually quite rich, and it represents a financial indicator, which determines the company's value in a comprehensive manner. The strength of EVA lies on its consideration of all possible financial parameters related to managing the organization, but incidentally, that is why its computation is quite cumbersome. Of course, nowadays spreadsheet programming, MATLAB, Maple, and the like do help the practitioners in this computational process. EVA includes net income (i.e., both the sales and expenses), interests, income tax, market value of firm's equity, assets, liabilities, and so on in the computational exercise. In short, *EVA* may be computed as follows:

$$EVA = NOPAT - (WACC * capital\ invested)$$

Where

NOPAT: Net operating profit after tax,
WACC: Weighted average cost of capital = Weighted average of cost of equity and cost of debt on the basis of percentage of capital sources (equity and debt)
Capital invested: Invested equity and long-term debt
or
Capital invested: Total assets – Current liabilities,

If *EVA* is positive, it indicates that the project generates wealth and is a good investment. If the *EVA* is negative, it is just opposite and it is a poor investment.

Use of this value-added concept may be further extended in capturing sustainability along with financial performance of the organization under study. However, translating all sustainability-related elements to financial terms in this value-added measurement is really a complex task. Figge and Hahn (2004) of Germany first developed an approach to quantify the sustainable value added, which was subsequently refined for making it an all-inclusive approach. This model has been developed considering certain premises/assumptions and principles while representing sustainable value added (SusVA) in financial or economic terms. These are summarized in the following.

1. Every business utilizes environmental and social resources to create certain level of economic output. So any company gets some economic output at the cost of environmental and social impacts. On the other hand, the company also earns some value over time because of proper management and utilization of all resources, facilities, and so on. **The sustainable value added is the excess value earned by the company, or in other words, the difference between its earned value or economic growth and the economic output or the value expected to be generated by environmental and social impacts on the basis of standard or established ecological and social efficiencies.**

2. As value addition is a continuous process, **SusVA is to be computed over a period in time scale**, such as *t0* (initial point of time) to *t1* (end point of time for measuring current sustainable value-added quantity).

3. **SusVA model includes primarily three parameters: value added (VA) or earned value, environmental and social impact added (EIA or SIA), and ecological and social efficiency.**

Ecological efficiency = *VA/EIA*
Social efficiency = *VA/SIA*
By EIA (or SIA) we mean the addition (consumption or procurement) of environmental (or social) resources along with their impacts. By ecological (or social) efficiency, we mean the VA earned by consumption of unit EIA (or SIA).

4. **It is proposed that the ecological (or social) efficiency will be computed as the efficiency of a benchmark business unit with the assumption that the company under study will also manage its activities with similar level of efficiency, but with different EIA or SIA.**

Keeping in view all the principles and assumptions mentioned here, let us explain the steps in SusVA computation.

1. Identify the benchmark entity and compute the following.

$$EE_b \, or \, SE_b = VA_b/EIA_b \, or \, VA_b/SIA_b$$

Where $EE_b \, or \, SE_b$: ecological or social efficiency of the benchmark unit
2. Economic growth during the period $t0$ to $t1$ created by the company under study or earned value by the company through business activities may be computed as follows:

$$EG = VA_{t1} - VA_{t1}$$

3. Then, $SusVA = EG - \sum_{i=1}^{m} EE_{i,b}(EIA_{i,t1} - EIA_{i,t0}) - \sum_{j=1}^{m} SE_{j,b}\left(SIA_{j,t1} - SIA_{j,t0}\right)$

Here, we consider i as the suffix for a particular environmental or social resource or impact, and there are m number of such resources utilized by the organization.

If SusVA ≥ 0, it indicates that the value created by the company through its business activities is in no way less than the value resulted from impacts on environment and society. Any positive value of SusVA is the excess value earned by the business unit.

It may further be noted by the readers that the SusVA and the popular model of Figge and Hahn follow the principle of weak sustainability only.

Ecological efficiency is the value addition or simply the value earned by unit impact on the environment—or in other words, unit consumption of environmental resource.

Social efficiency is the value addition or simply the value earned by unit impact on society—or in other words, unit consumption of social resource.

> *Sustainable value added by a company means the difference between the value earned, expressed in economic term, and the value expected to be generated by investing on environmental and social resources or at the cost of the impact on the environment and society.*

5.2.8.3   Direct and Indirect Cost-Based Approach

The third approach is primarily meant for cost-benefit analysis involving both the direct and indirect costs and benefits associated with environment and social dimensions and related to all business activities. Let us first focus on direct costs associated with maintaining sustainability.

The direct costs of sustainability may be divided into the following classes.

1. **Cost of impacts due to environmental damage:**

   - The loss or sufferings due to environmental pollutions
   - Ozone depletion and related skin diseases
   - Living with waste piling in urban areas or towns
   - Non-availability of scarce or non-renewable resources, when they are greatly needed
   - Scarcity of water
   - Direct impact on agricultural production because of lack of rain, acid rain, or drought

2. **Cost of mitigating or repairing the damage:**

   - Medical cost for treating the diseased persons staying closer to the polluted environment
   - Extra cost for fetching water to continue agricultural activities
   - Cost of water treatment due to the contamination of water
   - Recycling of water in order to handle the problem of water scarcity

3. **Cost of prevention and preparedness:**

   - Updating strategies, policies, and if necessary, structure of the organization to include sustainability at all levels and activities
   - Installation of new technology for arresting emissions
   - Improvement of energy efficiency by process re-engineering
   - Replacement of energy source by that of non-fossil-fuel origin or renewable sources
   - New product development using green design methodology
   - Implementation of green logistics

4. **Cost to society:**

   - Cost of rehabilitation and resettlement (to be borne by the owner of the factory, project, or business unit)
   - Cost of cultural conflicts due to resettlement and their impacts on the work environment

- Change of flora and fauna and its consequences, both physical and psychological
- Change of socio-cultural micro-environment after commissioning and operations of new facilities or factory
- Cost of loss in career, business activities, education, or other similar activities, to be borne by the project-affected families

The activities of environmental protection and restoration, in short, may simply be categorized under three heads—decarbonization by carbon footprint reduction, detoxication by reduction of emissions (of any type) or their impacts, and dematerialization by reduction of natural resource extraction and its further processing. Similarly, there may also be some direct benefits to society because of the new factory, project, or power plant in an otherwise rural and underdeveloped area or isolated hilly region. These are essential facilities for living, like hospitals or health centers, schools, colleges, shopping centers, and cinema halls, created in the region.

> *Direct costs of achieving sustainability or managing non-sustainable occurrences include cost of impacts due to environmental damage, cost of mitigating or repairing the damage, cost of prevention and preparedness, and cost to society*

Indirect costs of environment and society are primarily assessed using opportunity cost concept. However, another established approach of its assessment is the "shadow price" of the duality theorem. Let us try to measure the dimensions of sustainability in this direction.

Linear programming (LP) modeling is considered to be the most effective technique for taking optimal decision in planning for best use of existing resources, provided all the assumptions of LP application are fulfilled. If we plan for future production of a group of products of company A, the solution of LP problem will show the optimal production plan with maximum possible profit utilizing the available set of resources. Now, each LP problem is associated with its mirror image or counterpart, which is another LP problem, known as the dual LP problem. If the production of each product is a decision variable of the original (or primal) LP problem, each variable of the dual LP problem will represent the **worth** or **value** of a type of resource being used for the manufacturing of products. The relationships between the pair of LP problems is explained by a set of duality theorems. As detailed discussion on all these issues is beyond the scope of this book chapter, let us try to explain the economic interpretation of dual variables using some simple expressions. At the optimal stage, the values of objective function of both the LP problems are exactly the same, and we get the following:

Opt. objective functional value of primal LP = Opt. objective functional value of dual LP

$$Total\ optimal\ profit\ or\ financial\ gain = \sum_{i=1}^{m} Units\ of\ i\ th\ resource * Y_i$$

The left-hand side of the expression shows the total maximum profit or financial gain of the company A achieved from the optimal production plan. The right-hand side portion of the equation is the summation of some product values involving all the resources (total m in number here) being used for production. Each product value is a multiplication of quantity of ith resource and optimal value of the ith dual variable. So by dimensional analysis, we may infer the following:

$$Y_i = Per\ unit\ financial\ worth\ or\ value\ of\ ith\ resource\ in\ Indian\ Rupees\ per\ unit$$

It may be noted that this financial worth of a resource is the portion of optimal profit or financial gain, expected to be achieved by implementing the optimal production plan. So this worth or value is not the market price of the resource, and the value is also contextual in nature.

Now, suppose that an outside agency or company B requires any of the m resources, and the organization is ready to buy the resources at some reasonable price. How much price should be fixed by company A, which has already started using it (or is expected to use it), per optimal production plan? If the resource is fully utilized in its optimal production plan, the manager of company A will naturally reject the idea of selling it, as this will lead to substantial loss of profit or financial gain. In other words, it means that the **worth** or **value** of the resource will be very high, and this also means that the opportunity cost of its selling is huge loss of the profit of company A. The manager of company A will also have high bargaining opportunity in these circumstances. On the other hand, if it is underutilized in the production plan, the unutilized portion of the resource may be sold to company B at just throw away price or as a very cheap resource.

This actually happens in nature, when we extract the natural resources or when we go on degrading the nature and the physical environment. The extraction of scarce resources creates imbalance in nature among its set of related elements. It also can tolerate the degradation up to a certain level, beyond which it reacts adversely. These are cases of huge price of disturbing the state of equilibrium of nature or its optimized state of functioning. This is the indirect cost of environment to be paid in carrying our business activities, which can be understood as the shadow price of duality. The more pollution takes place, the higher the cost of emission, as the sink (from the concept of **source** and **sink**) of nature will be almost full. Similarly, the more we use the scarce or non-renewable resources (like fossil fuels, land, and even underground water), the higher the cost of the further use of natural resources or natural capital. On the other hand, if there are plenty of natural resources, like shrubs and bushes, which are not being fully utilized in nature's own growth or sustainability cycle, then this cost will be lower. So in those situations, impact on the environment is not much till a certain cutoff level of extraction, beyond which people are to pay high price for degrading nature. It is really difficult to compute that cutoff limit objectively. It is alleged that tribal people or people surviving in highly isolated zone or underdeveloped areas can read, listen, and understand the activities and underlying mechanism of nature without having any formal education.

Let us now explain the social costs (another set of dual variables or shadow price of social resources or impacts) with another simple example. Management of a factory is planning to start its manufacturing in a village that has rich and fertile land for agriculture. Most of the villagers are fully occupied in the paddies (or doing other agricultural activities). Once the factory and its related facilities are installed, the villagers are to be employed as the workers for manual labor. It may be noted that by social resources, here we mean the human resources available in the village. Now, in this case, what will be the cost of using the villagers in the factory? Would that be the market-dependent wage rate, somewhat more than their average earnings through agricultural work, or the highest rate among all possible avenues of earnings (applying opportunity cost concept)? Or is it the labor-law-based wage rate?

Till the entry of the company for building and commissioning the factory, society has been optimally using the villagers for state- or national-level paddy production. As ultimately this paddy output meets the demand of the country as a whole, this is directly reflected on India's GDP. So if the villagers leave the agricultural work and join the factory, the country will be deprived of its income, or there will be loss at the state/national level. This loss in GDP is surely more than the cost of market driven wage. Moreover, if the country wants to make up this loss of production of staple food for Indians by procuring the rice required from abroad, it will be more expensive. This amount is the shadow price of using social resources or the value of dual variables. Now, if the state/nation could not utilize the villager-hours maximally for agricultural production or if there is not sufficient demand of rice in the market, the social cost will decrease and the factory may engage free villager-hours at a lower rate. Thus, social costs are also contextual. If there are less free villager-hours available, the value of using the social resource will increase due to the impact of scarcity.

> *Indirect costs of using (or damaging) environmental and social resources may be expressed as shadow price, which depicts the non-market worth of the resources similar to opportunity cost concept.*

If the economic issues at the macro level are further explored, we may make use of the genuine savings index as the sustainability indicator at the national level (Moffatt, 2013).

Genuine savings index = gross national savings + education expenditure – consumption of fixed capital – depletion of energy resources – depletion of minerals – net depletion of forests – CO_2 damages – particulate pollution damages

This genuine savings index is often used by the World Bank. For human welfare, the human development index (HDI) is quite well known and popular. HDI is computed on the basis of three welfare indicators: health, education or literacy, and wealth (purchasing power parity).

Daly and Cobb attempted to capture both economic inequality and human welfare from the perspective of sustainability and developed the Index of Sustainable Economic Welfare (ISEW) using the following formula (Daly and Cobb, 1989).

ISEW = weighted personal consumption + net capital growth + health and education expenditure – depletion of non-renewable resources – pollution costs –

wetland and firm land loss – atmospheric damage (ozone depletion, global warming and other long-term environmental damages)

Key Learning

- Because of coercive and normative pressures, most of the industrial organizations nowadays use some popular and well-established performance measurement and/or management approaches incorporating sustainability along with the usual profitability-based indicators. As these are globally accepted and used quite often, transactions with all stakeholders and global reporting are easier. These are GRI, ISO 14000 certification of EMS, AQI, EIA, and so on.

- Global Reporting Initiative (GRI) is a universally acclaimed global reporting system to make the world aware of the business activities and their impact on environment and society. GRI norms are continuously updated capturing contemporary issues and geopolitical developments. Three GRI Standards—Foundation, General Disclosures, and Material Topics—of 2016 are the basic standards applicable for all organizations.

- The ISO 14000 family of certifications for environmental management systems (EMS) has already occupied the similar popularity in business community as enjoyed by the ISO 9000 series for TQM. These EMS standards are quite comprehensive and follow PDCA cycle for continuous development.

- Dow Jones Sustainability Indices are the set of globally popular indices that measure the performance of companies selected with ESG (environment, society, and governance) criteria using the best-in-class approach. The companies are recognized by the scores of its system of assessment.

- The air quality index (AQI) is the most recognized and popular component of Environmental Performance Index (EPI). Its limitation is the measurement of air quality only, not other environmental parameters. AQI is normally measured as the air quality of a region and its scores are categorized under six slabs—severe, very poor, poor, moderately polluted, satisfactory, and good.

- An environmental impact assessment (EIA) is a process of assessing both the beneficial and adverse impacts of business (project) activities on environment and society. Our Ministry of Environment, Forest, and Climate Change (formerly MoEF) demands EIA results for getting environmental clearance of projects. An environmental management plan (EMP) is prepared beforehand to mitigate the harmful impacts per the assessment of EIA. Similar to EIA and EMP, a strategic environmental assessment (SEA) refers to systematic analysis of environmental impacts in case of implementing various development policies, plans, programs, and other strategic actions.

- European Foundation for Quality Management (EFQM) is a model for continuous improvement of organizational performance. The EFQM excellence model uses the principle of RADAR cycle to maintain the continuity.

The components of this model are divided under two groups—"inputs" and "results." Recent EFQM versions, like those of 2013 and 2020, include sustainability, and thus, profitability and sustainability can be clubbed in representing organizational performance in preparing the road map for continuous improvement.

- The composite single index approach aggregates all indicators (both positive and negative) with their degrees of importance applying mathematical models for normalization, weighted averaging, and AHP.
- There are some finance-focused performance approaches, which measure sustainability in financial terms. ROI/ROA computation by the DuPont pyramid supports stage-wise analysis of all parameters, some of which affect sustainability operationally or financially. Economic value added (EVA) is considered to be a rich financial indicator of an organization's health. This EVA approach is further extended to capture the value-additive concept on sustainability, which is expressed in financial terms. The proposed model is well-known as sustainable value added (SusVA). Further, there may be direct cost computation of impact (or damage) of non-sustainable activities and also the cost of mitigating or repairing the damages. Indirect costs of impacts may be computed as the shadow price of the duality theorem. Most of the indirect environmental and social costs are expressed as opportunity costs only.

5.3 CARBON FOOTPRINT ANALYSIS AND LIFE CYCLE ANALYSIS AS THE TWO ESSENTIAL APPROACHES IN GREEN BUSINESS PROCESS MANAGEMENT

The approaches for measuring sustainability in operations discussed in the previous sections are mostly created by integrating various indicators related to sustainability, which meet the needs of stakeholders. Some of the approaches are meant for reporting the sustainability-related business activities to the world. Further, these approaches are primarily representing the single-index-based indicators of performance in terms of sustainability. The discussion in this section will be on the techniques, which are process-based and which demand thorough analysis of the business processes.

We know that sustainable development "meets the needs of the present without compromising the ability of future generation to meet their own needs" (WCED, 1987, p. 43). The concept drives the scientists and business practitioners to design the mechanisms for restricting the use of natural resources, some of which also accelerate the regeneration of renewable resources. Moreover, in the underlying principle of development, the present generation is shouldering the responsibility of handing over the earth in a healthy and living worthy condition to the next generation.

In order to achieve sustainable development in business functioning, the following four techniques seem to have gained world acceptance among the business community and policy-makers in assessing performance.

- Eco-efficiency
- Ecological footprints

- Carbon footprints
- Life cycle analysis

Let us briefly summarize the first two approaches and later on the last two approaches will be discussed in detail.

The World Business Council for Sustainable Development (WBCSD) describes *eco-efficiency* as the management strategy of doing more with less (with special focus on environmental issues). The "eco" part of this term is primarily meant for representing it as ecological efficiency or efficiency of making use of ecological system. But survival of any business entity lies on economic efficiency. However, as mentioned earlier at several occasions, higher economic efficiency leads to better environmental sustainability or ecological efficiency. A typical example in this context is energy efficiency. An energy-efficient process uses less energy for same quantity of work and thus naturally it helps in conservation of non-renewable natural resources like fossil fuels (very prominent in countries like India, China, and Russia with maximum share of thermal power source of energy). Additionally, for the same reason, less thermal power generation means less emission of GHGs and thus lower carbon footprint. There are three core objectives of *eco-efficiency*: increasing product/service value, optimizing the use of resources and reducing environmental impacts. There are some potential benefits from *eco-efficiency*:

1. Reduced costs: efficient use of materials and energy
2. Reduced risk and liability: avoidance of toxic or hazardous materials in product design
3. Increased revenue: scope for eco-efficient new product development and better brand image
4. Improved environmental performance: by reducing materials consumption and increasing recovery and reuse of waste materials
5. Optimal use of resources in all processes: eco-efficient strategies optimize the use of resources and energy in planning process for product design, manufacturing, purchasing, and logistics

Eco-efficiency is again discussed as a popular alternative to life cycle analysis in Japan at the end part of 5.3.2 sub-chapter (i.e. life cycle analysis).

Eco-efficiency aims at increasing product or service value with the minimum use of resources, particularly environmental ones. It reduces costs and risk and increases revenues and environmental performance, and like other efficiency measures, it optimizes the use of resources and energy.

Ecological footprints represent the demand of mankind on the earth's ecosystem. It includes the human need for consumptions and waste disposal. It is fulfilled by supplying the necessary resources for the human population and, by absorption of resulting wastes, making use of biologically productive lands and marine areas of the earth. There are certain terms associated with it. In the supply and demand perspective, if we treat the *ecological footprint* as the quantity of land and marine resources under the human demand, then the *bio-capacity* is the term meant for its supply. Bio-capacity measures the capacity of a given area to generate the natural resources (renewable ones) or to bear non-renewable resources and to absorb any waste generated by their consumption or use. In other words, *bio-capacity* is the capability of the area to support the human population living in the area. If the *ecological footprint* of a given population exceeds its bio-capacity, then the population has an *ecological deficit*. This results in shortage of resources and space for disposals, higher prices, and pollution due to industrial concentration to meet the extra demand. If the converse is true, that is population's *ecological footprint* is lower than its bio-capacity, the demand of the population will easily be met by its bio-capacity. It is the case of *ecological reserve*, and it reflects an extremely sustainable situation.

The *ecological footprint* is measured as the global hectare (gha) consumed and used up (for disposal of wastes) per annum per person. One gha is an area of biologically productive land, marine zones, and lands for disposal of 10,000 square meters (100 m × 100 m). Sometimes the ecological deficit is expressed as how many earth equivalents are required to make up this shortage. For example, in 2019, the bio-capacity of earth was measured as 1.6 gha, whereas the world average *ecological footprint* was expressed as 2.7 gha. It means that the world population needs more natural resources and space for consumption and waste disposal. So in short, we may say that a 1.75 earth equivalent (approximately 2.7/1.6) is required to meet the human demands of the world population.

It is well accepted that the ecological or natural resources of earth may be expressed in six bio-productive areas: agricultural lands or croplands for food and animal feed, grazing lands for animals, forest areas, marine fishing areas, built-up areas, and land for sequestering atmospheric carbon dioxide (Moffatt, 2013). In other words, these bio-productive areas cover the lands and resources for food, disposal, and sink for absorbing the carbon footprints. The following formula may be applicable for computing ecological footprints.

$$Ecological\,footprint = \sum_i [A_i * C * ((P + I - E)Y)]$$

A_i is the area of the land of ith type. C is an equivalence factor for a particular country. *(P + I − E)* is domestic production plus import minus export. Y is the global yield of the resources.

Ecological footprint is the term representing the quantity of land and marine resources under human demand, whereas bio-capacity is the term meant for its supply. Ecological deficit occurs in a population if the ecological footprint exceeds its bio-capacity.

Although *ecological footprint* covers almost all forms of resource consumption, carbon footprint has its specific and crucial role in creating global warming, and so carbon footprint analysis is important. There are also similar approaches like water footprint analysis, which are not discussed in this book. However, role of life cycle analysis is more on improving the sustainability-based performance of businesses and has importance in strategic decision-making. Here, the readers may note that attempts should always be made to decrease carbon footprint, to increase water footprint, and to create ecologically rich or reserve environment. Life cycle analysis, on the other hand, is meant for critical investigation of business processes on these issues and for facilitating the performance improvement.

5.3.1 Carbon Footprint Analysis

We understand that sustainable development emphasizes redesigning the development process by the people of a generation, who carry the responsibility of handing over the earth and its ecosystem to the next generation with least degradation and, if possible, in a better condition. This gives rise to global concern and worry on *global warming*, or the increase in the mean temperature of earth's near-surface air and ocean. Research has shown that greenhouse gases (GHGs) are primarily responsible for causing this global warming. These gases actually act like the glass that covers a greenhouse. The layer formed by the gases traps the sun's heat and does not allow it to leak away to outer space. This captured heat warms the earth and ocean. Studies have shown that from 1990 to 2019, because of this GHG generation by human activities, the earth's surface temperature increased by 45%. GHGs primarily include carbon dioxide, methane, nitrous oxide, and fluorinated gases (hydrofluorocarbons, perfluorocarbons, etc.), among which CO_2 carries the lion's share (70% to 80%). This CO_2 is produced mostly through human-triggered actions or anthropogenic actions, like deforestation, industrial pollutions (primarily paper, chemical, thermal power plants), use of automobiles in travels, use of AC, and several other actions. Business process management, popularly known as business process re-engineering (BPR), and other similar management techniques have already attempted to reduce emissions and resource consumption through better economic efficiency. These are lean manufacturing, inventory reduction or JIT, Six Sigma, digitization of data analysis for making these processes paperless, and so on. In this pursuit, *carbon footprint analysis* is a direct attempt of studying carbon dioxide generation during production and distribution of a product, during carrying out a process or during offering a service. This subsequently enables managers to pinpoint the process, items, or activities that require modification or improvement for the reduction of carbon footprint, if not for its complete removal.

CO_2 emission is universally measured by units meant for the weight of carbon as gm/kg/metric ton or tonne, shown as CO_2 equivalent (or tonne CO_2 eq). Based on the 2019 data, the top four carbon-generating countries are as follows:

1. China: 9.9 billion tonnes
2. USA: 4.7 billion tonnes
3. India: 2.3 billion tonnes
4. Russia: 1.6 billion tonnes

The main share of CO_2 generation in USA comes from transportation, industries, and power plants, whereas the other three countries burn fossil fuels for different reasons. It is quite relevant to reiterate the fact that nature has its mechanism to eliminate the CO_2 existing in space and atmosphere. If we consider that these man-initiated carbon-generating activities are the *sources* of carbon footprint, we should also know that there exist some *sinks* of nature to absorb it. The most prominent sinks are forests and oceans. Carbon footprints generated by sources are absorbed by sinks and the balance is maintained in nature. But in reality, sources are more active than the sinks, and all footprints cannot be absorbed by sinks. In this case, we call the region a *carbon-rich* (or *carbon-positive*) or most polluting state. The rarest situation is just the reverse, which is a *carbon-negative* (or *carbon-deficit*) state. Otherwise, a country is *carbon-neutral* if it has an equal capacity of being both a source and a sink. There exist three smallest countries in this world that are carbon-deficit. These are Bhutan, Suriname, and Panama, which have extra sink capacity even to absorb carbon footprints of other countries. Very often carbon neutrality may be achieved by *carbon offsetting*, which means that the pollution at one place may be compensated by carbon saving or creation of sinks somewhere else. For example, deforestation by a company for the expansion of its factory may be offset by afforestation and creation of lake somewhere else, which act as sinks.

> *Carbon footprint is the GHG generation by a process (or a set of processes) or in an area, measured as tonnes of CO_2 equivalent. Its main impact is global warming. It is generated by sources and absorbed by sinks. If the capacity of sources is more than that of sinks, it is carbon-positive state. If both the capacities are same, it is highly balanced situation, and we say that it is a carbon-neutral state. The rarest is a carbon-negative state, which means that the capacity of sinks exceeds that of the sources.*

Now, globally the average per capita carbon footprint is around 4 tonnes per annum. To have a better chance of avoiding 2-degree-centigrade rise in global temperature, the average global carbon footprint needs to drop to less than 2 tonnes per annum by 2050. Policy-makers, world leaders, and experts are trying utmost to incentivize people to curb footprint generation through various summits and during the Conferences of Parties (COPs) every year. The Kyoto Protocol of 1997 declared the intention of reducing the overall CO_2 emission to 5.2% below the 1990 level within 2008 and 2012. Unfortunately, we are yet to achieve it. But the Kyoto Protocol gave rise to a mechanism for carbon offsetting, known as *carbon trading*. This permits the companies to trade the polluting rights in a market under the control of some global regulatory body. The mechanism is called the *cap and trade* scheme, which allows a company to sell the polluting right, if its pollution level is lower than a prefixed cap, to another business unit, which is polluting more than the allowed cap level. Every country is given a cap, or allowable emission or polluting right, by the governing body of the scheme, and the country distributes it to all industrial units of the country as the cap for the company. So a company may continue polluting the environment by purchasing the polluting right from some other business unit, which could emit less than the cap, and that is carbon offsetting. This allowance trading acts like the trading in share market with price of certified emission reduction (CER) dynamically fluctuating on the basis of the demand and supply of CERs at a particular point of time. There are several drivers or enablers for applying carbon footprint analysis, like the following:

1. **Mandatory**: There are some legislative directives at national and/or global level which make the carbon footprint assessment mandatory for a business unit. European Union's Emission Trading Scheme (EUETS) functions as a *cap and trade* system and fixed 2.04 billion tonnes as "allowance" of cap for 2013. Here, this cap is the maximum limit of carbon emission for a year. Through ETS, companies may buy or sell the allowance emission in order to meet the specified target. Similarly, Indian companies do plan for carbon footprint analysis to meet the target fixed by the government. For example, the current target is the reduction of carbon footprint by 1 billion tonnes and of carbon intensity of the economy to less than 45% by 2030. By *carbon intensity*, we mean carbon emission in tonnes per unit of GDP of a country. Incidentally, this Indian policy is not mandatory based on any global directive, but the country on its own has created the target to be achieved by various industrial units of the country.
2. **Voluntary for corporate image**: Companies do not take any risk of tainting their brand image because of some of their activities that may be perceived by the society as unsustainable ones. Other than standard GRI reporting, most of the large corporations consider the declaration and announcement of goals with environmental sustainability as an essential marketing strategy for maintaining and uplifting corporate brand image. Declaration of emission reduction of 20 million tonnes of carbon by 2015 had surely contributed a lot to brand building of Walmart. So assessment and analysis of carbon footprints is necessary for these business units.
3. **Voluntary for seamless supply of essential raw materials**: Applying carbon footprint analysis the manufacturer can assess the carbon footprint content and any trace of hazardous items in its raw materials, which the manufacturer procures from the suppliers. So the manufacturer may discard certain input materials, look for their substitutes, and arrange for the training of suppliers on sustainability for the reduction of carbon footprint and removal of any risk of health hazards. Most of the Apple products are now free from lead, brominated flame-retardant, polyvinyl chloride, and mercury. Other computer-manufacturing giants also follow suit.
4. **Voluntary for better ecological and economic efficiency**: Reduction of material consumption in manufacturing means reduction of carbon footprint generation during production of those materials. Similarly, control on power consumption in manufacturing and distribution process reflects less CO_2 emission during power generation. Cost reduction may be achieved also by using recycled items, which is, in other words, a mode of avoiding carbon generation. So all these eco-efficient strategies need critical analysis of current carbon footprints on all products and processes

for getting a comprehensive result of economic, ecological, and carbon footprint reduction efficiency. Various organizations, including Wipro, introduced a take-back policy for collecting and recycling of e-wastes. Suzlon Energy (a wind turbine producer) uses wind energy in its own factory in Puducherry.

The carbon footprint assessment process may be explained by following three major stages.

Stage 1: Modeling of all the relevant processes applying tools and techniques of business process management (or business process re-engineering).
Stage 2: Measurement and critical analysis of carbon footprints on all these processes.
Stage 3: Application of corrective actions for reduction or if, possible, elimination of carbon footprints.

> *The Kyoto Protocol of 1997 had shown the global intention of reducing the CO_2 emission to 5.2% below the 1990 level within 2008 to 2012. The countries had been incentivized by making use of global trading of emission rights.*

The popular mode of depicting various processes in modeling is the use of standard symbols of basic process modeling for the elements like *operations*, *transport*, *delay*, *inventory*, and *inspection*. If the study is on a large supply chain, the methodology of supply chain operations reference (SCOR) model may be applied. This calls for using the generic functions like *plan*, *source*, *make*, *deliver*, and *return*. The process improvement by the reduction of carbon footprints is subsequently carried out by analyzing the *as-is* process network, keeping in view the *to-be* network as the target version of the process under study.

Normally, carbon footprints may be calculated via either of three approaches: bottom-up on the basis of process analysis, top-down on the basis of environmental input-output analysis, or by a combination of both. As mentioned earlier, GHGs will be assessed by CO_2 eq units. But GHGs must be measured following the guidelines of the GHG Protocol, which has been accepted as the best accounting tool to measure and manage the GHG emissions by businesses across the globe. The World Resource Institute (WRI) and the Word Business Council for Sustainable Development (WBCSD) jointly developed this protocol for creating a life cycle inventory as a mechanism for measuring, managing, and reporting carbon footprints. The GHG Protocol was developed in late 1997 (see www.ghgprotocol.org). It was first considered to be a corporate standard scheme for carbon footprint in 2001 after first edition publication of *The Greenhouse Gas Protocol: A Corporate Accounting and Reporting Standard*. In 2006, ISO adopted this corporate standard for its ISO 14064–1. The GHG Protocol divides the gas emissions under three types, primarily based on the sources of emission. These are known as scopes of emissions. Although we mentioned about the scope categories of carbon emissions earlier, here it requires little more detail because of their role in carbon footprint analysis.

Scope 1: **Direct emission**

Emissions that occur from the sources owned, operated, and controlled by the
company or the organization under study (i.e., from "within"). It includes
emissions because of using boilers or furnaces on-site or in the factory, generators in the factory, fugitive emissions (like air conditioner leaks), factory
material-handling units, and factory vehicles.

Scope 2: **Indirect emissions using procured energy**

These are all indirectly associated with operational activities, not directly
owned or controlled by the organization. For example, emissions due to the
production of electricity or steam, which are also used by the organization.
The organization purchases these items from outside. If the electricity is
produced only by thermal power plants, the emission generation will be
huge. Although the thermal power plant is directly responsible for emission,
the user organization cannot avoid the responsibility of this emission to the
environment. So indirectly the user organization may control this category
of emission by changing the source of electricity, by re-engineering the
production process for achieving more energy efficiency, or by technology
change for lower use of electricity.

Scope 3: **Other indirect emissions**

These emissions are also the results of the business activities of the company,
but they originate from the sources not directly owned, controlled, or operated by the company. However, the company does have indirect influence
on them. It includes the emissions associated with regular commuting
of the employees, business travels, or inefficient waste management by a
third party. These Scope 3 emissions also may occur due to the activities
of franchisees, retailers, customers, and suppliers. The organization under
study neither owns nor directly controls them. However, the organization
can manage them by selecting the right type of partners, by educating or
training them, or by extending supports for intensifying green activities at
their ends.

Computation of emission inventory for Scope 1 and Scope 2 is rather easy task,
whereas getting data and precise measurement of Scope 3 is difficult. Scope 3 emissions of one company very often represents Scope 1 and Scope 2 emissions of other
firms. The obvious difficulty in assessing Scope 3 emission is because of its data
collection from outside the company boundary, which is beyond its control or influence. But studies show that maximum GHG emissions associated with the product
or service produced by a company is often from supply chain (i.e., Scope 3), which is
much higher than company-level emissions (i.e., Scopes 1 and 2). So the assessment
of Scope 3 emissions is unavoidable and necessary.

> *Based on the GHG Protocol, the types of GHG emissions may be classified into three scopes—Scope 1 (direct emissions), Scope 2 (indirect emissions because of using procured energy), and Scope 3 (other indirect emissions).*

The critical assessment and analysis of carbon footprints may be conducted using one of the most effective tools of business process modeling, which is particularly quite popular in cost and operational control of business units. This activity-based costing (ABC) method assigns costs to operations and critically explores the multi-level process architecture. ABC is conducted using a critical parameter known as *cost driver*, which creates or attaches costs to each activity of the business unit. A similar tool, known as activity-based emission (ABE), has been designed and used in carbon footprint analysis (Recker et al., 2012) and brought out a good result in improving environmental sustainability.

ABE permits the calculation of CO_2 (representing CO_2 eq only) in each activity of a business process, by identifying the so-called *emission drivers* (similar to cost drivers) and by considering the impact of resources that facilitate the process execution. The emission drivers, along with the resources, generate the CO_2 emissions. The analyst should carefully identify the activities as the unitary components of business processes and distinguish between resources and emission drivers. The difference between the emission driver and resource can be clarified with an example. If we consider engine of a car or vehicle as the *emission driver* (i.e., creator of carbon emission), then fuel of the engine (e.g., petrol, diesel, ethanol) is a *resource*.

ABE may be executed by conducting the following six steps.

Step 1: Identify the product, service, or process to be analyzed. This identification is done on the basis of the business requirements, legislative pressure, study for achieving eco-efficiency, building market image, or any special directive for a process. Let us use the term *project* for such a product, service or process during this application of ABE method.

Step 2: Determine all the resources and activities involved in this project of carbon footprint analysis. It is similar to a modeling step with documentations of all possible linkages and data. It is further suggested that the actors responsible for all the activities are included in the documentation.

Step 3: Determine *emission drivers* for each activity. Here the sources of emission are important. So apply Scope 1, 2, or 3 on the basis of the GHG Protocol, as the case may be. Scope wise and responsibility or actor wise basic data are collected on emission drivers and resources.

Step 4: Calculate CO_2 eq emission for each activity, keeping in view the resources and emission drivers. Let us consider three common resources in business activities as an example and compute the emission. These are **fuel** (restricting only to business travels and employee travels to the company premises from their residence and back), **papers** used in the organization, and **electricity** consumption.

Fuel (Scope 3, as It Is Indirect Emissions Covering Only Traveling of Employees)

Carbon footprint (tonne) = Distance traveled (distance unit) × Emission factor incorporating standardized fuel efficiency in a given condition (efficiency per distance unit or tonne of emission/km, as an example)

Paper (Scope 3, as Indirect Emission by Paper Mills and Suppliers of Papers)

Carbon footprint (tonne) = Weight of papers used (weight unit; e.g., kg) × Emission factor during manufacturing and transport of papers and their use (efficiency per weight unit or tonne of emission/kg, as an example)

Electricity (Scope 2)

Carbon footprint (tonne) = kWh of electricity consumed × Emission factor per generation and distribution of kWh (tonne of emission/kWh)

Step 5: Calculate overall carbon emission of the project, also showing the emissions for each unit process or sub-process separately. This data for each process is required for pinpointing the process emissions to be improved for overall improvement of carbon emission.

Step 6: This is the most crucial step of taking corrective action for reduction of carbon footprint of the project as a whole. This step includes identifying processes, resources, and emission drivers, which contribute maximally to overall emission of the project. This often leads to process restructuring or redesigning, reduction or replacement of existing resources by some alternative ones (replacing diesel by electric battery), changing emission drivers (replacing ICE-based engine by battery-operated electric motor for electric vehicles), or making all administrative processes paperless.

> *Activity-based emission (ABE) is an effective tool for carbon footprint modeling, which is similar to activity-based costing (ABC) for cost and operational control.*

5.3.2 Life Cycle Assessment

Life cycle assessment or analysis (LCA) is another popular and internationally accepted concept for assessing sustainability using standardized methods and practices. LCA is somewhat similar to carbon footprint analysis, with larger scope of analysis. The analysis is not restricted to a specific project but to a long chain of processes. LCA is supposed to assess the environmental degradation issues and their impacts on all the processes (of the value chain) in the complete life cycle of the product.

LCA was primarily meant for environmental dimension of TBL. In the 1970s and 1980s some attempts were made on similar studies particularly on energy analysis and emission loadings from waste generation and packaging materials. In the 1990s the world witnessed many scientific and coordination activities on LCA practices and more on sustainability-focused research. Several workshops have been organized by the Society of Environmental Toxicology and Chemistry (SETAC), and a number of LCA guides and handbooks have been published. The years 1990–2000 may be treated as the decade of standardization. In fact, if we consider SETAC working groups are responsible for development and harmonization of LCA methods, the International Organization for Standardization (ISO) may be treated as the organization, which was intensely involved in standardization of LCA. The contribution of the ISO 14000 series (from the ISO website) is quite evident in environmental management—life cycle assessment principles and framework (ISO 14040), goal definition and inventory analysis (ISO 14041), life cycle impact assessment (ISO 14042), life cycle interpretation (ISO 14043), and requirements and guidelines (ISO 14044). It may further be noted that ISO never proposes any single and universal method for executing LCA; it shows guidelines, framework, and steps to be followed for carrying out LCA in a specific situation.

LCA is originally acronym of life cycle assessment, but sometimes, it is also called life cycle analysis or life cycle approach. It is also considered to be the synonym of cradle-to-grave analysis or eco-balance. In this idea or concept, we do not mean market-focused product life cycle, which includes stages like introduction, growth, maturity, and decline based on market demand. *The typical life cycle of LCA consists of a series of stages running from extraction of raw materials, through product design and formulation, processing, manufacturing, packaging, distribution, use, reuse, remanufacturing, and recycling, and ultimately disposal.* So although LCA was originally meant for cradle-to-grave situation, it is equally applicable for cradle-to-cradle-to-grave situation as well. Standard definition may be accessible in document of ISO 14040, where LCA is being recognized and treated as a technique. It is a *"technique for assessing the environmental aspects and potential impacts associated with a product, by: compiling an inventory of relevant inputs and outputs of a product system; evaluating the potential environmental impacts; and interpreting the results of the inventory analysis and impact assessment phases. LCA is often employed as an analytical decision support tool"* (Fava et al., 1993).

LCA is an interactive process and the boundaries of the assessment may be adjusted throughout the exercise. PAS 2050 is the most commonly used standard for assessing carbon footprint. For example, the threshold for significant assessment per PAS 2050 is 1% of the life cycle GHG emission of a product. Any emission lower than that is considered insignificant for assessment. In LCA, creating and managing both the input and output data is considered as inventory analysis. Interestingly, different standards show different ways of data allocation. Dias and Arroja (2012) had shown allocation of carbon footprints during LCA of office papers under three standards: ISO 14040/14044, PAS 2050: 2008, and the framework of the Confederation of European Paper Industries (CEPI). Among the three approaches, PAS considers maximum materials for accounting, then the ISO 14040 standard, and the CEPI framework accounts for relatively less percentage of GHG emission. The study of the researchers concludes that allocation of carbon footprint per tonne of office papers is 860 kg CO_2 eq (CEPI), 930 kg CO_2 eq (ISO 14040/14044), and 950 kg CO_2 eq (PAS

2050). This clearly justifies the importance of selecting the most suitable standard in carbon footprint allocation, depending on the goal and scope of a specific LCA exercise.

In case of any product, the sustainability aspects are assessed considering the four key ingredients that lead to environmental degradation: GHGs, water, packaging, and waste. Moreover, the life cycle of LCA comprises six main components of the supply chain—raw materials extractions/production and supplies, manufacturing, storage, transport, wholesale and retail, and use by end users and disposal (end-of-life). GHGs may be emitted from various sources (during activities under these six components), which are ultimately to be combined as the GHGs for the whole life cycle. For example, the embedded GHGs in bread comprise 45% from raw materials (i.e., wheat production); 23% in manufacturing (i.e., baking); 6% in logistics, distribution, and retail; 23% in consumer use; and 3% in recycling and disposal.

As mentioned earlier, the ISO-based definition states that LCA is a process of compilation and evaluation of the inputs, outputs, and potential environmental impacts of a product system throughout the life cycle. LCA framework takes care of various issues during this analysis. It may include the data item like bill of materials (BOM) of the product, local and global indices of LCA, and impact assessment factors for socio-economic analysis. In this context, let us refer to the content of Chapter 11, "Green Supply Chain Management: Product Life Cycle Approach," authored by Wang and Gupta (2011). The authors identified the most popular international standards for carbon footprint analysis, which are IPCC directives, ISO 14064 standards on GHG emissions (including all specifications of ISO 14064–1, 2, and 3), CNS 14040, CNS 14044, and PAS 2050. As a practical consequence of maintaining the Kyoto Protocol, LCA primarily focuses on GHG inventory for subsequently using them in impact assessment and analysis for emission reduction. Various software are now-a-days available on LCA execution, some of them are simple database management systems, and some others are for impact assessment and analysis. Some software is applicable for all situations, whereas others are meant for specific sectors and products. Wang and Gupta (2011) discussed the effective use of LCA software for estimating carbon emission of products and also analysis of recycling cost. The software output may even show the reuse, recycle, and recovery ratio per WEEE guidelines. Authors also claim that there exists software that supports the optimization of remanufacturing or recycling activities. This type of software is expected to compute recycling ratio, recycling cost, and recycling efficiency. Of course, it estimates the carbon emission generated during the whole process.

LCA is a systematic approach of assessing environmental impacts, and steps for implementing this approach are discussed further here.

Life cycle assessment (LCA) is a popular technique in assessing and analyzing environmental impacts during the life cycle of a product. It has been accepted by the business community as an effective tool in sustainability management. It primarily covers the critical activities like inventory analysis, impact assessment, and interpretation.

5.3.2.1   Identification of Purpose and Scope

Like every other study, LCA may be initiated with identification of its purpose, its boundaries or borderlines, and its limitations and constraints. Naturally, it starts with some assumptions. It is to be clearly decided at the outset that whether the full life line of the product to be studied or the study is to be restricted to production, transportation, or only a functional unit based on the purpose. The scope of the study may also be earmarked by the level of precision on data and its limits. According to ISO 14040, the purpose means the reason for carrying out LCA determined by specific stakeholders who are likely to make use of the study outcome. In this stage, even the specific impact categories and the required type of reports are to be finalized before the data collection. The type of product or customer category like B2B or B2C also influences consideration of stages of life cycle.

Before moving to the next stage (inventory analysis), inventory borderline is to be set. In other words, first the Product Category Rule (PCR) of ISO 14025 may be applied, and then the types of GHG emission are to be identified depending on the sources of emission on the basis of ISO 14064 GHG Protocol. These protocols are Category I or Scope 1 (direct emissions), Category II or Scope 2 (indirect emissions), and Category III or Scope 3 (other indirect emissions). These GHG protocols have been discussed earlier under carbon footprint analysis.

Among the three categories, Scope 3 seems to be the largest contributor of emissions. It even includes all emissions associated with extractions and processing of raw materials to be used by the suppliers for further processing to make them ready for supply to the manufacturer under study. The emissions under the three scopes are mutually exclusive. The items under Scope 3 are more related to processes included under the value chain of the manufacturer, not the manufacturer itself. This chain covers activities of distributors, wholesalers, retailers, and the like.

Mathews et al. (2008) show that Scope 3 emissions are primarily accounted for upstream activities, and they may be even around 74% of carbon footprint in industry. Huang et al. (2009) justify that in US these upstream emissions are primarily from top ten suppliers of companies. This information helps companies to identify the most emitting suppliers and their activities. Further, it has also been seen that business travels contribute immensely to carbon footprint. It is noticed that Scope 3 emissions widely vary from sector to sector. This causes difficulty in establishing a generic guideline on Scope 3 emissions data preparation for inventory analysis. Various carbon footprints generated by service providers, including consultants and designers, may also be included in Scope 3 emissions. Searching for the responsible stakeholders and their direct and indirect contribution to Scope 3 carbon footprint is really a herculean task, if done exhaustively. That is why a realistic LCA requires clear definition of the purpose and demarcation of the scope, as well as boundary of the analysis, depending on the type of product under study, sector, and strategic importance of the organization.

By purpose, we mean the main goal and ultimate use of LCA report. For example, it may mean exclusively carbon footprint computation, water footprint computation, total consumption of some scarce or non-renewable resource, requirement for resolving some litigation or legal conflict, image building of the corporate brand in the market, or meeting of some governmental legislative restriction. Thus, LCA may be conducted on the generation of a single output or all outputs due to all the relevant operations. Report writing may also be done per a standard or specific format or with special emphasis on certain issues and parameters.

By the scope of the LCA, we mean the operational units to be studied. For example, LCA may be conducted on inbound or outbound transportation, warehouses, and manufacturing units, not the whole chain of processes. There may be LCA for the whole supply chain or for the supply chain with restricted number of channels. The study may be carried out for a specific plant or for plants at various locations for the manufacturing organization with operations spread across the country or even across the globe. The scope of the LCA may also be for the manufacturing plants of the suppliers for making the whole supply chain a sustainable one.

This step thus maps the boundary and criticality of the issues for guiding the subsequent activities under LCA.

> *LCA may be initiated by clear identification of its goal or purpose and scope. This is practically a strategic decision involving the role of external stakeholders including governmental agencies, market, and society. It shows the direction, limits, and constraints in carrying out the whole LCA project.*

5.3.2.2 Inventory Analysis

Inventory analysis is the second phase of LCA. Here, by inventory, we mean the generation of stock of environmentally harmful items, particularly the carbon footprint due to the product production, distribution, use or consumption, and disposal. This step is actually collection of this data generation and assessment procedure. It is of significant interest, as the subsequent analysis, interpretation and corrective actions entirely depend on reliable data collection of harmful items that have been generated because of various direct or indirect processes associated with the entire life cycle of the product. The reliability of this data set also depends on its accuracy, relevance (with reference to the purpose of LCA), and whether it is of required quantity or not. Lack of required quantity of data or nonavailability of data may sometimes lead to the modification of the scope. Per ISO guidelines, this phase involves compilation and quantification of input and output data related to production and use of a product throughout its life cycle. The following data are primarily collected in inventory analysis of LCA.

- Energy requirements
- Requirements of raw materials or natural resources
- Atmospheric emissions (mostly gaseous wastes)
- Emission to land (mostly liquid wastes)
- Solid wastes
- Other releases to environment, which are expected to cause harm to the earth, the ecosystem, and the human body

Although it is advisable to collect data directly from the primary sources, sometimes it may be impossible, impracticable, time-consuming, or expensive. So analysts may make use of secondary sources of data like technical documents, commercially available databases, published governmental reports or scientific publications. However, the LCA reports should contain the list of these reports or websites as bibliography at the end of LCA reports. Each in-text citation should reflect the reference of secondary source in the report.

As data are collected on all components associated with each operation of the product life cycle, often the size of gathered data becomes huge. So it is advisable to fix some threshold value in order to ignore the items or components, which give rise to insignificant contributions with reference to the threshold value. For example, components contributing less than 5% of inputs may be ignored. Further non-quantifiable impacts (e.g., loss of beauty in landscape due to mining of an area for extraction of items like coal, iron ore, water, sand, and stones) may also be ignored because of the difficulty or impossibility of quantification. This also restricts the size of database and complexity in inventory analysis.

As the input and output data for LCA relate to processes involved in the life cycle of the product, the data collection and analysis require the following steps for carrying out the inventory analysis.

1. Development of a flow diagram of the processes within the defined boundary of LCA
2. Development of a data collection methodology
3. Collection of the relevant data
4. Evaluation and reporting of the results

In most of the manufacturing organizations, the main group of operations that are expected to be included in inventory analysis are the following.

1. Production, use, transportation, and treatment after use or disposal of the main or primary product being manufactured.
2. Production of ancillary or secondary products or materials, such as packaging materials, and making it ready for packaging the primary product (both industrial and consumer packaging). Sometimes, because of the uniqueness of the manufacturing process, some co-products are also produced along with the primary product.
3. Energy production (Scope 2) required for the previous two types of operations.

A complete system of production-use-disposal of a product, or in other words, its life cycle, is divided into various subsystems. Figure 5.6 exhibits a short snap of the flow diagram of product life cycle.

Most common set of subsystems are as follows:

- Raw materials extraction and processing to convert them to the required input materials for the manufacturing the primary product
- Manufacturing, fabrication, assembling, and packaging

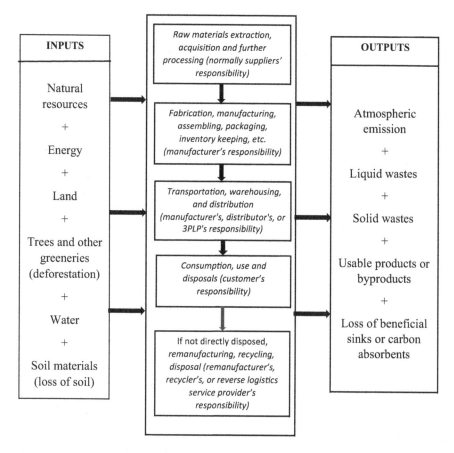

INPUTS		OUTPUTS

Raw materials extraction, acquisition and further processing (normally suppliers' responsibility)

Natural resources

+

Energy

+

Land

+

Trees and other greeneries (deforestation)

+

Water

+

Soil materials (loss of soil)

Fabrication, manufacturing, assembling, packaging, inventory keeping, etc. (manufacturer's responsibility)

Transportation, warehousing, and distribution (manufacturer's, distributor's, or 3PLP's responsibility)

Consumption, use and disposals (customer's responsibility)

If not directly disposed, remanufacturing, recycling, disposal (remanufacturer's, recycler's, or reverse logistics service provider's responsibility)

Atmospheric emission

+

Liquid wastes

+

Solid wastes

+

Usable products or byproducts

+

Loss of beneficial sinks or carbon absorbents

FIGURE 5.6 System flow diagram for inventory analysis of a product life cycle.

- Transportation and distribution
- Consumption, use, and disposal at the consumer's or customer's end
- Remanufacturing, refurbishing, or recycling

Let us focus on some uniqueness in inventory analysis for each of the subsystems.

5.3.2.2.1 Raw Materials Extraction and Processing

This subsystem covers operations involving extraction of raw materials from nature (e.g., mining, oil exploration, extraction, mineral processing), transportation, material-handling, and conversion of raw materials to the input materials required for manufacturing of the main product (and/co-products). However, the selection of right materials as inputs to the manufacturing process precedes the planning for carrying out the operations in this subsystem. It is the outcome of a cost-benefit analysis considering environmental cleanliness. A package made from recycled material is surely more environment-friendly. But because of its decreased strength (through recycling), its thickness increases, so a greater amount of packaging material is required. So, its waste generation will surely be more at the end of the life cycle of the primary product than that of packages made from virgin materials. Non-renewable and renewable raw materials are to be segregated during this data collection exercise. Energy and water are two common inputs to be treated separately. Energy may be used directly as input material, like solid or liquid fuels, in both manufacturing operations and transportation. But if electricity is considered as input material (used in almost all machine-driven operations), the data collection may be done at one step backwards—that is, including all carbon emissions at the level of electricity generation (Scope 2 GHG Protocol). Moreover, direct inputs like diesel also require consideration of backward analysis. Here, the crude exploration is subjected to refining operations, which is also not a very clean production process. Of course, in case of imported crude the data may only be considered from refinery stage. This again depends on the *scope of analysis* defined and clearly mentioned in the first step of LCA.

5.3.2.2.2 *Fabrication, Manufacturing, and Assembling*

Maximum waste generation, emissions and use of materials, energy, and water are expected to occur during operations of this subsystem. However, the analysis of this subsystem demands consideration of some specific issues. First, although the process of data collection will be same for all cycles of production, in case of some variations of material use or some change in production process itself or even machine efficiency, the data collection is to be carried out exclusively for this manufacturing plant (or manufacturing subsystem with that changed materials/process/technology) only. Second, there will be repetition of data collection, if new co-products are to be produced by the same production process. The allocation of materials and energy needs to be recalculated in this case. Third, scrap generated from production process may be reused as inputs (e.g., sectors involving steel, paper, and glass). Input-output relations are to be exclusively analyzed in other similar cases as well, like if energy (electricity) is required for generation of energy itself (in case of any power plant). Fourth, while using industry average data (secondary data source), it is to be noted that there may be differences in technology level, utilization of technology, and interventions of machines in place of human beings among the organizations under the same industry. The higher the intervention of human elements in lieu of machines, the lesser the chance of maintaining a standard set of data across the industry. This results in the creation of inventory data for a technology driven business are quite different from that of businesses intensive with human elements or with less technology intensive in the same industry. The difference will be not only in use of materials and energy but also in the generation of unwanted wastes and emissions.

5.3.2.2.3 *Transportation/Distribution*

Normally, the standardized data on emissions may be obtained for each mode of transportation on the basis of average distance traveled and load it is carrying. If the distribution network includes multi-level or multi-channel transportation, then the inventory data not only include inputs and outputs for each mode at each level but are also expected to cover inputs and emissions at each warehouse in the network. In case of the multi-level distribution network, data collection at the hub or warehouse level will be the additional task. Moreover, the use of refrigerators in vehicles (cold supply chain) and/or warehouses may require some special treatment for inventory generation and analysis. However, it entirely depends on the scope of LCA.

5.3.2.2.4 Customer Use and Disposal

This subsystem covers input/output data relating to all the activities for consumption and use of the product under study by the customers. These include its consumption (in case of consumer good), storage, further processing (in case of industrial good), maintenance, and reuse (by some other customer; e.g., secondhand car). Disposal may take place after the completion of its usable life. The data under this subsystem are affected by the consumer's behavior, work environment, and biases during the use. On the other hand, after disposal, it may be not only reused by some other customer (in case of durable items) but also recycled or composted. Some products may be remanufactured for further use (as good as new), used for landfilling, or incinerated at the end of its life. Each generates different types of inventory data with different data sources and different modes of data collection.

5.3.2.2.5 Recycling/Remanufacturing

This refers to the possible value recovery activities on discarded products after use of the primary customers. This has been discussed extensively in Chapter 4 of this book. If the recovery of products leads to recycling or cannibalization (per Thierry's classification of product recovery options or process), then the output will be a completely different product or material, which is expected to be used by other customers or to be used as inputs in other production processes. In this case, the data collection will continue till the production of the recycled items only. The consumption of the recycled item or its use as input material for another production process is normally expected to be beyond the scope of the current LCA. If the product recovery is done through remanufacturing of used products, then there may be two situations. If the remanufacturing is conducted by the OEM itself, LCA may continue (depending on predefined scope and goal) till the production of the remanufactured products, which are subsequently sold to secondary markets. In this case, the period of the life cycle may be further extended by the inclusion of reverse logistics process, remanufacturing, and distribution of remanufactured products to secondary users or customers. Assessment and analysis of sustainability parameters are done on all those additional processes. This may be applicable for car, big machineries, earth moving heavy machineries, aircraft, etc. In this case, this LCA is meant for the cradle-to-cradle situation, unlike the popular cradle-to-grave LCA studies. On the other hand, if the remanufacturing is done by some service provider or outside remanufacturer, then the situation will be slightly different. Here the OEM is only responsible for supply of the used products collected from its customers to those external remanufacturers. The inventory data collection for LCA may end after the assessment of the processes like, collection, reverse logistics, and delivery of used products to that outside agency. However, if any component or part is remanufactured and the remanufactured item is reused in the main manufacturing process of OEM, then it is a closed-loop cycle. In this case, the data of various environmental impacts on the remanufacturing process may be collected as the data of input materials (in inventory of LCA) for the manufacturing of the product under study. As a matter of fact, most of the manufacturing processes nowadays are carried out using recycled input materials. The use of recycled or remanufactured components/materials replaces virgin inputs and eliminates unwanted emissions due to pre-processing of natural resources for conversion to the desired components/parts. If an automobile company uses remanufactured engines (with same period of warranty for remanufactured component as offered in newly manufactured cars) in place of a new one for car assembly, then this results to conservation of raw materials required for engine manufacturing and avoids unwanted emissions in making the new engines. It may further be noted that any remanufacturing or recycling process once again generates some emissions, although it saves carbon footprints for the disposal of primary used products. However, the scope of LCA directs whether this additional emission will be added to the LCA of the parent product or the LCA of the product that requires the recycled items as inputs to its production process.

As mentioned earlier, the first step is drawing the flow diagram of the system, keeping in mind the scope, goal, and the boundary of the study. The flow diagram should contain the essence of the processes(s) and all possible inputs and outputs. If the flow diagram is quite detailed, you may expect greater accuracy and multiple use of the results. But this makes the flow diagram more complex and the data collection and analysis often demand more time and resources. For convenience in data gathering, a large system is often divided into a series of connected subsystems. For each subsystem, the analyst should separately analyze and collect materials and energy requirement data, and all forms of environmental emissions, along with the processes involved in the subsystem itself.

Prior to data collection, the quality of data is also to be ensured per the guidelines framed at the beginning of the study. Criteria for checking the data quality primarily include the precision, completeness, representativeness, consistency, and reproducibility. There may also be several categories of data, like operations- or process-specific data, aggregated data, industry-average data (primarily for benchmarking), and generic data. It is advisable to include the facts like purpose of inventory creation, scope of analysis, data category, and source of data in the spreadsheet modeling while designing the data file. In order to properly collect the requisite data, first the main product and co-products are to be defined. Then, the uses of all the resources, materials, and energy are to be appropriately apportioned among the products or co-products. The excellent research article by Babu (2006) provides the guide lines of International Standards Organization (ISO) in this context, which are discussed in the following sections.

Step 1: Instead of direct allocation of resources, the LCA analyst should judiciously identify and prioritize the processes or functions on the basis of their contributions to production and other value-additive activities of all the products (including co-products). Subsequently the resource allocations may be estimated as the allocation of input data for LCA of products.

Step 2: In this pursuit, the underlying physical relations and transformation of inputs may be quantitatively established between inputs/outputs and products through the processes.

Step 3: The physical relationships may also be represented in terms of economic values during this allocation.

The data for inputs are to be created considering the scope of study and significance of the inputs in generating pollutants beyond some threshold values. Normally energy inputs are assessed on each process, transportation, or any other material-handling equipment based on certain standard rate of consumption. Energy inputs may also be computed by a one-step-backward approach, like fuel inputs in generating electricity, which is subsequently used for production of the products under study. This refers to the Scope 2 protocol of environmental study. Analysts make use of standard and published data set on the requirement of energy for carrying out each activity, which is further equated with production of the product. There may be various energy sources (coal, petroleum, hydropower, nuclear, solar, wind, solid waste, etc.). Most often the cost of the conversion of energy sources to electricity per kilowatt-hour and availability of these sources determine the most economic electricity generation process. This process and the energy sources are once again environment-polluting activity and consuming a lot of scarce natural resources.

Water is also another input item, which requires special consideration due to environment-polluting pumping operations, electricity consumption, and conservation issues because of the scarcity of water resources. However, its data collection will be little more complex, if it is not the simple flow of fresh water, whenever required. What we try to mean is the use of recycled water along with fresh water. It needs a different treatment, as fresh and recycled water may be used in different processes; the generation rate of recycled water depends on the rate of recycling, and recycling itself is an environment-polluting process.

Of course, it is advisable to produce energy and water inputs from renewable resources, although in such situations, we should remember that **it is only valid if the rate of regeneration of renewable resources is higher than the rate of consumption of energy and water**.

> *Inventory analysis is a crucial phase of LCA for the collection and storage of relevant input and output data for subsequent environmental impact analysis. The data are collected from activities and processes in all stages of product life cycle, like raw material processing, manufacturing, transportation, use and disposal, and recycling of used products before ultimate disposal.*

5.3.2.3 Impact Assessment

Impact assessment, or sometimes known as life cycle impact assessment (LCIA), is the stage of LCA for understanding and evaluating the magnitude and significance of the potential environmental impacts for the production, use, and disposal phases of a product throughout its life cycle.

This impact assessment is carried out analyzing the inventory table or database developed in the preceding stage of LCA. Two situations often arise at this stage— too large a database and too small a database. If the data table is huge and much detailed and covers critical facts demanding in-depth domain knowledge, then the

LCA analyst is to consult the relevant domain experts for assessing, analyzing, and making out meaningful conclusions from the inventory data. It is thus little time-consuming and expensive. On the other hand, handling a small inventory database is relatively easy, less time-consuming, and relatively cheaper. But this may end up with relatively less-detailed analysis, the interpretation and conclusion may be incomplete, or the recommendations may be ineffective.

The aim of this stage is to convert (preferably by quantitative analysis) the inventory of data to various types of impact on well-being of earth and mankind. The impacts are often classified on the basis of certain categories to facilitate the process of mitigating the environmental degradation, improving the work environment and quality of the product, and taking proactive corrective actions for making the product life cycle sustainable.

Perhaps, the broad classes of environmental impacts will be the global warming, health hazards through toxic chemicals, and conservation of natural resources, particularly the non-renewable fossil fuels. Among these three classes of environmental sustainability issues, climate change (or global warming) seems to be the most crucial one because it is the concern of almost all global citizens.

Let us illustrate the main idea of environmental impact in LCA. The relevant data (particularly the output) in inventory database or table are supposed to contain CO_2, CH_4, N_2O, and other GHGs. These are directly contributing to global warming and thus climate change. It may be noted that climate change occurs as a result of a chain of causes and their links. The following is a small chain of causal links, which may lead to climate change.

1. GHG emission causes changes in the composition of the atmosphere.
2. Change in atmospheric composition causes change in radiation balance.
3. Change in radiation balance causes change in temperature distribution.
4. Improper temperature distribution causes climate change, changes in the ecosystem, biodiversity, and so on.

Moreover, high GHG emissions trap the heat like greenhouses and ultimately causes global warming, glacier melting, and floods.

The next step is to select the appropriate impact indicator(s) and to find the mechanism for converting emission data to impact indicator(s). Experts in biology, chemistry, meteorology, and environmental science have already developed models depicting the causal relations (conversion) of each link in the multistage model of sustainability impact. Experts in the Inter-Governmental Panel on Climate Change (IPCC) tried to capture all these technical models to quantitatively depict the conversion functions, along with the impact formulae of GHGs in causing climate change and global warming. These are known as global warming potentials (GWPs) (Guinee and Heijungs, 2017).

In LCIA the difficult task is to determine the *indicator* variables corresponding to each *impact category*. Measured values of indicator variables will reflect the environmental impact of any process or business activity. So what is most important is to find out the conversion function:

$$Impact = f\ (indicator\ variables)$$

We may also explore inverse function as value of *Indicator variable = F (Impact)*, as the cause-effect linkage.

GWPs may help in this pursuit. Moreover, because of the chain of causal linkages, there may be midpoint or in-between indicators (indicators manifested within the chain, or before the completion of the chain of linkages) and also the end-point indicator within a particular impact category. Table 5.7 exhibits an illustrative example of seven impact categories.

We are not to forget that there may be more than one indicator for an impact category, and also an indicator may represent more than one impact categories.

For better understanding of impact analysis or assessment, let us consider a hypothetical example of inventory data set. It shows 150 kg of CO_2 emissions, 1.2 kg of CH_4 emissions, and 1.4 kg of SO_2 emissions. Now we are to refer to standardized mode of conversion (or equivalence) like GWP to represent all the three types of emission to a single type, CO_2 emission, which is the universally accepted as standard unit of GHG emissions. Based on the GWP table, if the CO_2 equivalent emission is considered to be 1 unit (1 kg), then CH_4 is equivalent to 25 kg CO_2 equivalent. SO_2 does not have much impact on climate change, so let us ignore it. Thus, the conversion will lead to the following.

$$1*150+25*1.2 = 150+30 = 180 \text{ kg } CO_2 \text{ equivalent}$$

TABLE 5.7
Examples of Indicator Variables Corresponding to Impact Categories

Impact Category	In-Between Indicator Variable	End-Point Indicator Variable
Global warming	Effect on infrared radiation and GHG coverage	Loss of life, disappearance of species
Increase of carbon emissions	Worsening of the AQI rating and increase of CO_2 in the atmosphere	Bronchial issues
Ozone layer depletion	Change in tropospheric ozone concentration	Increase in the number of cancer patients and loss of life
Depletion of coal, petroleum, and other hydrocarbon resources	Increase in energy demand and thus meeting it by economically efficient fossil fuel exploitation at a higher rate	Increase of cost of production because of deeper mining and exhaustion of fossil fuel in the near future
Depletion of natural resources	Searching for alternative material resources or attempt for its artificial substitutes	Huge investment for production facility of alternative materials or scarcity or dearth of material sources
Land use impact	Amount of biologically productive land occupied, used for some purpose, and transformed to something else	Disappearance of trees and other species
Deforestation	Reduction of sinks, greeneries, source of fresh vegetables, and other essential resources	No natural sinks for CO_2, natural calamities, and so on

Mathematically, its generalized version may be expressed as the global warming score (GW) of the project in CO_2 equivalent:

$$GW = \sum_i (GWP_i * m_i)$$

GWP_i is the potential of ith GHG component in global warming and m_i is its emitted amount per the inventory database. GW provides the influence of a project on global warming measured quantitatively in a unified scale, which can be compared with other projects for benchmarking or for taking up any control measure.

Now, the equation is meant for assessment of global warming effect by the GW as one of the impact categories (similar to climate change) because of GHG emissions. But each component of GHG or the polluting substance generates impact under other impact categories as well. So in order to generalize the impact assessment, the following two equations may be considered to represent a meaningful environmental impact model.

$$Imp_j = \sum_i (F_{ji} * m_i)$$

Total environmental impact due to project operations or process operations in the product life cycle =

$$\sum_j (W_j * Imp_j)$$

Imp_j is the environmental impact under j^{th} impact category. F_{ji} is the functional relation or the characterization that links i^{th} indicator or component to the impact of jth category. W_j is the weight, relative importance, or criticality of the jth impact category.

In the impact analysis phase of LCA, the analyst computes the total overall environmental impact due to various processes of the product life cycle. The environmental impact is classified under various impact categories, and the impacts are manifested through various indicators.

5.3.2.4   Interpretation/Analysis

This is the final phase of LCA, which primarily covers conclusions and further recommendations after critical analysis of inventory data and subsequent assessment of environmental impact. In this phase, the LCA analyst also checks whether the goal of LCA has been achieved and the study has been made within its prescribed scope. If the reader happens to come across ISO document on LCA, they may refer to some narration on this phase of LCA. It includes the identification of significant issues, critical analysis of inventory data along with their impacts, and sensitivity of the impacts to all known and partially known situations. Pinpointing the most critical or significant issue is really a difficult task. Various models or techniques, like the fish bone diagram or Pareto analysis, may be applied as the supporting tools in this exercise.

As concluding remarks on LCA, let us briefly comment on the following two aspects, which may represent the possible extensions of LCA.

5.3.2.5 Alternative to LCA

This popular alternative technique is also a measure of environmental sustainability, although having a focus on economic achievements at the cost of some environmental load or degradation. So it is an efficiency-oriented measure known as environmental efficiency or, more popularly, eco-efficiency. It was first proposed during the Earth Summit in Rio by the World Business Council for Sustainable Development (WBCSD) in 1992. Eco-efficiency (or socio-efficiency) has been discussed earlier, but it is repeated here because of its role as an alternative to LCA in some cases.

It is the ratio of the value of a product or service produced along with resulting environmental load—or in other words, the value of the product or service generated by unit environmental load. The Japan Management Association, because of its orientation toward quality and productivity, expressed its interest more toward this tool than the conventional LCA. The environmental efficiency may be perceived as a **productivity measure** showing generation of economic output at the cost of unit environmental damage. Alternatively, it may be perceived as a weak sustainability measure, reflecting best possible capital substitution (man-made capital substituting natural capital), which means trade-off between profit and planet dimensions in TBL/3Ps. The following formulae may be used for evaluation of eco-efficiency. The second formula is for comparison or benchmarking.

$$Eco\text{-}efficiency = (Functional\ value\ of\ the\ product)/(Environmental\ impact$$
$$during\ product\ life\ cycle)$$
$$Performance\ w.r.t.\ Environmental\ efficiency\ in\ a\ particular\ year =$$
$$(Eco\text{-}efficiency\ of\ the\ product\ under\ study)/(Eco\text{-}efficiency\ of\ the\ standard\ product)$$

It may further be noted that in similar line social efficiency, or socio-efficiency, may be measured, like the following:

$$Socio\text{-}efficiency = (Functional\ value\ of\ the\ product)/(Societal\ impact\ during$$
$$product\ life\ cycle)$$

So we may even conclude that through this technique, we are addressing weak sustainability involving the three bottom lines, as the value of product is primarily assessed in economic terms.

5.3.2.6 Extension of LCA

Extension of conventional LCA methodology has been done by incorporating the other two dimensions of TBL in the scope of LCA. It means that the extended LCA includes all three dimensions of sustainability. So this life cycle sustainability assessment (LCSA) may be treated as the completeness of LCA at least in conceptual form (Klopffer, 2008).

$$LCSA = LCA + LCC + SLCA$$

Here, by LCC and SLCA we mean life cycle costing and social or societal life cycle assessment, respectively.

Moreover, please note that the right-hand side of the equation is not the algebraic summation of the three components. It only indicates that the completeness of LCSA may be achieved if we can capture all the three dimensions of sustainability during our analysis: environmental impact, cost issues resulting from penalty or management of unsustainable practices, and social impact.

The LCA and LCC were developed in the 1980s and 1990s, respectively. SLCA got the acceptance of global experts only in the first decade of the 21st century. Although LCSA in totality has been understood by researchers, practitioners, and policy-makers as a meaningful concept measuring the three-dimensional impact of any project or business endeavor, it is yet to be put into practice because of some difficulty and ambiguity encountered by the analysts while applying it in industrial activities.

There lies some difference between the conventional cost accounting and LCC. Costs involved in the sustainable use of products and in waste removal or recycling generally do not show up in conventional cost accounts. LCC includes the use and end-of-life treatment (cradle-to-grave or even cradle, as in LCA), unlike the usual cost accounting-based computation of cost of goods sold (cradle-to-gate or cradle-to-point-of-sale). LCC is meant for economic assessment of the impact, not for preparation of documents in standard format or reporting for cost control. Besides, LCC includes accounting for hidden or less-tangible costs associated with environmental protection and other practices for maintaining sustainability. All sorts of such interventions are to be reflected in the report or document of LCC. However, in this process the analyst faces difficulties. For example, external costs due to environmental damage are the expenditure of the society or even the future generation which is supposed to suffer after two to three decades. So the difficulty lies in capturing the data from the societal representatives and/or because of the time lag between the occurrence of the cause and the result of the damage.

SLCA is yet to be developed as a full-fledged technique for assessing societal impact of a product during its life cycle. Unlike LCA (environmental), SLCA parameters cannot be quantified so easily. Even in LCA, an impact category like biodiversity can hardly be quantified with enough of objectivity. The following, in fact, are the unsolved issues in SLCA.

- There is hardly any complete and universally accepted list of social indicators, which may be easily measured and applied in social impact assessment.
- There is no exhaustive list of health measures, impact of toxicity, and measures of well-being of mankind. These impacts and measures are also dependent on the human body, lifestyle of the individual, and several other factors, which restrict us to consider these measures as standard ones. Further, most of societal impacts do have time lag for their full-blown result.
- It is difficult to include social impacts in the life cycle inventory database.
- Getting the universally applicable data set on societal impacts is really a difficult task. Let us consider worker's satisfaction as the societal impact due to business activities during a product life cycle. Worker's satisfaction

apparently seems to be a qualitative parameter, and this requires surrogate parameters for measurement. For its measurement we need values like income per hour or day, number of working hours per day, expectation level in meeting social and physiological needs (education, healthcare, etc.), need for savings for future security, and so on. But there is no universally acceptable standard scale available even for measuring all these parameters. Values often change with the type of process, location, or demographic characteristics of individuals.

LCA may be extended by the use of eco-efficiency as its alternative and for making conventional LCA concept more inclusive LCA may also be extended by including other two dimensions as life cycle sustainability assessment (LCSA).

LCSA = LCA + LCC (life cycle costing) + SLCA (social life cycle assessment)

If we try to summarize the initiatives in measuring sustainability-related performance, it may be classified in a set of concepts, measures, and mechanisms like the following.

Concepts: sustainable development, triple bottom line (TBL or 3Ps)

Measures: carbon footprint (environmental air pollution), water footprint (water conservation), ecological footprint (conservation of land and natural resources), eco-efficiency (economic value generation at the cost of environmental damage)

Initiatives: Annual summits or COPs by UNFCCC and IPCC leading to carbon credit and carbon trading on the basis of the mechanisms like cap-and-trade, clean development mechanism, carbon tax, or extended producer responsibility (EPR)

In this context, it may be proposed that analysis and assessment based on the product life cycle concept originates from life cycle thinking, and slowly it gets matured over time. The final or fifth stage of development is LCSA. Figure 5.7 is a snapshot of the chronological stages, which represents how the life-cycle-based thinking of sustainable development can slowly be transformed into its most comprehensive form (LCSA).

Nowadays many software are commercially available for carrying out computational activities in LCA. LCA software is primarily covering the computational part, which is mostly simulation-based, and also the database management part in managing the inventory data set. Although the list of available LCA software in the market has been expanding over time, the following are some of the popular LCA tools that demand mentioning in this context.

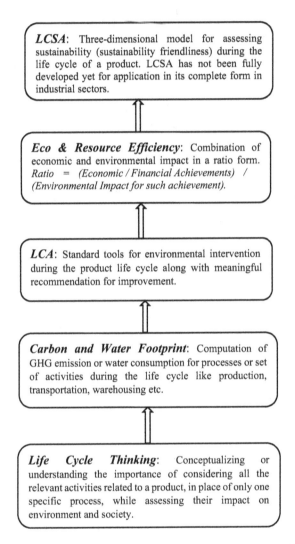

FIGURE 5.7 Snapshot of stages showing development of contemporary LCSA.

Ecochain is an environmental system platform. Its support is primarily in reflecting carbon footprint covering the whole organization (www.ecochain.com).

OpenLCA is an open-source LCA solution, available to everybody (www.openlca.org).

Mobius has been developed with the primary focus on product design. Alternative scenarios may be developed during design process and environmental impacts may be assessed (www.ecochain.com/mobius).

SimaPro was developed around three decades ago. It is a popular LCA tool in the market, primarily among academicians and LCA consultants (www.pre-sustainability.com/all-about-simapro). It was developed in the Netherlands.

GaBi Software was developed some time in mid-'90s and soon gained the popularity in domestic market of Germany. Its primary clients are LCA consultants (www.gabi-software.com).

One Click LCA is an LCA solution tool, which is specifically meant for LCA in construction industry (www.oneclicklca.com).

Athena is the software developed in Canada as Eco Calculator. It is more popular in the academic community than in the commercial circle (www.athenaSMI.ca)

ECO-it originates from the Netherlands as a software tool. It is popular in the commercial or business community, with a special focus on LCA for building design (www.pre.nl).

BRE is the tool developed in UK and is publicly available. The LCA is primarily meant for the whole assessment (www.bre.co.uk).

Ecoinvent was created in the Netherlands primarily as a database software for inventory data of LCA. It is more popular as a tool for comparison of sustainability friendliness among products (www.pre.nl/ecoinvent).

LISA is an Australian LCA tool publicly available and primarily meant for building design (www.lisa.au.com).

Key Learning

- The moderately to highly popular techniques for assessing organizational performance, including sustainability-focused (primarily on environmental dimension) performance, are eco-efficiency, ecological footprint, carbon footprint, and life cycle analysis.
- Eco-efficiency measures the value creation by the process, project, or business unit, which also generates environmental impact. In other words, this reflects the efficiency of utilizing the environmental resources. The similar definition is applicable for energy efficiency and social efficiency.
- In a populated area (a town, district, state, or country or even the whole earth), ecological footprint shows the demand on land and marine resources for consumption and disposal by the population. Bio-capacity represents the availability of these natural resources. Ecological footprint, on the other hand, is measured in terms of hectare (gha), or 10,000 square meters, which contains the natural resources demanded by the population.
- The Kyoto Protocol of 1997 declared the intention of reducing CO_2 emission to 5.2% below the 1990 level within 2008 to 2012 so as to restrict the ongoing global warming. Carbon trading was proposed as the mechanism to enable it. Like a share market, it is a global trading process of certified emission reduction (CER) or right-to-emission applying the cap and trade scheme.
- Carbon footprint analysis is a step-by-step approach for improving a process in order to reduce carbon footprint.
- During process analysis for assessing carbon emission, it is necessary to divide the types of GHG emissions on the basis of the GHG Protocol— Scope 1 (direct emissions), Scope 2 (indirect emissions because of procured energy, steam and the like), and Scope 3 (other indirect emissions).

- Carbon footprint analysis may be conducted using the activity-based emission (ABE) method, which is similar to the ABC method, which is often used for cost and operational control in an organization.
- Life cycle analysis (LCA) is a popular technique used by various organizational units and governmental agencies for assessing and analyzing the environmental impacts caused by a long chain of processes and for the complete life cycle of a product.
- LCA ideally covers analysis of all the processes of product life cycle, ranging from extraction of raw materials to final disposal. This technique includes activities like identification of purpose and scope, inventory analysis, impact assessment, and interpretation for further recommendation.
- LCA has been further enriched with the inclusion of other two dimensions of sustainability by extending it to life cycle sustainability assessment (LCSA), which combines LCA, LCC (life cycle costing), and LCSA (social life cycle assessment).

5.4 SCORECARD-BASED TECHNIQUES FOR INTEGRATED BUSINESS PERFORMANCE ANALYSIS

A scorecard-based framework for performance analysis is not essentially a mechanism for computation of any single index, but rather it is a matrix-based and multidimensional management framework. The scorecard represents separate existence of more than one dimensions, which are linked among themselves and all contribute to mission, vision, and strategic goal of the strategic business unit or a corporation.

Kaplan and Norton developed the balanced scorecard (BSC) approach (Kaplan and Norton, 2009), which is primarily meant for integrating needs of all stakeholders, which culminating to the strategic goal. BSC can be effectively applied for dynamic performance analysis of a business unit using the closed-loop cybernetic concept—plan-do-check-act—for this dynamism. The following shows a simple example of leading and lagging indicators as a chain of cause-and-effect relations in BSC modeling.

Know-how of employees → Quality and efficiency of the process → On-time delivery → Customer retention → Return on assets employed or ROI

These five parameters making the unidirectional flow of cause-and-effect relation represent the four dimensions or **perspectives** of BSC model like the following.

- "Know-how of employees" is the parameter of the **learning and development perspective**.
- "Quality and efficiency of the process" and "On-time delivery" are the parameters of the **internal process perspective**.
- "Customer retention" is the parameter of the **customer perspective**.
- "Return on assets employed or ROI" is the parameter of the **financial perspective**.

So the BSC model can make us understand how an intangible asset like "know-how of employees" leads ultimately to improvement of ROI of the company.

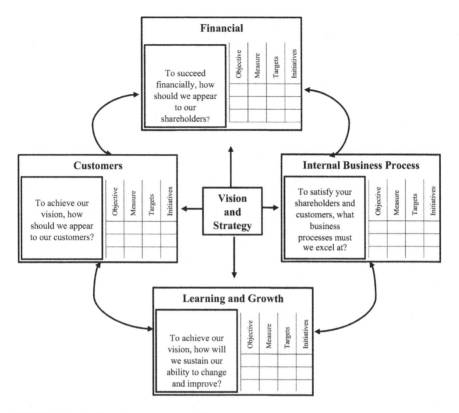

FIGURE 5.8   Balanced scorecard model.

Figure 5.8 displays the basic and popular model of Kaplan & Norton's BSC. It is quite evident that all managerial activities originate from the vision of the organization, which is subsequently translated into goals and then strategies.

This scorecard model proposes representing the apex-level strategies in four perspectives: *customers, financial, internal processes,* and *learning and growth or development.* Each of the perspectives is operationalized by a scorecard with its own *goal or objectives, measurements, targets,* and *specific initiatives.* The business activities start at meeting the demand of customers, and the performance is reflected as the financial outcomes. These four perspectives are closely related and linked. Moreover, the integration is also extended to all organizational levels. In other words, the BSC model enables an organization to create a suitable path of developing operational initiatives to meet its mission, vision, and strategic goals, bridging the gap between mission/vision and operational activities, simultaneously satisfying its stakeholders. It is thus emphasized that BSC is a model that uses an organization that can manage the demands of the relevant stakeholders and translate strategies to action. The four perspectives of BSC correspond to three stakeholders of the business. These stakeholders are shareholders (financial outcome), customers, and employees (internal processes and benefit of learning for growth). However, it is also noted that BSC is open enough to integrate little distant stakeholders, like environment and society. This, in fact, led to creation of the idea of including sustainability-related factors in the BSC model. Learning and development, along with creativity, is a critical perspective, which keeps almost all other perspectives updated and re-engineered continuously. This triggers the flow of activities that ultimately results to improvements at all perspectives or dimensions and finally to the financial outcome like ROI. BSC is balanced, as it brings in equilibrium in at least four ways.

First, it addresses both *long-term* and *short-term goals* and creates a linkage between them. Second, BSC interconnects both *internal* (internal processes, organizational development, HRM, etc.) and *external* (stakeholders like shareholders and customers) issues. Third, the implementation of this model means simultaneous consideration of both *quantitative* (financial) data and *qualitative* parameters like customer satisfaction, knowledge, and creativity. Last and most importantly, this model clearly identifies the *enablers* (*leading variables* or causes) and *results* (*lagging variables* or effects) and helps in depicting the inter-perspective and intra-perspective causal relationships.

5.4.1 Sustainability in the BSC Model

The original BSC model, in fact, ignored to include sustainability or related issues (ethics, human safety, etc.) as essential pillars or perspective in the scorecard. BSC is practically a finance-focused strategic instrument for an organization. So the inclusion of sustainability in BSC is a real challenge to business organizations. The integration of sustainability in BSC aims at improving corporate performance in all the three sustainability dimensions: 3Ps or TBL. If the improvement takes care of prioritized achievement on environmental and social dimensions, it may be termed as achievement of *strong sustainability* (very rare occasion); otherwise, achievement through judicious trade-offs will be the example of *weak sustainability*, which is quite common in business community due to its ease of implementation. Interestingly, there exist a couple of counter-arguments against the popular belief that achievement in economic dimension is always in conflict with achievements in other two non-economic dimensions of sustainability, as described in the following.

- If the economic condition of a business organization is quite sound, the organization may tend to spend the money from its reserve for better image building and for activities having no expectation for future financial gain. This automatically enables the company contribute toward environmental and social activities without any conflict with financial achievement. There are many such examples in business community in India and abroad.
- More focus on environmental and social issues improves brand image of the corporation (large conglomerates like Tata Group). This surely leads to better revenue and market share. Again, there is no conflict between the financial gain by revenue generation and the contribution toward environmental and social issues.

Because of the highly effective and comprehensive form of the BSC model, it is quite logical to explore the possibilities of including sustainability as another perspective in performance management. But the question is, how do we do that? The very nature of environmental and social issues is creating the challenges for the experts who intend to capture this dimension in BSC model. For ease of expression, let us use the term sustainability-related (or simply sustainability) BSC, or SBSC, to represent this newly developed model, and in fact, this term is already gaining popularity among the concerned members of the business and expert communities.

In this pursuit, Kalender and Vayvay (2016) propose three options or variants of integrating sustainability in the BSC model or for developing the SBSC model.

Variant I: Under this variant, environmental and social aspects are integrated in the four perspectives of BSC through respective strategic elements or performance drivers for which both the cause-and-effect indicators and also targets and measures are identified. Environmental and social aspects subsequently become the intrinsic parts of the conventional scorecard. As these two dimensions of sustainability are now in-built in strategy formulation, the measurement in financial perspective (e.g., ROI), influenced by shareholders' expectations, will surely have the required modifications with compromises for contributing to sustainability. Similarly, green-conscious customers will take part in the market share, and green processing will take place instead of conventional internal processes. Further, the learning, training, and knowledge gathering of employees will occur in their professional growth and creation of a system for organizational learning on greenness and sustainability.

Variant II: The second option is the addition of sustainability as the fifth perspective. Of course, this logically matches with the inherent openness of the BSC model. Further, the consideration of sustainability as an additional perspective in performance declaration often adds value to corporate brand. Here the sustainability-oriented BSC may be developed by adding two perspectives (six in total), like environmental and social perspectives. Each of the added ones will have same parameters—objectives, measures, targets, and initiatives—like other perspectives. Sustainability thus maintains its footprints in both cause factors and effect factors and balance them with those of other perspectives. The primary challenge in adding the fifth perspective of environment or social aspects (or both as fifth and sixth) to the conventional BSC is its non-market-oriented characteristics. Fundamentally, these two dimensions of sustainability have originated as social constructs, having least or no value through market transaction. The four perspectives of the conventional BSC model are showing the stakeholder needs and their linkages are ultimately aiming at achieving the ROI invested, which is determined by the current market exchange process or pricing mechanism. There lies the conflict in fitting the fifth one in the conventional BSC model. Actually, the necessity of this fifth and sixth non-market perspectives of sustainability arises, when the role of environment and social perspective becomes strategically relevant to all the other four standard perspectives. The addition of sustainability perspective also becomes important, when its strategic contribution because of the drivers from non-market supports the organization in achieving the success in a competitive market.

Variant III: The third option is the creation of a separate scorecard exclusively on environmental and social aspects. This may help measure the performance in terms of non-economic sustainability dimensions separately. This is also practiced in industries. But effectiveness of this approach is quite questionable, as sustainability issues are somewhat detached from the essential integration process of business activities. On the other hand, Figge and Hahn (2004) indicate that this third approach of creating a separate scorecard for sustainability is, in fact, not a standalone, separate approach. A derived separate scorecard is not an independent alternative but an extension of the previous two approaches of SBSC. This variant of SBSC represents a scorecard, which keeps a coordinated control on strategically relevant environmental and social aspects of the whole BSC system.

Let us have some snapshots (two business cases) on how businesses implement the BSC system, which includes sustainability issues (Epstein and Wisner, 2001).

5.4.1.1 Bristol-Myers Squibb

Bristol-Myers Squibb (BMS) is a multinational pharmaceutical corporation. It integrated the sustainability aspects in the conventional BSC model. This is a typical example of **Variant I** of integrating sustainability for development of SBSC. So sustainability factors are included in all four perspectives as the measures of performance. The following shows the environmental and social performance measures under each of the four perspectives: operational and environmental, health, and safety (EH&S) criteria (adopted from Exhibit 3 of Epstein and Wisner, 2001),

Learning and growth/development perspective: employee practices (training hours, ergonomic reviews, diversity) and transfer of best practices (ISO 14001 certification, product life cycle reviews)

Internal business processes perspective: environmental performance (water consumption, packaging reduction, percentage of solvent recycled, energy consumption, generation of hazardous waste materials, number of suppliers reviewed on environmental performance and fines charged, workers' exposure), employee performance (number of workdays lost, workplace related sickness or injuries)

Customer perspective: external customer support (product safety, post-customer-use waste recycled, customer education toward sustainability), good citizenship (number of awards related to social or philanthropic contribution, donations as philanthropic or social work)

Financial perspective: cost savings (savings from accident reduction, savings from PLC reviews), investment (fund earmarked on EH&S or environment, health & safety capital projects, expenditure on preventive and corrective actions against environmental and human health damages, community improvement programs), revenues (sales of socially and environmentally positioned products)

The corporate strategy of the business unit essentially includes the goal of social and environmental responsibility. BMS ensures implementation of all these measures as the performance metrics.

5.4.1.2 Severn Trent

Severn Trent in the UK is an international provider of waste, water, and utility services. The scorecard developed by this organization is similar to **Variant III** of the SBSC model—that is, a separate scorecard is created for sustainability-related performance. In this pursuit, it first identified four sustainability-focused strategic goals and subsequently corresponding 14 indicators. The indicators or measures are discussed in the following list, along with the strategic goals for creating the required separate scorecard. The strategic goals on the LHS are in italics, and the measures are in parentheses.

1. *Economic and employment growth* (economic growth as sales turnover, social investment, and employment as number of employees)
2. *Meaningful social progress* (health issues as services related to water supply and waste disposal, education and training, and housing quality)
3. *Environment protection* (climate change as GHG emissions, other air pollution, transport emission, water supply, wildlife as afforestation or plantation, and restoration of land)
4. *Sensible and careful use of natural resources* (consumption of natural resources and waste generation and disposal using land)

Thus, any business organization may identify its most suitable SBSC model for a balanced translation of its corporate strategy to actions. Care should be taken in selecting the list of measures that covers all aspects of perspectives and at the same time that is not too large a list causing enough of operational complexity. However, the measurement mix should be a combination of leading/lagging, external/internal, input/output, and financial/operational measures. It may further be noted that most of the businesses are feeling more comfortable in using Variant I of SBSC. In other words, Kaplan and Norton's balanced scorecard is kept intact with its four-perspective framework and their interactions, whereas the existence of sustainability measures is quite prominent under each perspective. Out of the three dimensions of sustainability, the conventional BSC model already includes financial/economic achievement. So only environmental and social issues are included as the measures of sustainability in this SBSC model. Like the conventional BSC, this model also maintains all required linkages for balancing and connecting between strategic phenomenon and operational actions. Figure 5.9 exhibits such a SBSC model as an illustrative example.

5.4.2 Sustainable Supply Chain (SSC) Scorecard

In any business scorecard, there is always existence of various stakeholders looking for satisfaction of their individual areas of interest. This means the existence of different strategic and operational goals, some of which are essentially conflicting in

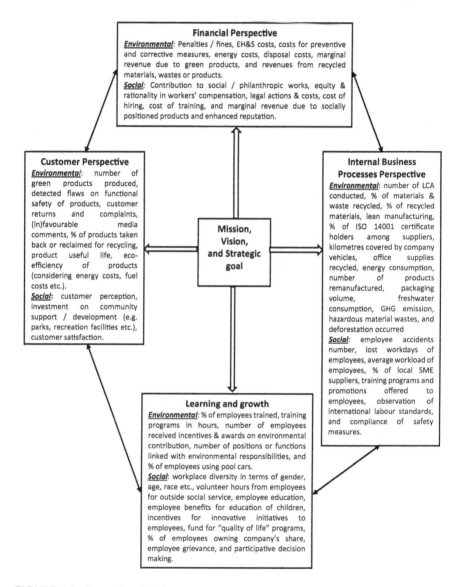

Financial Perspective
Environmental: Penalties / fines, EH&S costs, costs for preventive and corrective measures, energy costs, disposal costs, marginal revenue due to green products, and revenues from recycled materials, wastes or products.
Social: Contribution to social / philanthropic works, equity & rationality in workers' compensation, legal actions & costs, cost of hiring, cost of training, and marginal revenue due to socially positioned products and enhanced reputation.

Customer Perspective
Environmental: number of green products produced, detected flaws on functional safety of products, customer returns and complaints, (in)favourable media comments, % of products taken back or reclaimed for recycling, product useful life, eco-efficiency of products (considering energy costs, fuel costs etc.).
Social: customer perception, investment on community support / development (e.g. parks, recreation facilities etc.), customer satisfaction.

Mission, Vision, and Strategic goal

Internal Business Processes Perspective
Environmental: number of LCA conducted, % of materials & waste recycled, % of recycled materials, lean manufacturing, % of ISO 14001 certificate holders among suppliers, kilometres covered by company vehicles, office supplies recycled, energy consumption, number of products remanufactured, packaging volume, freshwater consumption, GHG emission, hazardous material wastes, and deforestation occurred
Social: employee accidents number, lost workdays of employees, average workload of employees, % of local SME suppliers, training programs and promotions offered to employees, observation of international labour standards, and compliance of safety measures.

Learning and growth
Environmental: % of employees trained, training programs in hours, number of employees received incentives & awards on environmental contribution, number of positions or functions linked with environmental responsibilities, and % of employees using pool cars.
Social: workplace diversity in terms of gender, age, race etc., volunteer hours from employees for outside social service, employee education, employee benefits for education of children, incentives for innovative initiatives to employees, fund for "quality of life" programs, % of employees owning company's share, employee grievance, and participative decision making.

FIGURE 5.9 Exemplary SBSC mode.

nature. This is a challenge in designing and creating a scorecard. A decent scorecard also enables a business to translate the strategic elements into a structured KPI system explaining the cause-and-effect relationship.

Section 5.4.1 shows the mechanism to be followed in businesses while implementing the sustainability-focused BSC or SBSC model. But that was for a single business entity. Let us now extend the concept to a supply chain, which is a set of business entities. The scorecard for sustainable supply chain represents a modified framework (Cetinkaya, 2011).

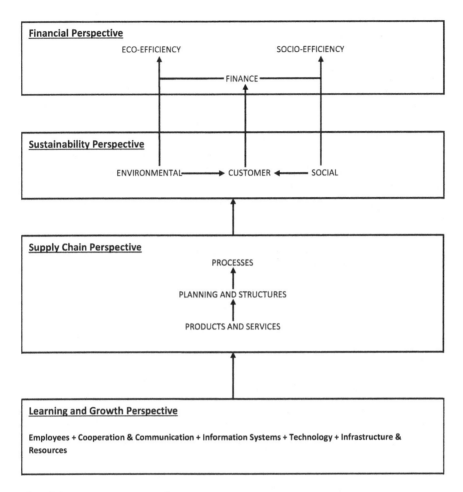

FIGURE 5.10   SSC scorecard framework.

(Source: Chapter 2, page 46, Cetinkaya, 2011)

The framework for sustainable supply chain (SSC) scorecard is similar to conventional BSC with some changes in defining perspectives. The SSC scorecard also takes into consideration four perspectives. However, these perspectives are somewhat different—*financial, supply chain, learning and growth,* and *sustainability.* The coverage of each of these four perspectives does include all the three dimensions of sustainability. The financial and the learning and growth perspectives of BSC are remaining in this SSC scorecard as the two perspectives with some changes in coverage. The supply chain perspective in this scorecard takes care of the internal processes in the BSC model and other activities involving tangible items and resources. On the other hand, the sustainability perspective includes the customer perspective of BSC and linked to financial dimension and revenue generation in SSC scorecard. Figure 5.10 exhibits the framework of this scorecard.

The following summarizes the role, content, and actions relevant to each of the perspectives in the SSC scorecard.

5.4.2.1 Financial Perspective

Unlike the financial perspective of BSC, this perspective of the SSC scorecard includes eco-efficiency and socio-efficiency measures incorporating the role of two sustainability dimensions along with conventional financial efficiencies. The primary idea in this perspective is to assess the sustainability-related activities in terms of both the cost of activities and the financial benefits (direct and indirect) received because of taking up green ventures in the sustainable supply chain. Actually, all dimensions of sustainability are expressed in financial or related terms in this perspective. The two efficiencies of sustainability measure the ratio between the value generation (primarily in financial terms) from the product or service and the environmental and social impact or cost due to their damages. For example, we may consider the generation of sales revenue from products per unit energy consumption or per unit carbon emission in tonnes in production or transportation as the current eco-efficiency of the company.

5.4.2.2 Sustainability Perspective

As mentioned earlier, this perspective in the SSC score involves customer satisfaction in the product or service, including the issues related to sustainability. This perspective is linked with the financial perspective, reflecting the achievement of customer needs as a result of the value propositions delivered by the supply chain. As customer satisfaction is linked with the sustainability dimension, this serves as the measure of supply chain impact on the environment and society. If the customers are retail stores or if the customers are users of industrial products meant for the production of some other item, linkage between customers and sustainability will represent the responsibility of customers toward two aspects or dimensions of sustainability. The existence of sustainability in the financial and supply chain perspective in this scorecard enables measuring the value proposition in environmental, social, and economic terms.

5.4.2.3 Supply Chain Perspective

In the sustainable supply chain, the supply chain perspective emphasizes the importance of the factors that have maximum impact on customers and also on environmental and social goals and thus ultimately the defined financial goals. Moreover, this perspective in the SSC scorecard includes both internal and external influence factors in carrying out all the internal processes. In other words, the internal process perspective of conventional BSC is included in this perspective. The external influence factors originate from other supply chain entities, like suppliers, distributors, and logistics service providers. These factors affect the process capabilities and reliability of processes of the company under study. They are also responsible for the green performance of the whole supply chain. Performance of a sustainable supply chain is measured by various factors, some of which are conflicting in nature, which are primarily affected by operational processes of the supply chain. The examples of these performance measures may be quality of products or services (in terms of meeting the desired specifications or standards, avoiding late deliveries, etc.), environmental damage (GHG emissions during production and transport, water consumption, energy efficiency, etc.), or social aspects (resettlement of displaced families, traffic accidents, etc.). Strategies of product and process design, planning for suitable distribution network, and so on enable the supply chain perform in a balanced manner.

5.4.2.4   Learning and Growth Perspective

This perspective seems to be the original driving force, which activates all the external and internal factors of supply chain perspective to achieve the ultimate financial goals (including eco- and socio-efficiencies) through customer satisfaction and environmentally and socially responsible performance. The categories comprising the capability of employees and information system, along with proper cooperation, are the essential drivers for activating elements in other perspectives. Besides, the technology-oriented capability, infrastructure, energy, and resources are also necessary to enrich learning and growth process and to translate all elements to effective performance of a sustainable supply chain. In other words, these drive the ultimate achievement of financial and sustainability goals by the sustainable supply chain. In addition to production facilities, transportation, warehousing, packaging, and material-handling operations are increasingly getting impacted by technological development, sometimes even by adopting imported technology. Modern technologies are more sustainability-friendly. These often include the mechanisms for environmental protection, control of resource consumption, and workplace safety. The use of alternative fuels, digital distance maintenance for trucks, and renewable energy in factory and warehouse facilities are some of the common characteristics of the new technologies. The SSC scorecard is also an excellent tool for monitoring infrastructure (transport and logistics) and energy resource data, using transport costs (including cost of fuel consumption, salary of driver and helper, and toll costs) as an appropriate measure. This parameter (as the cost of infrastructure) does have impact on the objectives of other three perspectives of the scorecard. The learning and growth perspective feeds its output directly to supply chain perspective, activating internal and external parameters. Consequently, the supply chain processes, collaboration and coordination with suppliers and/or distributors, and the resultant business model may be adjusted according to the capabilities and resources available. Further, the processes are enriched by parameters representing sustainability issues through the green-conscious market or customers. The ultimate achievement of financial gain reflects this endeavor.

5.4.3   Challenges in Scorecard Implementation

The implementation of any scorecard-based performance framework is a challenging task because of its multidimensional characteristics linking with all processes, functions, and roles or responsibilities of the organization. It would be little more complex if the objective is to create a well-balanced scorecard and once the sustainability issue is also added to it.

The SSC scorecard is a template and generic model, not a one-for-all type model, which may be used in any possible business environment. The measures, priorities, and targets for each perspective vary, as most of them are company-specific. The level of importance of stakeholders and perspectives also changes sector-wise. For example, the supply chain perspective for the manufacturing sector seems to have more importance than that in the e-commerce sector. Sectors with a highly competitive market like electronic sector (cellphone, television, etc.) are expected to prioritize sustainability perspective because of their customer-centric strategies incorporating enough importance of brand image maintenance (environment and society friendliness). So the SSC scorecard is to be adjusted to fit it to the industry, company, and region under study. The SSC scorecard is to be implemented at the strategic level. Any change in the socio-economic environment, product or process design, or technology adoption will surely mean changes in the SSC scorecard framework or elements.

While implementing scorecards, you may note that BSC is the tool through which the mission or strategic/corporate goal at the top-most level of the organization is converted to goals at all relevant responsibility centers. This is like a top-down approach, and it happens because of linking characteristics of BSC between the strategic and action levels. The objectives identified at various responsibility centers (as their goals) are correlated with the corporate or strategic goal of the organization. In fact, the BSC model is quite effective in establishing the linking both across all the perspectives and also across all the strategic business units (SBUs), divisions, or departments. The strategic goal cascades into objectives in all perspectives and relevant activity centers. In each of these cases, this is further converted to measures, targets, and initiatives. This issue has been identified and reported by Epstein and Wisner (2001) very clearly and emphatically. The goal cascading also includes various support functions of the organization. Most of the operational and local challenges are captured in goals/objectives/KPIs at SBUs or functional or departmental levels. The example of Unilever is cited by experts. The corporate level of Unilever management announced the reduction of environmental imprints. Now, some of its SBUs are under threat because of the quality and availability of water. So the strategic goal of Unilever is translated to the goal/objective of minimum use of water and water recycling for these SBUs. For some other SBUs the same strategic goal may be treated as reduction of packaging materials, GHG emissions, and reduction of use of fossil fuel to represent their SBU-level objectives for the BSC modeling. These objectives may also be termed as the sustainability sub-goals of the strategic sustainability goal (at the corporate level) in the BSC.

The following are some examples of cascading the strategic sustainability goal of maintaining 3Rs down to goals of different departments or activity centers.

Corporate/Strategic Sustainability Goal: Triple Rs (Reduce, Reuse, and Recycle)

Sustainability Goal of the Manufacturing Division

- Cost minimization
- Reduction of wastage, wastes, non-value-additive activities, and inventory
- Creation of remanufacturing facilities
- Reduction of packaging
- Recycling of water and effluents

Sustainability Goal of the Procurement/Purchase Division

- Preference for recycled materials
- Preference for recyclable materials

Sustainability Goal of the Marketing and Distribution Division

- Reduction of empty vehicle runs
- Reduction of kilometers traveled by vehicles
- Preference for EVs or hybrid vehicles
- Sales promotions for remanufactured products

Sustainability Goal in Supplier Selection (Supply Chain Perspective)

- Preference for suppliers with ISO 14000 certification
- Preference for green suppliers using recycled items as their inputs for manufacturing

Sustainability Goal for Retailers (Supply Chain Perspective)

- Maximum retailers to be engaged in take-back activities and collection of used products from primary customers/users for their remanufacturing

While implementing sustainability-focused scorecards created on the basis of Kaplan and Norton's BSC, the following two aspects are to be given utmost importance.

- Activities in each responsibility center—profit center, cost center, investment center, or revenue center—are influenced by their own requirements, limitations on resources and facilities, and other constraints. These are to be taken into consideration while identifying its suitable sub-goal, which is essentially going to contribute to achievement of the strategic goal of the organization.
- A scorecard with its necessary perspectives and framework is to be designed for each responsibility center (SBU, department, etc.). These scorecards are all linked among themselves and collectively linked with the strategic scorecard of the corporation. So any failure on achieving target or initiatives in a department ultimately leads to non-achievement of strategic goal of the organization.

Thus, some suggestions on effective implementation of scorecards for achieving sustainability may be outlined as the following.

- The possible conflicts among the sub-goals of responsibility centers must be identified beforehand. Take appropriate steps, if possible, for resolving them and attaining goal congruence. If not possible, the concerned persons should be aware of this conflict well in time.

- There should be continuous monitoring of meeting the satisfaction level of stakeholders, input resources, and risk factors arising from time to time.
- Before implementing a sustainability-based scorecard for an organization or supply chain, if necessary, some restructuring of organization may be conducted for ease and effectiveness of the scorecard.
- Data collection and feedback systems may be enabled for better monitoring and control.
- Employees should be made aware and trained before implementing the scorecard model.
- The effectiveness of scorecard implementation may only be perceived if the financial measures are linked with customer needs and the link is further extended by environmental and social aspects.
- The organization or the supply chain should always consider sustainability as an opportunity, not a cost or risk.

5.4.4 EXPECTED BENEFITS FROM SBSC OR SSC SCORECARDS

On the basis of these discussions on various related issues, we may summarize the benefits or advantages of implementing a BSC with sustainability as an effective tool for performance management. This is outlined in the following list.

1. These scorecards provide the logically effective linkages between corporate strategies and performance indicators at various levels of the organization. Consequently, this creates the guidelines for the employees on how to contribute not only to departmental performance but also to the corporate financial and sustainability performance.
2. The resultant links also bind the social and environmental strategies with the corporate values and its broad image.
3. A successful SBSC is expected to increase employee satisfaction, reduce operational costs, improve productivity, increase market opportunity through the green-conscious customer community, uplift corporate image and reputation, increase stock market premiums, and so on.
4. The inclusion of environmental and social aspects as performance measures in balanced scorecards promotes the role of sustainability strategy as a key element in the corporate strategy basket. This surely increases the likelihood of success in achieving the company's strategic objectives.
5. As the criteria showing environmental and social contributions are explicitly included in performance metrics, the implementation of these scorecards automatically enhances the environmental and social accountability of employees and other supply chain members. Moreover, the scorecards clearly recognize the interconnection of sustainability goals with corporate objectives.
6. The company may be repositioned with new image of enhanced corporate responsibility toward the well-being of the outside world (environment and society) other than its commercially focused limited world of the organization itself.

7. Once the team is engaged in developing the SBSC, the members get the deep knowledge and insights about the core meaning of the mission, strategic goals, and issues related to performance measures through the process of translating the corporate mission to a set of manageable performance indicators at various levels of organizational activities.

8. The key strength of balanced scorecards lies on integrating the financial measures with other non-financial ones, each of which reflects the satisfaction level of a stakeholder, establishing simultaneously strong linkages with the corporate objectives and mission. So on the one hand, it generates the desired profitability, and on the other hand, it meets the demands of customers, employees, outside community, and policy-makers at the national and UN levels.

5.4.5 FEASIBILITY ANALYSIS FOR IMPLEMENTING CORPORATE SUSTAINABILITY STRATEGY—NET PRESENT SUSTAINABLE VALUE (NPSV) APPROACH

The decision on whether to implement a strategy or not is surely an important issue in strategic management. In this context, it may be noted that unlike other corporate strategies, corporate sustainable strategy aims at improving all the financial, environmental, and social performance simultaneously. Thus, the conventional techniques for investment appraisal or financial feasibility analysis no longer remain valid in implementation decision of corporate sustainability strategy. If a company decides to implement a sustainable strategy, it is not exclusively aspiring for efficient allocation of monetary resources, but it is also efficiently investing for environmental and social resources.

Most of the discounted cash flow (DCF) methods (NPV, IRR, etc.) have been successfully applied to analyze financial or economic feasibility of any investment strategy. Here, let us discuss an appraisal technique, known as the net present sustainable value (NPSV) approach, which is exclusively meant for feasibility analysis of implementing any sustainability strategy. Liesen et al. (2013) very clearly described the theoretical background, formulation, and practical implications of applying NPSV for implementation of this strategy. NPSV is primarily based on the concept of *capital substitution*. In this context, we may consider four types of capital—*man-made capital or economic capital* (represented by produced or manufactured goods with financial payoffs through market transactions), *human capital* (mostly intangible but the enablers of generating man-made capital with examples like various skills, knowledge, judgment, etc.), *natural capital* (natural resources from the earth), and *social capital* (another intangible but quite rich capital in business processes, with examples like groups, relationships among individuals and institutions, various networks, communities, etc.). It may be noted that investment or use of environmental and social resources or capital also means impacts (mostly negative) of the organizational or business activities on the environment and society. The main challenge in this endeavor is the creation of a single tool for appraising the worth of creating impacts on the environment and society with its corresponding financial gains or value equivalence.

This requires the application of the well-established concept of opportunity cost. In fact, we use the same opportunity cost while applying the conventional net present value (NPV) method. It justifies the worth of any investment proposal by generating at least as much financial return by investing economic capital, which any alternative use of this capital could have generated. Now let us extend this concept little further. Experts consider that maintenance of constant level of environmental and social capital is the key prerequisite for sustainable development. Moreover, societal development requires the continuous creation of economic or man-made capital, which is only possible by making use or, often, overuse of other types of capital. It is thus an accepted case of substitution of capital for the development of society or the mankind. Unlike strong sustainability, capital substitution is emphasized and addressed by weak sustainability. However, weak sustainability may only be applicable in an industrial situation, if the substitution is logical and acceptable by the management, stakeholders, and society.

Opportunity cost is the cost or penalty of missing the opportunity of earning the best possible return or value considering some other alternative.

The sustainable value approach takes into consideration substitution and opportunity cost. The sustainable value approach was originally developed by Prof. F. Figge and Prof. T. Hahn and was published first time in 2001. "Sustainable value is created whenever the return that is achieved through the use of financial, environmental and social resources exceeds the opportunity cost of resource use" (Liesen et al., 2013). The sustainable value added (SusVA) approach has been discussed at length in Section 5.2 of this book. In NPSV, we address almost similar issues, only with consideration of time value of money, which means that money generates its value over the time incorporating the possible returns by its investment. The NPSV model incorporates all the three concepts: NPV, capital substitution, and opportunity cost.

Thus, in NPSV, we compute present value of future financial returns using some discount rate like NPV and one type of capital is substituted by another (environmental and social capitals are substituted by financial return; i.e., economic capital). Opportunity cost is also incorporated by comparing this value with the amount expected, if the corporate sustainable strategy goals are exactly fulfilled.

The net present sustainable value (NPSV) may be defined as the present value of the future returns generated by using environmental and social resources in excess of the returns targeted by corporate sustainability strategy,

---

The following two parameters are the essential inputs for NPSV computation.

5.4.5.1 Minimum Rate of Return

This is the rate of return of using a resource, or in other words, it is the financial gain or value creation by using or exploiting unit resource. Its value is fixed as the target of achievement based on the corporate sustainability goal of the company. It is applicable to environmental resources (natural capital) and social resources (social capital). For example, suppose that the sustainability strategy of a company fixed Rs 50,000 per tonne of CO_2 emission as its goal of rate of return or sustainability efficiency (or eco-efficiency). Then it means that at least Rs 50,000 is to be earned as the compensation of one tonne of CO_2 emissions from business or project activities. Here, let us reiterate the fact that the use of environmental resources does not simply mean the consumption of natural resources or capital; it also means degradation of natural assets or environmental degradation like GHG emissions. So this Rs 50,000 per tonne of CO_2 emission is treated as the strategic goal of environmental efficiency. The similar understanding is applicable for social efficiency (rate of return) in case of social capital or social resources use or degradation. Sustainability-conscious corporations are supposed to aspire for earning as much financial returns as possible to compensate the use of environmental and social resources. However, so as to maintain being a sustainability-oriented business unit, the earnings should never be less than the rate of return or efficiency earmarked as the strategic goal. Moreover, as it is a ratio measure, the attempt for reduction of using environmental resources for same financial gain also will lead to same consequence. If the company is highly sustainability-conscious and plan for continuously improving the sustainability value, then the strategic target of rate of return should be dynamic with some rate of increment as the target for every year into the future. This will surely go on creating harder challenges to be faced in every year during a planning period (e.g., five years into the future).

Let us assume that the company is planning to increase its profit by 5% each year by investing for a completely new production technology and by other related operational restructuring. The technology, being a contemporary one, is also expected to support the green strategy of the company. It is envisaging 4% reduction of CO_2 emissions. So the company is increasing its rate of return or eco-efficiency by $1.05/0.96 = 9.375\%$ every year dynamically. The whole process justifies the substitution of natural capital (that is, the use of environmental resources or environmental degradation) by creating economic capital (that is, the generation of financial gain or value). By using the dynamic target of rate of return and meeting the target every year, the company actually improves its contribution to sustainability.

5.4.5.2 Discount Rate

In the NPSV model, it is required to capture the fluctuating future cash flows (values expressed in financial terms) to their present value equivalence like the discounted cash flow analysis techniques (e.g., NPV) in conventional financial appraisals. So we are to consider a discount rate for NPSV computation. To date the researchers had been working on identification of the most meaningful discount rate for annual consumption of environmental and social resources. It is really a difficult and unsolved problem, particularly in selecting the right type of discount rate for the future use of social resources. This is simplified by using the same discount rate for the time value of money as used in conventional investment appraisal.

5.4.5.3 NPSV Model Formulation

NPSV model analyzes and tests feasibility of an investment proposal or project considering the objective of corporate sustainability strategy. NPSV determines whether or not the value created or the financial gain achieved by investing the financial, environmental, and social capital is more efficient than the efficiency objective fixed by the sustainability strategy. The following are the proposed steps for NPSV computation, considering **one resource (sustainability-related) only**.

Step 1: Let E_t be the target efficiency of using the resource to be achieved in t^{th} period (or year) (like eco-efficiency or socio-efficiency). This is the corporate sustainability target to be achieved or maintained by the organization.

So E_t = (Value creation or financial gain)/(Resource use or consumption)

If it is static for all future years, then it will be E_t only. But if it is dynamic with expected annual increment of d, then target efficiency in year or period $t = E_t (1+d)^t$. We may consider $d = 0$ for static target of annual efficiency.

Step 2: Forecast or predict the use or consumption of the resource in period or year t as C_t, which is the result of the organizational activities. These economic activities give rise to anticipated financial gain or return G_t in year t. So the anticipated efficiency of the resource use for value creation by the company is G_t/C_t in period t.

Step 3: Thus, the anticipated resource efficiency in excess of the target efficiency fixed by sustainability strategic goal in the period $t = G_t/C_t - E_t (1+d)^t$.

Step 4: The net value creation or financial gain achieved in period t in excess of the gain according to the sustainability strategic goal $= (G_t/C_t - E_t (1+d)^t)*C_t$.

Step 5: Finally, NPSV for a particular resource during the whole project life

$$= \sum_{t=1}^{T} \frac{\left(G_t/C_t - E_t (1+d)^t\right)*C_t}{(1+i)^t}$$

where i is the discount rate, and T is the total life span of the project in number of periods or years.

Once we consider all the resources, we get the total NPSV of the project. If NPSV ≥ 0, then it means that the sustainability target or the strategic goal is fulfilled; otherwise, we should not accept the project.

5.4.5.4 Example of NPSV Computation with Only GHG Emissions as the Environmental Impact

Let us take up an illustrative example for NPSV computation. A company has selected a project after carrying out complete feasibility study including the financial feasibility. However, company is yet to appraise its feasibility in terms of sustainability, as the project is to fulfill the corporate strategic goal of sustainability during its tenure of operations. Let the life span of the project be only five years.

The anticipated financial net cashflows or financial gains from the project is Rs 10,00,000 per annum, and let that be constant for ease in calculation. However, the project operations cause environmental damage by emitting 5 tonnes of CO_2 in each year. It means that the expected eco-efficiency generated by the project management is Rs10,00,000/5 = Rs 2,00,000 per tonne of CO_2 emissions per annum. The corporate strategic goal for maintaining sustainability by the company is expressed by the minimum rate of return of Rs 1,50,000 per tonne of CO_2 emissions for each year. In other words, this is the target resource efficiency for environmental resource (CO_2 emissions) to be achieved in each year through the project operations. Let us also assume that this target is the static rate, and it remains constant during the whole five-year planning horizon.

So the excess rate of return generated in each year = Rs 2,00,000 − 1,50,000 = Rs 50,000 per tonne of CO_2 emissions. The net value created in each year = Rs 50,000 × 5 = Rs 2,50,000.

If we consider 10% as the discount rate, then the annual discounted value contribution by the project for next five years even after environmental degradation by GHG emission will be as follows:

1st year: 2,50,000/(1 + 0.1) = Rs 2,27,273
2nd year: 2,27,273/(1 + 0.1) = Rs 2,06,612
3rd year: 2,06,612/(1 + 0.1) = Rs 1,87,829
4th year: 1,87,829/(1 + 0.1) = Rs 1,70,754
5th year: 1,70,754/(1 + 0.1) = Rs 1,55,231

Thus, we may conclude that the NPSV generated by the project, considering the sustainability strategic goal of the company on eco-efficiency for GHG emissions during the five-year life span of the project, is Rs 9,47,699.

Key Learning

- A balanced scorecard, developed by Kaplan and Norton, is an established and all-encompassing performance measurement, planning, and control tool, which simultaneously satisfy all the stakeholders of any corporation. An attempt is made to express sustainability-focused performance management system through a scorecard-like framework that originates from the corporate sustainability strategy.
- The most effective scorecard approach in sustainability is consideration of sustainability in balanced scorecard, or SBSC. Sustainability may be integrated to BSC in three different ways: by enriching each existing perspective with sustainability, by adding a fifth perspective, and by creating a separate scorecard for sustainability. Two case studies show how the companies are actually embracing sustainability in managing the organization.
- A generalized scorecard has been proposed for sustainable supply chains. It also has four perspectives like BSC, but with some differences, like the learning and growth perspective, the supply chain perspective, the sustainability perspective, and the financial perspective.
- Any development process accomplished by mankind may be explained by interplay of four types of capital—man-made capital (economic), natural capital, human capital, and social capital. Environmental resources or

capital is represented by the natural capital itself. The environmental and social impacts or degradation because of any economic or business activity will be treated as the use of natural and social capital.

- The feasibility of implementing corporate sustainability strategy in any investment project may be analyzed and appraised by the net present sustainable value (NPSV) approach. It is the net present value of sustainable value addition as a result of investment of economic, environmental, and social capital for a project and the future value creation.
- The NPSV model incorporates conventional NPV, capital substitution, and opportunity cost.
- Both the eco-efficiency and socio-efficiency contribute to the NPSV modeling as the rate of return against investing environmental and social capital.
- The essential input parameters in NPSV computation are minimum rate of return (static or dynamic) and discount rate.

Prior to the discussion session, it is expected that student groups will be formed. Now each of these questions may be discussed among the group members. The objective of the discussion session is to encourage students to think threadbare and explore all related issues, not arriving at the answer or solution to the problem,

Discussion Questions

1. Various viewpoints show various sets of indicators for measuring sustainability performance. Classify them showing the implications of each viewpoint. Is there any specific class focusing on industry specific mode of classification?
2. If LCA shows high carbon emission by other members of the supply chain, like the supplier or distributor, then how do we convince or incentivize them to lower their carbon emissions, given that the market is quite competitive?
3. Let us assume that some remanufactured parts from the used products of the OEM are being used as input materials for another manufacturer. This manufacturer and OEM is initiating LCA separately in its organization. Now we also know that ideally the scope of any LCA includes the environmental impact of extraction and processing for producing input materials in any manufacturing process, and the life cycle may also be extended to product reuse and recovery process (cradle-to-cradle). Then should the environmental impacts during remanufacturing be include in scope of LCA for OEM or for the downstream manufacturer? Give your comments with proper logic.
4. Carbon footprint analysis ultimately aims at improving processes in order to reduce carbon emissions. If redesigning a process for lower emissions affects the performance or efficiency level of a critical process, then how do we resolve the conflict? Readers are requested to validate their answers with some industrial examples.

5. A manufacturer invested in a high-capacity production technology with high expected eco-efficiency on carbon emissions. So the company can now get high value or high financial return with low carbon emissions. Apparently, it is a good proposition. Suppose that the company is envisaging an upsurge in demand, and the new production technology has the capacity to expand production. So it is planning to intensify production levels. The result is, of course, high financial gain, but with considerably higher magnitude of emissions because of expanded production level, or over use of machines, and the like. Is it a sustainable and acceptable proposition?

Special Question as Food for Thought

The ISO 14000 series is an acceptable certification for environmental sustainability, like the ISO 9000 series is for quality assurance. In that case, what is the equivalent concept or philosophy for TQM? Is it TSM (total sustainability management) or something else? Can the "right first time, every time" of Prof. Philip Crosby be applicable in sustainability as well as a step toward adopting TSM? How do we implement this concept of Prof. Crosby in our industry?

REFERENCES

Babu, B. V. (2006). Life cycle inventory analysis (LCIA). www.researchgate.net/publication/252225470_Life_Cycle_Inventory_Analysis_LCIA. Accessed on 14 September 2021.

Cetinkaya, B. (2011). Developing a sustainable supply chain strategy. In: Tyssen, C., Cetinkaya, B., Cuthbertson, R., Ewer, G., Claas-Wissing, T., Piotrowicz, W., & Tyssen, C. (Eds.). *Sustainable Supply Chain Management*. Berlin Heidelberg: Springer, Chapter 2, 17–55.

Choudhury, A., Mondal, S., & Mukherjee, K. (2017). Analysis of critical factors influencing the management of green supply chain practice in small and medium enterprises. *International Journal of Logistics Systems and Management*, 28(2), 200–224.

Daly, H. E., & Cobb Jr., J. B. (1989). *For the Common Good: Redirecting the Economy Toward Community, the Environment, and a Sustainable Future*. Boston, MA: Beacon Press.

Dias, A. C., & Arroja, I. (2012). Comparison of methodologies for estimating the carbon footprint—case study of office paper. *Journal of Cleaner Production*, 24, 30–35.

Drohomeretski, E., da Costa, S. G., & de Lima, E. P. (2014). Green supply chain management—drivers, barriers and practices within the Brazilian automotive industry. *Journal of Manufacturing Technology Management*, 25(8), 1106–1134.

Epstein, M. J., & Wisner, P. S. (2001, Winter). Using a balanced scorecard to implement sustainability. *Environmental Quality Management*, 1–10.

Fava, J. A., Consoli, F., Deusion, R., Dickson, K., Mohin, T., & Vigon, B. (1993). A conceptual framework for life cycle impact assessment. Society of Environmental Toxicology and Chemistry and SETAC Foundation for Environmental Education, Inc. Workshop Report, Sandestin, FL.

Figge, F., & Hahn, T. (2004). Sustainable value added: Measuring corporate contributions to sustainability beyond eco-efficiency. *Ecological Economics*, 48, 173–187.

Guinee, J., & Heijungs, R. (2017). Introduction to life cycle assessment. In: Bouchery, Y., Corbett, C. J., Fransoo, J. C., & Tan, T. (Eds.). *Sustainable Supply Chains—a Research-Based Text Book on Operations and Strategy*. Cham: Springer Series in Supply Chain Management, Chapter 2, 15–41.

Hervani, A. A., Helms, M. M., & Sarkis, J. (2005). Performance measurement for green supply chain management. *Benchmarking: An International Journal, 12*(4), 330–353.

Hristov, I., & Chirico, A. (2019). The role of sustainability Key Performance Indicators (KPIs) in implementing strategy. *Sustainability, 11*, 1–19, 5742. doi:10.3390/su11205742.

Huang, Y. A., Weber, C. L., & Mathews, H. S. (2009). Categorization of Scope 3 emissions for streamlined enterprise carbon footprint. *Environmental Science and Technology, 43*(22), 8509–8515.

Jain, V. K., & Sharma, S. (2012). Green supply chain management practices in automobile industry: An empirical study. *Journal of Supply Chain Management System, 1*(3), 20–26.

Jankalova, M., & Jankal, R. (2020). How to characterize business excellence and determine the relations between business excellence and sustainability. *Sustainability, 12*, 1–25, 6198. doi: 10.3390/su12156198.

Kalender, Z. T., & Vayvay, O. (2016). The fifth pillar of the balanced scorecard: Sustainability. 12th International strategic management conference, ISMC 2016. *Procedia—Social and Behavioral Sciences, 235*, 76–83.

Kaplan, R., & Norton, D. (2009). *The Balanced Scorecard: Translating Strategy to Action.* Boston: Harvard Business Press.

Kazancoglu, Y., Kazancoglu, I., & Sagnak, M. (2018). A new holistic conceptual framework for green supply chain management performance assessment based on circular economy. *Journal of Cleaner Production, 195*, 1282–1299.

Klopffer, W. (2008). Life cycle sustainability assessment. *International Journal of Life Cycle Assessment, 13*(2), 89–95.

Krajnc, D., & Glavic, P. (2005). How to compare companies on relevant dimensions of sustainability. *Ecological Economics, 55*(4), 551–563.

Latif, B., Mahmood, Z., Said, R. M., & Bakhsh, A. (2020). Coercive, normative and mimetic pressures as drivers of environmental management accounting adoption. *Sustainability, 12*, 1–14. doi:10.3390/su12114506.

Liesen, A., Figge, F., & Hahn, T. (2013). Net present sustainable value: A new approach to sustainable investment appraisal. *Strategic Change, 22*, 175–189.

Mathews, H. S., Hendrickson, C. T., & Weber, C. L. (2008). The importance of carbon footprint estimation boundary. *Environmental Science Technology, 42*(16), 5839–5842.

Moffatt, I. (2013). Measuring sustainable development. In: Quaddus, M. A., & Siddique, M. A. B. (Eds.). *Handbook of Sustainable Development—Studies in Modelling and Decision Support.* Cheltenham, UK and Northampton, USA: Edward Elger, 39–60.

Mukherjee, K. (2011). House of Sustainability (HOS): An innovative approach to achieve sustainability in the Indian coal sector. In: Quaddus, M. A., & Siddique, M. A. B. (Eds.). *Handbook of Corporate Sustainability.* Cheltenham, UK: Edward Elger, 57–76.

Olaru, M., Stolerin, G., & Sandru, I. M. D. (2011, February). Social responsibility of SMEs in Romania from the perspective of the EFQM European excellence model. *Amfiteatru Economic, 13*(29), 56–71.

Recker, J., Rosemann, M., Hijalmarsson, A., & Lind, M. (2012). Modelling and analyzing the carbon footprint of business processes. In: vom Brocke, J., et al. (Eds.). *Green Business Process Management.* Berlin Heidelberg: Springer-Verlag, 93–109.

Sadler, B., & Verheem, R. (1996). *Strategic Environmental Assessment: Status, Challenges and Future Direction.* Zoetermeer, The Netherlands: Ministry of Housing, Spatial Planning and the Environment, The Netherlands and the International Study of Effectiveness of Environmental Assessment and the EIA Commission of thjhe Netherlands.

Singh, R., & Chauhan, A. K. S. (2020, July). Impact of lockdown on air quality in India during Covid 19 pandemic. *Air Quality Atmosphere & Health, 13*, 921–928.

Uygur, A., & Sumerli, S. (2013). EFQM excellence model. *International Review of Management and Business Research, 2*(4), 980–993.

Walton, S. V., Handfield, R. B., & Melnyk, S. A., (1998). The green supply chain integrating suppliers into environmental management processes. *The Journal of Supply Chain Management, 34*(2), 2–11.

Wang, H. F., & Gupta, S. M. (2011). *Green Supply Chain Management.* New York: McGraw Hill Companies Inc.

Wendling, Z. A., Emerson, J. W., deSherbinin, A., Esty, D. C., et al. (2020). *2020 Environmental Performance Index.* New Haven, CT: Yale Center for Environmental Law & Policy. www.epi.yale.edu.

World Commission on Environment and Development (WCED). (1987). *Our Common Future.* Oxford, England: Oxford University Press.

Zhu, Q., Gang, Y., Fujita, T., & Hashimoto, S. (2010). Green supply chain management in leading manufacturers—case studies in Japanese large companies. *Management Research Review, 33*(4), 380–392.

Zhu, Q., Sarkis, J., Lai, K. H., & Geng, Y. (2008). The organizational size in the adoption of GSCM practices in China. *Journal of Corporate Social Responsibility and Environmental Management, 15*(6), 322–337.

Epilogue or Appendix

SUSTAINABILITY IN PRACTICE BY SOME CORPORATIONS

FORD COMPANY IS ACHIEVING SUSTAINABILITY BY RECYCLING

- Ford has the history of successful implementation of sustainability in various activities of the organization. Keeping in view of achieving this goal of sustainability, Ford focused on recycling along with increase in fuel efficiency and efficient use of materials. Ford uses the recycled plastic bottles for underbody shields of cars, trucks, and SUVs, nearly consuming 1.2 billion recycled plastic bottles per annum, with approximately 250 bottles per vehicle manufactured. So its green strategy is reflected clearly on 3Rs (reduce, reuse, and recycle). Sheet disposed plastic bottles are turned into plastic fibers, which are subsequently converted to a sheet of materials by a textile process. This sheet of materials is further processed to automotive parts.
- Ford is also aspiring for achieving its long-term sustainability goal of complete elimination of plastic use by 2030, managing all its globally located plants by local sourcing using renewable energy by 2035, and meeting goal of carbon neutrality by 2050.
- Ford aims to meet the target of 40% of sales volume as EVs (electric vehicles) by 2030.
- By 1999 Ford could get its 73 North American plants certified under ISO 14001, and thereafter, the plants synchronize their management systems with the environmental policy of the company.
- Ford Taurus of 2010 includes eco-friendly renewable and recyclable materials for seat cushions and seatbacks.
- More than 85% by weight of a Ford vehicle is made up of recyclable materials. Because of this, it could significantly save costs of materials.

(Source: corporate.ford.com/articles/sustainability; cnbc.com/2019/05/31)

HEWLETT-PACKARD—THE FIRST IT COMPANY TO DECLARE CARBON FOOTPRINTS

- In 2021, Hewlett-Packard (HP) expressed its aim of becoming the world's most sustainable and technology-rich company by 2030.
- Environmental stewardship was one of its few commitments, which HP has been maintaining since its inception. It is the first global IT company to publish its full carbon footprints and set carbon emission reduction goals for the total value chain. HP always enriches its corporate culture by sustainability, accountability, and transparency. In its corporate initiative, HP follows and includes the UN Sustainable Development Goals (SDGs), Global Reporting Initiative (GRI), and norms of the Sustainability Accounting Standards Board.

- HP addresses sustainability by taking initiatives in three different directions—climate action, human rights, and digital equity. In terms of climate action, HP strives for a net zero carbon and circular economy. It is reducing climatic impact on entire supply chain and related activities, opting for renewable sources for electricity generation, and maintaining transparency in reporting all related impacts.
- Since 1991, HP has been active in recycling and remanufacturing programs under circular economy. It manufactured cartridges using recycled plastics and recycled/remanufactured cartridges. Its future targets include renewable electricity, circularity of products and packaging, and also the restoration of forests.
- HP maintains the values of human rights, diversity, equity, and inclusions. It invests in employees' career growth and supports. Similarly its aim of digital equity supports many to have access to education, healthcare, and economic opportunities needed for their survival and growth.

(*Source:* hp.com/us-en/shop/tech-takes; hp.com/us-en/hp-information/sustainable-impact.html)

WALT DISNEY IMAGINEERING MADE USE OF ITS CREATIVITY IN ACHIEVING SUSTAINABILITY

- Being a corporate giant in entertainment industry, some reports shows that the Scope 1 and Scope 2 emissions of Walt Disney is approximately 1.3 million tons in 2020. But the company has been systematically trying hard to reduce this emission, and it is told that at least 28% has been reduced since 2018.
- Walt Disney formulated and declared its sustainability goals in 2020, which was targeted to be achieved in 2030. These goals are divided into five areas: greenhouse gas emissions, water, waste, materials, and sustainable design.
- In order to meet sustainability goals in all five areas, the following targets are to be fulfilled.

  - Achievement of net-zero direct GHG emission by Disney's activities (Scope 1) and zero-carbon electricity use (Scope 2) by 2030
  - Focus on water stewardship in each operation
  - Use of low-impact materials
  - Zero waste in all operations
  - Achievement of targets on Scope 3 emissions, developed scientifically and analytically (i.e., science-based goals)

- Walt Disney Imagineering deals with various facilities and avenues that contribute to Scope 3 emissions. Scope 3 of Disney includes the emissions generated by goods/services, materials, manufacturing, transport, warehousing, and other aspects of theme parks, hotels, office buildings, and other similar facilities. Science-based emission targets are finalized on the basis of facts and realistic and scientific analysis of their manufacturing process, constraints, materials, and technology being used.

(*Source:* app.impaakt.com/analyses, written by Meenakshi N., 4 December 2022; thewaltdisneycompany.com).

JOHNSON AND JOHNSON ACHIEVED SUSTAINABILITY THROUGH ITS ENVIRONMENT-FRIENDLY PRODUCTS

- Johnson and Johnson claims that around 1.2 billion people of the world practically rely on its various products (e.g., life-changing medicines, baby lotion, and non-medicinal items like medical devices to adhesive bandages). So it already maintains highest-quality products for its main clients, like patients, doctors, and other customers, as these are related to human healthcare.
- The company is also engaged in manufacturing sustainable products and packaging keeping the following items in priority.

  - Regularity compliance on environmental products and packaging
  - Sustainability in product development process
  - Product life cycle assessment
  - Green chemistry and engineering principles in both design and manufacturing of products and packaging
  - End-of-life impacts of products and packages
  - Both intra-industry and inter-industry collaborative attitude in management of the company
  - Reduction of plastic uses
  - More reusability and recyclability

(Source: jnj.com/about-jnj/policies-and-positions)

VEDANTA COULD ACHIEVE SIGNIFICANT REDUCTION OF GHG EMISSIONS

- Vedanta Limited is a large corporation engaged in production of oil and gas and various important metals and its aluminum business is managed by Vedanta Aluminium. This aluminum-producing corporate giant is the largest producer of aluminum in India, manufacturing more than half of India's aluminum (per data of FY22). Vedanta had always been market leader in value-additive and inclusive manufacturing of aluminum. Its strategic intent toward sustainable development practices is clearly reflected on its fourth rank in DJSI 2021.
- Based on its press release, it was declared that in FY 2022 it could reduce the carbon footprint by 12% with respect to that of the previous financial year, whereas the production volume increased by 20%. On World Environment Day, Vedanta Aluminium expressed its commitment of GHG reduction by 25% by end of 2030 with the baseline FY 2021.
- Its strategy of dealing with climate change issue includes the following:

  - It developed low-carbon aluminum, named as Restora, with carbon footprint of almost half of global threshold level.
  - It is the largest consumer of renewable energy, amounting to almost three billion units in FY 2022.
  - It intended to save carbon footprint in its large smelter at Jharsaguda, Orisha, by replacing its diesel-fueled forklifts by electric ones (e-forklifts). It has estimated that by replacing 23 such forklifts, Vedanta can reduce

consumption of 250,000 liters of diesel per annum, which results in almost 690 tonnes of GHG savings. Actually, in FY 2022 Vedanta replaced 27 forklifts with their electric versions powered by lithium-ion batteries. It collaborated with GEAR India for these EVs equipped with cutting-edge technology similar to Industry 4.0.

- Its plan to plant 4 lakh of saplings added significantly to carbon sink base of the region and this country as a whole.
- Operations efficiency and energy conservation resulted in savings of 23 lakh gigajoules in FY 2022.

(Source: https://vedantaaluminium.com/media/press-release, press release of 6 June 2022, New Delhi)

TATA CONGLOMERATE IS PRACTICING SUSTAINABILITY VERY INTENSIVELY

- Creation of Tata Sustainability Group (TSG) in 2014 is a burning example of the seriousness of Tata Group in implementing sustainability practices in its corporations. Being a part of Tata Sons, TSG is supposed to collaborate with all Tata Group companies for guiding, supporting, and exchanging thoughts in embedding sustainability in their business strategies, reflecting their essential responsibility toward the environment and society. TSG provides its guidance and supports overall sustainability (sustainability goals and KPIs, governance and reporting, and framework for sustainable supply chain), environment (climate change issues, water stewardship, and waste management), and CSR (strategies and program assessment for community development).
- The Tata Sustainability Policy framed by TSG emphasizes some minimum sustainability agenda on the following thrust areas.

  - Investors and regulators—global reporting
  - Environment—climate change mitigation and adaptation, energy, and water and solid waste
  - Community—CSR and disaster response
  - Employees—volunteering, diversity and inclusion, and safety
  - Value chain partners—business and human rights and sustainable supply chain
  - Customers—product/service stewardship

(Source: tatasustainability.com/AboutUs/TataSustainabilityGroup; tatasustainability.com/AboutUs/OurApproach)

SCHNEIDER ELECTRIC FOCUSED ON CARBON FOOTPRINT IN ORDER TO ACHIEVE ITS TOUGH TARGET OF CARBON NEUTRALITY

- Schneider Electric (SE) is a European multinational corporation headquartered in France. It has approximately 1.5 lakh workers spread around 100 countries. The company apparently moved its focus from steel and industrial products to electricity and energy sometime in the 1970s. It has now

specialized in digital automation and energy management addressing various facilities, like buildings, data centers, and infrastructures integrating
energy management with automation, software, and services.

- SE is also paving its road of excellence in sustainability. It bagged the recognition of best global sustainable supply chain in 2021. Now SE has a
stringent target of achieving carbon neutrality by 2025 and net zero emission by 2030. For its longer end-to-end supply chain these targets are to be
fulfilled in 2040 and 2050. SE has already reduced supply chain's footprints
by 100,000 tonnes some years ago. SE involved its suppliers in achieving
the targets on reduction of the GHG emissions of the supply chain as a
whole. Various Industry 4.0–focused IoT-enabled tools are being used to
management of the sustainability issues on the supply chain.
- Its primary targets of 60% energy consumption using renewable energy,
85% recycling of wastewater and 93% recycling of solid waste significantly
contributed to the reduction of carbon emissions.
- Most of its products have been developed based on eco-design approach,
and customers are also getting benefits of easy repairability, upgrade, and
ease in dismantling or disassembly for end-of-life treatment of the products.

(*Source:* www.thehindubusinessline.com/specials/clean-tech/paving-a-road-to-turn-
supply-chains-sustainable/article35522120.ece)

HERO MOTOCORP LTD INITIATED ITS JOURNEY TOWARD
EXCELLENCE IN SUSTAINABILITY BY GREEN BUILDING DESIGN

- Hero MotoCorp Ltd constructed its new plants on the basis of green building concept. Latest plants like Neemrana, Global Parts Centre (GPC), and
CIT Jaipur are LEED IGBC certified. These reflect the responsibility of
the company toward water and energy consumption. The optimization of
energy consumption is achieved by using green roofs of 1,16,500 square
meters in all plants of Hero MotoCorp, with green roofs of 25,000 square
meters only in the Vadodara plant commissioned in FY 2017.
- The company uses hydroponics technology in its plants, which drastically cuts
water consumption. In this case, it consumes only 2% water. It also recycles
carbon dioxide into a greenhouse, which activates and enhances the photosynthesis of the plants and small trees inside. Simultaneously, the green walls
generate oxygen, and that gives back the fresh oxygen to the work environment. The green walls have been built in the Neemrana and Vadodara plants.
- In this pursuit of achieving the excellence in sustainability, Hero MotoCorp
used the zero waste to landfill (ZWL) approach. The GPC of Neemrana,
Dharuhera, and Gurgaon plants achieved ZWL certification in 2019–2020.
- Hero MotoCorp also intended to spread the sustainability practice upstream
and downstream by converting the other members of supply chain as green
organizations. Hero MotoCorp collaborated with CII and came up with
two Green Partner Development Programs (GPDP) involving suppliers
and dealers separately. It has been organizing the programs for more than

13 years, engaging nearly 200 supply chain partners. In this process, Hero MotoCorp and the whole supply chain got benefitted enormously.

(Source: www.heromotocorp.com/en-in/digital-annual-report-2019-20)

SOME INDUSTRIAL SNAPSHOTS

- **eBay** developed the green strategy primarily focusing on product recovery process. Its aim was emphasizing exchange and reuse instead of disposal and creating environmental degradation. So eBay maintains cradle-to-cradle as the strategy of sustainability-focused performance. It also sells used furniture.
- Big coffee corporations like **Starbucks** manage their businesses sustainably by constructing and maintaining green buildings with LEED certification for their stores.
- **Google** measures its sustainability performance by considering all relevant criteria used in GSCM and by using renewable energy for its facilities.
- Supply chain partners of **Tata Chemicals** are valued as business associates of the company and integral stakeholders of this big corporation. The goal of the company is not simply its own success but the success of the whole supply chain and thus the value chain partners. It is taking an enterprise-wide risk management initiative for all the partners in the chain.
- **Walmart** maintains its sustainability-focused performance by orienting its activities toward three goals—create zero waste, operate with 100% renewable energy, and sell products that sustain Walmart's resources and environment. In this pursuit it involves all its supply chain partners. Walmart is already running 46% of its operations power from renewable energy sources, and it is planning to harvest from the nonconventional sources like wind and solar more intensively for achieving the target of 100% renewable energy by 2035. By 2040, it is expected that Walmart will only use the zero-emission vehicles for its transportation activities.
- **Nestle** seems to be one of the forerunners in using recyclable plastic bottles, which can be used in other industries in future. By 2025, its target is to use 100% recyclable plastic bottles, which can be recycled over and over again.
- The FMCG giant **Unilever** implemented its sustainability policy at East African plants in choice and maintenance of suppliers. Per the responsible sourcing policy, the suppliers are to ensure that all the natural raw materials and naturally derived ingredients are not linked with deforestation or exploitation of indigenous peoples, employees, and local communities. FMCG is active in converting food waste to animal feeds in East Africa, using biogas in situ, and composting waste as fertilizers.
- **Mitsubishi** is engaged in the production of clean and green energy. The innovation of technology for reuse of EV batteries and Mitsubishi Corporation Energy Solutions Ltd (MCES) are reducing the environmental impacts. The rooftop solar panels (approximately 3.3 MW) were installed by MCES at the Mitsubishi Motors Okazaki Plant, and the power generated is supplied to the facility itself. This reduces around 1,600 tons of carbon emission per annum.

Index

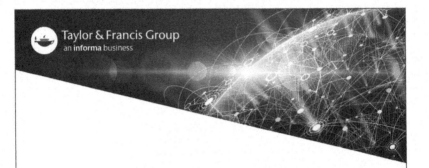